Biological
and
Medical Sensor

TECHNOLOGIES

Devices, Circuits, and Systems

Series Editor

Krzysztof Iniewski

CMOS Emerging Technologies Inc., Vancouver, British Columbia, Canada

Biological and Medical Sensor
TECHNOLOGIES

Edited by

KRZYSZTOF INIEWSKI

CRC Press
Taylor & Francis Group
Boca Raton London New York

CRC Press is an imprint of the
Taylor & Francis Group, an **informa** business

CRC Press
Taylor & Francis Group
6000 Broken Sound Parkway NW, Suite 300
Boca Raton, FL 33487-2742

First issued in paperback 2017

© 2012 by Taylor & Francis Group, LLC
CRC Press is an imprint of Taylor & Francis Group, an Informa business

No claim to original U.S. Government works

Version Date: 20111118

ISBN 13: 978-1-4398-8267-2 (hbk)
ISBN 13: 978-1-138-07321-0 (pbk)

Visit the Taylor & Francis Web site at
http://www.taylorandfrancis.com

and the CRC Press Web site at
http://www.crcpress.com

Contents

PART I   Sensors for Biological Applications

PART II   Sensors for Medical Applications

Preface

Sensor technologies are a rapidly growing topic in science and product design, embracing developments in electronics, photonics, mechanics, chemistry, and biology. Their presence in monitoring biology and in medical applications has become widespread. The demand for portable and lightweight sensors is relentless, filling the needs of many applications, some of which we are just starting to dream of.

The book is divided into two parts. Part I deals with biological sensors. Dr. Abhisek Ukil, from ABB Corporate Research, starts with a chapter on advanced sensing and communication in the biological world. This is followed by a chapter on DNA-derivative architectures for long-wavelength bio-sensing by Dr. Dwight Woolard of the U.S. Army and Dr. Alexei Bykhovski from North Carolina State University. Researchers from the University of Washington follow up with a review of label-free silicon photonic, while researchers from Ankara University discuss quartz crystal microbalance-based biosensors. Dr. Scarpa Giuseppe from the Technical University of Munich provides an in-depth study of lab-on-chip technologies for cell-sensing applications, while Dr. Yue Cui of Princeton University elaborates on enzyme biosensors. Part I concludes with researchers from the University of California at Davis discussing future directions for breath sensors and with a chapter on solid-state gas sensors for clinical diagnosis by Dr. Giovanni Neri of the University of Messina.

Part II deals with medical sensors. Researchers from Northeastern University start this part with a chapter on bio-sensing and human behavior measurements, while Dr. Pietro Salvo from the University of Ghent describes sweat rate wearable sensors. The remaining chapters deal with various aspects of medical imaging. Dr. Mark Nadeski and Dr. Gene Frantz from Texas Instruments provide an introduction to the future of medical imaging, Dr. Bjorn Heismann of Siemens describes spatial and spectral resolution aspects of semiconductor detectors in medical imaging, while Dr. Jim Christian of RMD writes about CMOS SSPM detectors. These chapters are followed by an in-depth study of CdTe detectors and their applications to gamma-ray imaging by Dr. Tadayuki Takahashi of JAXA. The concluding chapter discusses positron emission tomography (PET) based on research carried out by researchers at Stanford University.

With such a wide variety of topics covered, I am hoping that the reader will find something stimulating to read. I also hope that they will discover the field of sensor technologies to be both exciting and useful in science and in everyday life. Books like this would not be possible without many creative individuals meeting together at one place to exchange thoughts and ideas in a relaxed atmosphere. I would like to take this opportunity to invite you to attend CMOS Emerging Technologies events that are held annually in British Columbia, Canada, where many topics covered in this book are discussed (see http://www.cmoset.com for presentation slides from the previous meeting and announcements about future ones). If you have any suggestions or comments about the book please mail me at kris.iniewski@gmail.com

Krzysztof Iniewski
Vancouver, British Columbia, Canada

Editor

Krzysztof (Kris) Iniewski manages R&D at Redlen Technologies Inc., a start-up company in Vancouver, Canada. Redlen's revolutionary production process for advanced semiconductor materials enables a new generation of more accurate, all-digital, radiation-based imaging solutions. He is also a president of CMOS Emerging Technologies (www.cmoset.com), an organization that covers high-tech events concerning communications, microsystems, optoelectronics, and sensors.

Dr. Iniewski has held numerous faculty and management positions at the University of Toronto, the University of Alberta, SP, and PMC-Sierra Inc. during the course of his career. He has published over 100 research papers in international journals and conferences. He holds 18 international patents granted in the United States, Canada, France, Germany, and Japan. He is a frequently invited speaker and has consulted for multiple organizations internationally. He has also written and edited several books for IEEE Press, Wiley, CRC Press, McGraw Hill, Artech House, and Springer. Dr. Iniewski's personal goal is to contribute to healthy living and sustainability through innovative engineering solutions. In his leisurely time, he can be found hiking, sailing, skiing, or biking in beautiful British Columbia, Canada. He can be reached at kris.iniewski@gmail.com.

Contributors

Martin Brischwein
Heinz Nixdorf-Lehrstuhl für
    Medizinische Elektronik
Technische Universität München
Munich, Germany

Alexei Bykhovski
Electrical Communication Engineering
    Department
North Carolina State University
Raleigh, North Carolina

Jeffrey W. Chamberlain
Department of Bioengineering
University of Washington
Seattle, Washington

James F. Christian
Radiation Monitoring Devices Inc.
Watertown, Massachusetts

Yue Cui
Department of Biological Engineering
Utah State University,
Logan, UT

Cristina E. Davis
Department of Mechanical and
    Aerospace Engineering
University of California, Davis
Davis, California

Jean-Pierre Delplanque
Department of Mechanical and
    Aerospace Engineering
University of California, Davis
Davis, California

Y. Murat Elçin
Faculty of Science
Tissue Engineering, Biomaterials and
    Nanobiotechnology Laboratory
Stem Cell Institute
Ankara University
Ankara, Turkey

Gene Frantz
Texas Instruments Inc.
Houston, Texas

Helmut Grothe
Heinz Nixdorf-Lehrstuhl für
    Medizinische Elektronik
Technische Universität München
Munich, Germany

Björn Heismann
Pattern Recognition Lab
Friedrich-Alexander University
Earlangen-Nuremberg
Erlagen, Germany

Shin-nosuke Ishikawa
Japan Aerospace Exploration Agency
Institute of Space and Astronautical
    Science
Sagamihara, Japan

and

Department of Physics
University of Tokyo
Tokyo, Japan

Craig S. Levin
Department of Radiology
Stanford University
Stanford, California

Yingzi Lin
Department of Mechanical and
    Industrial Engineering
Northeastern University
Boston, Massachusetts

Mark Nadeski
Texas Instruments Inc.
Houston, Texas

Giovanni Neri
Department of Industrial Chemistry and
    Materials Engineering
University of Messina
Messina, Italy

Arunima Panigrahy
Department of Mechanical and
    Aerospace Engineering
University of California, Davis
Davis, California

Daniel M. Ratner
Department of Bioengineering
University of Washington
Seattle, Washington

Pietro Salvo
Centre for Microsystems Technology
and
Interuniversity Microelectronics Center
University of Ghent
Ghent, Belgium

Goro Sato
Japan Aerospace Exploration Agency
Institute of Space and Astronautical
    Science
Sagamihara, Japan

Giuseppe Scarpa
Institute for Nanoelectronics
Technische Universität München
Munich, Germany

David Schmidt
Department of Mechanical and
    Industrial Engineering
Northeastern University
Boston, Massachusetts

Şükran Şeker
Faculty of Science
Tissue Engineering, Biomaterials and
    Nanobiotechnology Laboratory
Stem Cell Institute
Ankara University
Ankara, Turkey

Kanai S. Shah
Radiation Monitoring Devices Inc.
Watertown, Massachusetts

Michael R. Squillante
Radiation Monitoring Devices Inc.
Watertown, Massachusetts

Farhad Taghibakhsh
Department of Radiology
Stanford University
Stanford, California

Tadayuki Takahashi
Japan Aerospace Exploration Agency
Institute of Space and Astronautical
    Science
Sagamihara, Japan

and

Department of Physics
University of Tokyo
Tokyo, Japan

Shin'ichiro Takeda
Japan Aerospace Exploration Agency
Institute of Space and Astronautical
    Science
Sagamihara, Japan

Stefan Thalhammer
German Research Center for
    Environmental Health
Institute of Radiation Protection
Helmholtz Zentrum München
Neuherberg, Germany

Abhisek Ukil
ABB Corporate Research
Integrated Sensor Systems Group
Baden-Dättwil, Switzerland

Shin Watanabe
Japan Aerospace Exploration Agency
Institute of Space and Astronautical
    Science
Sagamihara, Japan

and

Department of Physics
University of Tokyo
Tokyo, Japan

Bernhard Wolf
Heinz Nixdorf-Lehrstuhl für
    Medizinische Elektronik
Technische Universität München
Munich, Germany

Dwight Woolard
Electrical Communication Engineering
    Department
North Carolina State University
Raleigh, North Carolina

and

U.S. Army Research Office
Durham, North Carolina

Part I

Sensors for Biological
Applications

1 Advanced Sensing and Communication in Biological World

Abhisek Ukil

CONTENTS

1.1   INTRODUCTION

The animal kingdom is vast and varied. The entire animal kingdom, ranging from the smallest insects to the biggest animals, relies on various sensory systems to interact with nature and live. The types of biological sensors found in nature exhibit enormous variation, like advanced sonar, infrared (IR) thermal sensors, advanced gas sensors, image sensor, vibration detector, chemical sensor, and so forth. Besides having advanced sensory systems, many creatures possess extraordinary communication capability to facilitate proper utilization of the sensory signals and transmit the important information to other fellow creatures. The sensors, associated signal processing, communication, sensor fusion often surpass the state-of-the-art man-made engineering systems. This chapter would provide an explorative study of the different biological sensory systems with the associated signal processing and the possible engineering adaptation of the relevant systems. However, this cannot in anyway be an encyclopedia of the vast amount of advanced sensory science abundant in nature. Some of the best studied creatures and their sensory mechanisms are described in the following sections.

The remainder of the chapter is organized as follows: In Section 1.2, echolocation capabilities of bats are discussed. Section 1.3 presents the amazing acoustical defense mechanisms of nocturnal insects, against the attacks of the bats. IR thermography in certain type of snakes is discussed in Section 1.4. This is followed by a discussion on the waggle dance communication of the honeybees, a Nobel prize winning discovery, in Section 1.5. Advanced sonar capabilities of the dolphins and the whales are discussed in Section 1.6. Finally, Section 1.7 presents brief discussions on sensory systems of several other species, including electrolocation by fish, polarized light sensing by shrimps, bioluminescence in marine lives, acoustic mimicry in insects, advanced gas sensing in beetles, flexible mechanical structures of cockroaches, chemical sensing and communication in ants, etc.

1.2 BATS: ECHOLOCATION

1.2.1 INTRODUCTION

Bats are one of the oldest mammals, the oldest known bat fossil is approximately 52 million year old. Over centuries, people had had curiosities for bats. They fly like birds, bite like beasts, hide by day, and see in the dark. In the folklores of different cultures, we can find instances of bats, for example, in the Greek storyteller Aesop, in the Latin culture, some Nigerian tribes and the North American Cherokee Indians, Chinese, Buddhists, Mayan, etc. In modern days, we also see big influence of bats on good (e.g., Batman) and bad (e.g., vampire) symbolizations.

Bats are the only mammal that can fly and has bidirectional arteries [1]. The forelimbs of all bats are developed as wings; that is why bats have scientific classification order as "Chiroptera" "cheir" (hand), and "pteron" (wing). There are about 1100 species of bats worldwide. About 70% of bats are insectivorous; most of the rest are frugivorous, with a few species being carnivorous [2]. Bats range in size from Kitti's hog-nosed bat measuring 29–33 mm (1.14–1.30 in.) in length and 2 g (0.07 oz) in mass [4] to the giant golden-crowned flying-fox, which has a wing span of 1.5 m (4 ft 11 in.) and weighs approximately 1.2 kg (3 lb). Bats are classified in many different ways; most prominently they are classified as "megabats" and "microbats" [3]. Main characteristics of these two classes are mentioned as follows:

Megabats:

- Usually bigger in size (but not always)
- Frugivores
- Visual cortex and good vision
- Poor auditory system
- No echolocation [4]

Microbats:

- Usually smaller in size
- Insectivores or carnivores
- Practically no vision
- Advanced auditory system: using echolocation to move and hunt [4,5]
- Different ear and cochlea structure [6]

From the aforesaid classifications, it can be noticed that microbats employ an interesting mechanism called "echolocation" [7,8] to forge and hunt in almost darkness.

It is to be noted that they have usually very poor vision; therefore, they rely on their auditory sensors to navigate.

Lazzaro Spallanzani made first experiments on bats' navigation and location methods in 1793. He noted that bats were fully able to navigate with their eyes covered and in total darkness. Spallanzani conducted series of experiments in 1799, eventually concluding that even though bats did not use their eyes, covering or damaging their ears would prove invariably fatal to them, as they would stumble on any and all obstacles and be unable to catch any prey.

1.2.2 Echolocation

Bat echolocation is a perceptual system where ultrasonic sounds are emitted specifically to produce echoes. By comparing the outgoing pulse with the returning echoes, the brain and the auditory nervous system can produce detailed images of the bat's surroundings. This allows bats to detect, localize, and even classify their prey in complete darkness [8,9]. Bats emit echolocation sounds in pulses. The pulses vary in properties depending on the species and can be correlated with different hunting strategies and mechanisms of information processing [1], as shown in Figure 1.1. Bat uses timing, frequency content, duration, and intensity of the echo pulses in an adaptive way to catch the prey that is also moving.

Echolocating bats use frequency modulated (FM) or constant frequency (CF) signals for orientation and foraging. Many types of bats use FM pulses with downward sound that sweeps through about an octave. Others use CF and FM signal combinations. The CF component occurs around 27 kHz, with a duration of about 20–200 ms. The FM component sweeps down from 24 to 12 kHz, with a prominent second harmonic from 40 to 22 kHz [10]. Evolutionary developments of the FM or

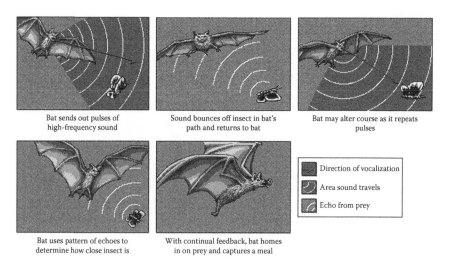

Bat sends out pulses of
high-frequency sound

Sound bounces off insect in bat's
path and returns to bat

Bat may alter course as it repeats
pulses

Bat uses pattern of echoes to
determine how close insect is

With continual feedback, bat homes
in on prey and captures a meal

Direction of vocalization

Area sound travels

Echo from prey

FIGURE 1.1 Echolocation mechanism of bats to catch prey in real time using ultrasound signals in an adaptive manner.

CF/FM type echolocation capability in bats have been motivated by several factors like nature of structural anatomy [11], habitat [12], foraging behavior [13], etc.

The CF and FM signals serve different purposes. For example, the narrowband CF signals are used to localize targets and to determine target velocity and direction [10]. In comparison, FM signals, with broader bandwidth, are used to form a multi-dimensional acoustic image to identify targets [10].

Bats that use mixed signal forms usually hunt in cluttered environments where prey detection is harder for bats that use only FM signals. The mixed signals are of three basic types: FM/short-CF, short-CF/FM, and long-CF/FM sounds. Even though most of the species use only one type of sound [10]. The difference in foraging behaviors of the FM bat and CF/FM bat is shown in Figure 1.2.

Main characteristics of the CF/FM and FM type bats are mentioned as follows:
CF/FM bat

- Uses mixed signal with longer CF part with short narrowband FM part
- Lives mostly at dense places, caves, etc.
- Has Doppler shift compensation (DSC) to compensate own wing flattering and increase resolution accounting for insects' wing flattering [8]
- Does not have advanced delay time processing
- Has higher sensitivity and frequency selectivity, adaptive motion control [13]

FM bat

- Uses shorter duration broadband signals
- Lives mostly in open forest
- Has no DSC
- Can differentiate delay time less than 60 μs [8]
- Has better target localization

Evolutionary developments of echolocating capabilities in different bat species are described in Ref. [11]. For example, *Cynopterus brachyotis* shows no echolocation,

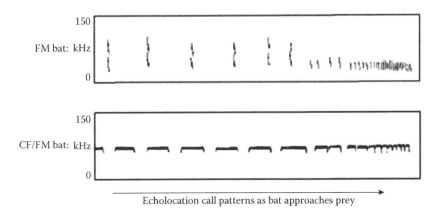

Echolocation call patterns as bat approaches prey

FIGURE 1.2 Different echolocating calls: FM and CF/FM bats.

Rousettus aegyptiacus shows brief, broadband tongue clicks, *Lasiurus borealis* uses narrowband, fundamental dominated pulses, narrowband multiharmonic pulses are used by *Rhinopoma hardwickii* and *Taphozous melanopogon*, short broadband multiharmonics are used by *Megaderma lyra* and *Mystacina tuberculata*, and long broadband multiharmonics are used by *Myzopoda auritam*, etc. [11].

1.2.3 NEUROBIOLOGY AND PHYSICS OF ECHOLOCATION

The anatomical structures of the bat ear, vis-á-vis human ear, are shown in Figure 1.3. In general, for echolocation, bats have bigger pinna [6]. They also have specialized muscles like tensor tympani and stapedius, which are used to control the signal

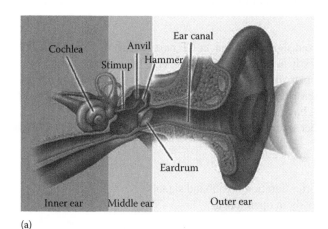

(a)

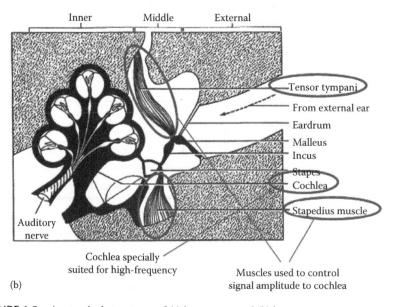

(b)

FIGURE 1.3   Anatomical structures of (a) human ear and (b) bat ear.

amplitude to the cochlea part in the inner ear [6]. Bat's external ear produces changes in the spectrum of incoming echoes, which creates patterns of interference that are used by the bat to estimate target elevation. The cochlea is the auditory portion of the inner ear. The cochlea is filled with a watery liquid, which moves in response to the vibrations coming from the middle ear via the oval window. As the fluid moves, thousands of "hair cells" are set in motion and convert that motion to electrical signals that are communicated via neurotransmitters to many thousands of nerve cells. The cochlea of the bat is specially suited for high-frequency sounds.

The echolocator enables bat to detect small airborne prey and determine the direction (azimuth and elevation) and range of the prey. If the acoustic power transmitted by the bat is P, it is shown in Ref. [9] that the signal power S received by the bat can be expressed as

$$S \approx \frac{PGA\sigma}{16\pi^2 R^4} \exp(-2\beta R), \tag{1.1}$$

where
  G is the gain of the transmitter, relative to an isotropic beam
  A is the received antenna area, that is, the size of the bat ears
  R is the range of the prey
  σ is the acoustic cross section (echo area) of the target
  β is the atmospheric attenuation factor which depends on frequency ($\beta=0.16\ m^{-1}$ at 30 kHz, $\beta=0.69\ m^{-1}$ at 100 kHz)

In general, σ depends on the shape of the insect. Equation 1.1 is known as the range equation, from which the typical power ratio comes as $S/P \approx 10^{-12}$ [9]. This indicates that the hearing sensitivity of bat must be high enough to detect the small reflected signal power from the target. The CF bats overcome this problem by physiological adaptations to the ear and the brain. These include filters that greatly reduce their sensitivity to a very narrow band of frequencies near the transmitted frequency [9].

The sound-transforming features of the head and the outer ear (pinna, see Figure 1.3) provide animals and humans with important cues for localizing a sound source in space. This can be expressed in the head-related transfer function (HRTF). Bats use spectral cues and interaural intensity differences (IIDs) for sound localization because of their small head size and the ultrasonic echolocation calls [14].

Bats estimate the azimuthal direction from the difference in received power in their two ears, and the elevation similarly or perhaps from the spectral shift (the frequency difference between transmitted and received signals, due to the Doppler effect) [9]. This is demonstrated in Figure 1.4. In Figure 1.4, the left and right ears of the bat, separated by a distance $d \ll R$, the range of the prey, receive reflected signals with different power as a function of the direction and the elevation.

Field studies using bat species showed that bats are capable of detecting the direction of signal source within approximately 1° in azimuth and 0°–7° in elevation [15].

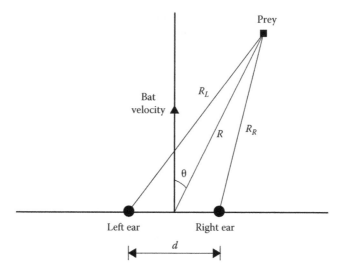

FIGURE 1.4   Directional detection in bats hearing.

1.2.4   UTILIZATION OF ECHOLOCATING CAPABILITY

Bats use their amazing auditory capabilities to perform daunting tasks. Among these, finding and selecting proper caves for hibernation and mating grounds is quite impressive. The study by Glover and Altringham [16] shows finding of proper caves by bats in different weather and day/night conditions, considering factors such as the number of chambers in the caves, entrance orientation, hydrology, sex ratio and age compensation of the bat swarms, etc.

It was already mentioned that bats use their echolocating capabilities for identifying prey. The study [17] shows that the fringe-lipped bat, *Trachops cirrhosus*, has a unique ability among bats to prey on frogs by listening to the advertisement calls male frogs produce to attract their mates. So, the bats intercept the frog mating calls to launch their attack. The study [17] also shows that *T. cirrhosus* hunt in groups, being social. This gives them unique opportunity to learn about any unpalatable prey, for example, a poisonous toad, very quickly. Experiments show that inexperienced bats quickly learn from more experienced "tutor" *T. cirrhosus*.

Another study by Page and Ryan [18] shows that bats (*T. cirrhosus*) are capable of localizing frog calls amid complex environments, for example, with obstacles, other auditory noise, etc.

1.2.5   ENGINEERING ADAPTATION

The closest engineering adaptation of the echolocation is the radio detection and ranging (RADAR) [19] systems, found in airports, ships, national securities, etc. RADAR (shown in Figure 1.5) uses electromagnetic (EM) waves to identify the range, altitude, direction, or speed of both moving and fixed objects. A radar system

FIGURE 1.5   RADAR systems.

has a transmitter that emits either microwaves or radio waves that are reflected by a target and detected by a receiver, depending on the fact that EM waves reflect (scatter) from any large change in the dielectric or diamagnetic constants, for example, solid objects in air.

Not only bats but many human beings, especially those who lost their vision in childhood or so, are reported to demonstrate echolocating capability as well [20–22].

The system described in the study by Waters and Abulula [23] allows free-ranging humans to locate a virtual sound location using active sonar, much like the bats. An emitted pulse, centered on the users' head, serves as an intensity and time marker. The return pulse is rendered at the virtual target location and emitted after a time delay corresponding to the two-way path from sender to target and back again [23].

Studies by Carmena and Hallam [24,25] present biologically inspired engineering on the use of narrowband sonar in mobile robotics. Motivated by the CF-FM bats' capability to exploit Doppler shifts in dynamic conditions, the study focuses on implementing the Doppler shift mechanisms in an experimental robotics systems.

Barshan and Kuc [26] had developed a sonar system capable of making obstacle avoidance using bat-like mechanisms.

1.3 INSECTS: ACOUSTICAL DEFENSE

1.3.1 INTRODUCTION

In the previous section, we have noticed how bats hunt their prey such as nocturnal insects using the advanced echolocation capability. In response to that, many insects, especially moths, have developed an evolutionary defense mechanism to avoid hunting bats by listening to their echolocation calls and taking evasive maneuvers to escape predation [27].

Many of these insects use ultrasonic waves for their intraspecific communication. Thus, it is possible for these insects to detect the ultrasound echolocating calls by the hunting bats. Other species have probably evolved ultrasound hearing in order to detect the predators. Typically, frequency range of insect hearing lies between 10 and 80 kHz, which covers the frequency range of the hunting bats as well. Roughly, there are three defense mechanism hypotheses of the insects based on the auditory clicks [28,29].

1. Aposematism, that is, the sounds remind the bat of the moths' noxious qualities (e.g., bad taste) [30].
2. Jamming, that is, the sounds confuse the bat by generating phantom echoes, causing to abandon the hunt [29].
3. Startle, that is, simply surprising the bat with the sound click, causing a momentary halt which would allow enough time for the insect to fly away.

A key question asked by several researchers is at what point during a bat's attack does an arctiid emit its clicks? The postulated answer would be if the sounds are aposematic, the moth should emit them as early as possible in the bat's attack echolocation sequence to allow the bat enough time to react to them. If, however, the sounds disorient the bat, they should be emitted later in the attack to maximize their confusion effects by denying the bat enough time to adjust [29].

1.3.2 APOSEMATISM

Bates and Fenton [31] reported possible aposematic characteristics of arctiid clicks. The bat recognizes a certain type of prey by the clicks. It primarily detects whether the prey is palatable or noxious. The study shows that certain insects adopt this as a defense mechanism to emit clicks that have a frequency content resembling that of a noxious prey, thus befooling the bat to abandon the hunt.

1.3.3 SIGNAL JAMMING

Studies using the tiger moths (*Cycnia tenera*) and big brown bat (*Eptesicus fuscus*) show that these moths emit high-frequency clicks to an attacking bat, to disorient the bat by interrupting the normal flow of echo information required to complete a successful capture, causing a signal jamming effect [29]. The study shows that *C. tenera* does not respond to approach calls but waits until the terminal phase of the attack before emitting its clicks [32]. This can cause severe signal jamming to the bat, ultimately leading to abandoning the hunt or getting misguided. Otherwise, the bat can get startled momentarily which would provide enough time for the moth to fly away.

Lepidoptera class insects have fine, delicate, round, angular, wavy, or pointed scales, which are organized cross or longwise. As the single scale has a tree-like shape in it cross section, the incoming sound waves are refracted and reflected several times. This can cause severe interference to the signal, allowing no signal or wrong signal fed back to the echolocating bat.

1.3.4 INSECT AUDITORY MECHANISM

Bats employ advanced echolocating calls. For a moth, as the hunting bat approaches, the incident amplitude of the call increases and its duration shortens. This provides a significant cue to the moth despite its anatomically simple ears. A typical moth's ear consists of only three mechanosensory neurones: two auditory A cells with similar frequency tuning and one stretch-sensitive B cell [27,33,34].

The insects can create sound clicks either by a stridulatory organ or a tymbal organ [28,35]. Study by Miller and Surlykke [28] shows that in insects ultrasound clicks are produced by scraping hard surfaces covered with spikes or small ribs against solid edges or surfaces. This is called stridulatory organ, for example, many insects use hind legs to scrape against wings or head. Tymbal organ consists of air filled bags covered with cuticular plates from the abdominal segment. The bags collapse as the muscle contracts. This causes the plate of top of the bag to buckle. As the plate is elastic, it jumps back on its own, producing sound clicks [28,35]. Mechanical model to simulate the tymbal dynamic auditory tuning in the moths has been presented in the study by Frederick et al. [33].

1.3.5 ENGINEERING ADAPTATION

The Lockheed F-117 Nighthawk [36] is a stealth ground attack aircraft formerly operated by the United States Air Force, shown in Figure 1.6. As the name suggests, it incorporates features to hide from or jam enemy radar. The F-117 was born after a combat experience in the Vietnam War when increasingly sophisticated Soviet surface-to-air missiles brought down heavy bomber flights. The comparison of the insects' ability to react to acoustical sonar to that F-117's EM waves is a qualitative one.

The F-117 has got broadband and spot radio jammers. Besides the electronic jammers, the F-117 is shaped to deflect radar signals, owing to different composite materials that scatter and refract the incoming radar waves. This is quite similar to the scales of the moths described earlier.

1.4 SNAKE: INFRARED THERMOGRAPHY

1.4.1 INTRODUCTION

Sensory systems in the animal world are often seen to be optimized with respect to the survival needs as well as biological and habitual characteristics of certain

FIGURE 1.6 F-117 Nighthawk stealth bomber aircraft. (From Wikipedia resources, F-117 Nighthawk.)

species. Analogously, as an evolutionary gift to the cold-blooded reptiles, some snakes, mainly of two groups, pit vipers and boids [37,38], have extraordinary IR radiation detectors. The IR detector system enables the snakes to perceive a two-dimensional image of the heat distribution, enabling them to form a thermal image of their prey or predators, helping to hunt or survive [39], even in complete darkness.

1.4.2 ANATOMICAL STRUCTURE

The IR sensing is possible due to a set of cavities known as the pit organs [40]. In pit viper (*Crotalus horridus*), the pit organ is located on each side of the snake's head near the eyes [41]. In rattlesnake (*Crotalus atrox*), the IR detection is done by some special loreal pit organs located between the eye and nostril on either side of the viper's face, with a thin membrane suspended between these chambers acting as an IR antenna [40]. The snakes can detect a temperature difference in the range of milliKelvins. Nonvenomous pythons (e.g., *Python regius*) and boas have about 5–10 times less sensitivity than the pit vipers [40]. The pit organs of the rattlesnake and the python are shown in Figure 1.7.

In pit vipers, behind the heat-sensing membrane, an air-filled chamber provides air contact on either side of the membrane. The mitochondria-enriched, highly vascular pit organ is heavily innervated with numerous heat-sensitive receptors formed from terminal masses of the trigeminal nerve fibers (terminal nerve masses, or TNMs). The fibers convey IR signals from the pit organ to the optic tectum of the brain [39,40]. Biological tissues like pit organ with high water concentration are expectedly highly absorptive in the mid-IR region of the EM spectrum. As revealed by spectral analysis of the pit organ epidermis of pit viper (*Crotalus adamanteus*), two major regions of IR transmission capabilities are detected, namely, 1700–2900 cm^{-1} (3.4–6.0 μm) and 700–1200 cm^{-1} (8.3–14 μm) wavelengths [43].

FIGURE 1.7   Pit organs of pythons (a) and rattlesnake (b). (From Wikipedia resources, Infrared sensing in snakes.)

1.4.3   IR Detection Principle

The snakes detect IR signals by radiant heating of the pit organ than photochemical transduction, utilizing the transient receptor potential (TRP) channels [40]. The different anatomy and sensitivity of the IR detection in different snake species most likely demonstrate molecular differences in the IR sensing. The study [40] used transcriptome profiling to investigate about the "wasabi receptor" TRPA1 as IR detector. The study reported snakes TRPA1 as heat-activated channel, demonstrating that the rattlesnake TRPA1 was inactive at room temperature, but active above $28.0°C \pm 2.5°C$ [40].

1.4.4   Neurological Interpretation

The pit organ in the snakes allows them to estimate the temperature difference of a prey and a predator and make a thermal image of them. The principle of the IR detection is quite similar to the pinhole camera. The study [41] presents a transformational model of this two-dimensional image projection in the snake brain based on the three-dimensional spatial heat distribution detection. Experiments like presenting discretized versions of images to pit organs and measuring the heat distribution in the membrane assuming realistic aperture and number of heat receptors, and then reconstructing back the images, show significant improvements as might happen in the simple processing unit in the snake brain.

1.4.5 Engineering Adaptations

IR thermography [44] is very commonly used in modern temperature measurement systems, which basically can be divided into two groups: contact and noncontact type. Conventional contact temperature measurement methods include thermocouples, resistance temperature detectors (RTDs), and thermistors, while IR is a method of noncontact measurement. Discovery of the IR measurement is credited to Sir Frederick William Hershel. The noncontact IR thermography has several usages, for example, in biophysics, communication, remote sensing, medical imaging, security, astrophysics, engineering, etc. The engineering applications include

- Remote temperature measurement.
- Temperature measurement in harsh environments.
- Temperature profiling of electrical and mechanical devices in power generation and transmission.
- Temperature profiling and safety in nuclear power plants.
- Thermal monitoring of boilers and chimneys.
- Thermal investigations and hot-spot detection in electronic printed circuit boards.
- Temperature monitoring in glass fiber production.
- Hot metal monitoring, for example, in steel production, flare detection, and slag monitoring.
- Inspection and quality assessment in car manufacturing.
- Outdoor fire detection in coal mines.
- Fire detection in waste recycling applications.
- Quality control on refrigeration systems and refrigerators.
- Process control on calendering machines in the plastics industry, home automation, building and fire safety, etc. In defense of the recent outbreaks of the SARS virus and swine flu epidemic, many airports employed IR noncontact detectors to screen infected passengers [45]. The different applications are shown in Figure 1.8.

Thermal IR detectors convert incoming radiation into heat, raising the temperature of the thermal detector. This change in the temperature is then converted into an electrical signal, which can be displayed and amplified. There are three types of IR detectors:

1. Thermal detectors
2. Photon detectors
3. Pyroelectric detectors

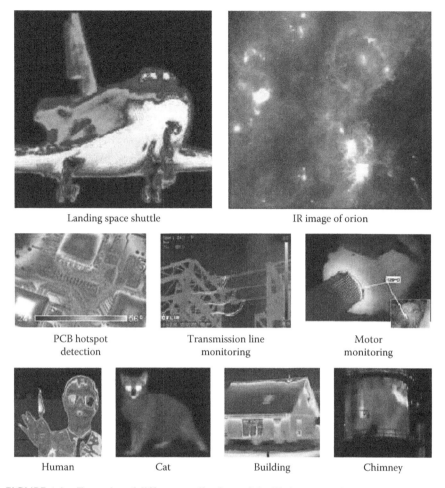

Landing space shuttle   IR image of orion

PCB hotspot   Transmission line   Motor
detection   monitoring   monitoring

Human   Cat   Building   Chimney

FIGURE 1.8   Examples of different applications of the IR thermography.

The thermal IR detectors comprise the following types:

- Bolometer, in which resistance varies with received radiation.
- Thermopile consists of multiple thermocouples in series whose voltage output varies with the received radiation.

Photon-type detectors react to the photons emitted by the object. The IR radiation causes changes in the electrical properties of photon-type detectors. Two main types of photon IR detectors are

1. Photoconductive, which shows increased conductivity with received radiation
2. Photovoltaic, which converts received radiation into electric current

In the pyroelectric detectors, surface charge varies in response to the received radiation.

There are several commercially available industrial IR thermal cameras, for example, "-i, -P, -T series" from FLIR [46], "Ti- series" from Fluke [47], etc.

1.5   HONEYBEE: WAGGLE DANCE COMMUNICATION

1.5.1   INTRODUCTION

Olfactory sensors are common in the insect world. However, for the tiny insects, communicating important messages is particularly difficult with their limited biological organs. Nevertheless, honeybees (*Apis mellifera*) [48,49] possess a unique, outstanding capability to perform a dance in order to convey specific coded message about new food source, its direction, and distance from the beehive. Karl von Frisch made the Nobel prize winning discovery [49] to reveal the sophisticated non-primate communication language, which he termed as "Tanzsprache" in his native tongue, meaning "dance language" [48]. In beekeeping and ethology, the dance of the honeybees is known as "waggle dance" [50,51].

1.5.2   WAGGLE DANCE PATTERN

In the waggle dance, a dancing bee typically executes fast and short forward movements straight ahead on the comb surface, returns in a semicircle in the opposite direction, and starts the cycle again in regular alternation. This is shown in Figure 1.9.

Each waggle dance consists of several of these cycles [48]. In the straight portion of the dance, the bee executes a single stride by lateral waggling motions of the abdomen, known as the "waggle run" [52]. Experiments by Frisch [48] revealed that forager bees which locate food source at a long distance (more than about 50 m)

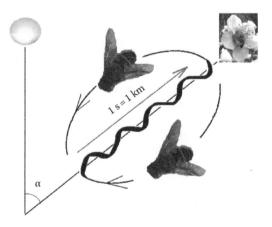

FIGURE 1.9 Waggle dance by honeybees to convey information about food source. (From Wikipedia resources, Waggle dance.)

usually perform the tail-wagging dance, while the bees locating food at a nearby place usually perform a round dance.

1.5.3 INFORMATION ENCODING IN WAGGLE DANCE

In the course of the waggle dance, the honeybees encode information about the food source in various ways. First, the forager bee allows other hive-mates to taste a small sample of the food, in order to convey the type and goodness of the food source. Experiments [48] have shown that if the food source is not good, for example, in synthetic experiments the sugar syrup is diluted, either the forager bee does not at all perform the waggle dance or performs it with less intensity.

Apart from the round or tail-wagging type of waggle dance, the forager bees convey the distance information usually by the duration of the tail wagging in the waggle run. For shorter distances, the sharply marked tail-wagging dance movements with a buzzing sound typically last for about 0.5–1 s. This increases from about 1–4 s for distances of 200–4500 m [48,49].

It is very important for the bees to inform other mates about the exact location of the food source. Thus, the tail-wagging runs are oriented in different directions by forager bees coming from different direction. For example, the waggle run of the "figure-eight" type [48,51] dance might be downward or upward depending on whether the food source is in the direction of the sun or the opposite from the beehive. The angle with respect to the vertical line provides the azimuth of the food source. As day advances, the direction of the dance changes by same angle as that traversed by the sun, but in a opposite direction. This is depicted in Figure 1.10.

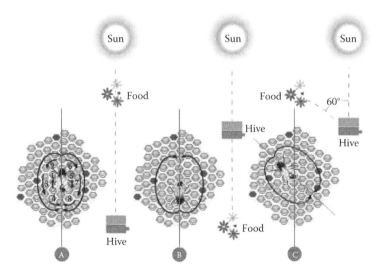

FIGURE 1.10 Direction indication of the food source with respect to the sun in the waggle dance.

1.5.4 Controversy around Waggle Dance

Inspite of series of successful experimental results by von Frisch, there were some scepticism [53–55] about the distance and direction information encoding in the waggle dance. However, with experimental demonstrations using harmonic radars [51], odometers [56], von Frisch's discovery is now widely accepted in the biology community. However, even though the waggle dance is highly effective, bees following the dancing instructions must use their odor and visual cues to successfully find the food source [51]. This also raises some doubt about the overall effectiveness of the waggle dance, with some experiments showing that many bees fail to follow the waggle dance or some need to watch more than 50 runs to follow it [57]. Some experiments [58] showed that most followers ignored the waggle dance for resource direction and about 93% foragers returned to areas previously known to them.

1.5.5 Other Information in Waggle Dance

Karl von Frisch's experiments [48,49] already indicated that bees are able to perceive polarized light. Besides the sun, the bees use the polarization pattern of the sky to guide themselves better. For example, the bees can continue to recognize sun's position after sunset or if it is obscured by mountains, etc. [49]. Other species like crayfish [99], spiders, ants, and octopuses can also detect polarized light.

Apart from the direction, distance, and relative profitability of the food source, it is also hypothesized that the waggle dance of *A. mellifera* could possibly also indicate potential dangers in the food source. Studies like Ref. [59] show that honeybees avoid flowers where they experienced predator attack, like spiders, flowers containing dead conspecifics, etc. Experimental results [59] suggest that honeybees returning from dangerous flowers are less likely to perform the waggle dance or engage in fewer waggle runs than foragers returning from equally rewarding, safe flowers.

Klein et al. [60] reported that sleep-deprived bees perform the waggle dance less proficiently, with significant deterioration in communication. Similarly, sleep deprivation in humans often cause imprecise and irrational communication.

Previous experiments had shown that different species of honeybees employ different version of waggle dance [49]. A study by Su et al. [61] demonstrates that a mixed colony of Asiatic (*Apis cerana*) and European (*A. mellifera*) honeybees could gradually understand each other's "dialect" of waggle dance.

1.5.6 Engineering Adaptations

In computer science, particle swarm optimization (PSO) is a computational method that optimizes a problem by iteratively trying to improve a candidate solution with regard to a given measure of quality, inspired by the behavior of social insects and animals such as fish, birds, ants, etc. [62].

Similarly, there are ongoing research activities on developing efficient fault-tolerant routing, inspired by bee waggle dance [63].

Imitating the honeybees, in operations research, a population-based search algorithm, "the bee algorithm" [64], was developed in 2005.

Another bee-inspired algorithm in the operations research domain is the "artificial bee colony algorithm (ABC)" [65]. In ABC system, artificial bees fly around in a multidimensional search space and some (employed and onlooker bees) choose food sources depending on the experience of themselves and their nest mates, and adjust their positions [65].

1.6 DOLPHIN, WHALE: SONAR

1.6.1 Introduction

Dolphins and whales can be found in oceans and seas across the world. They belong to the mammalian order Cetacea. They vary considerably in size, for example, from the blue whale (*Balaenoptera musculus*), the largest living mammal, to small harbor porpoise (*Phocoena phocoena*), the latter typically being about a meter long [66].

As cetaceans live underwater where visibility is typically limited from one (dirty water) to tens (best condition) of meters, lack of light at depths, acoustics is very important. Also, acoustic energy propagates better than any other form of energy underwater. Synonymously, the cetaceans have evolved an advanced acoustical system, enabling them to live in the aquatic environment, helping to navigate, forage, find mating partners, survive from predators, etc.

Dolphins and whales use wide frequency range for acoustics, for example, large blue whales use about 15 Hz frequency range, while many odontocetes use over 100 kHz, Table 1.1 in Ref. [66] summarizes the different frequency bands used by different species. The range of frequencies used tends to be governed by the anatomical constraints, for example, larger animals use lower frequency range while smaller animals the higher ones. Distance factor also influences the frequency band usage. For example, frequencies below 5 kHz would have negligible losses for ranges more than 100 km, at about 20°C temperature [66], thus suiting the larger animals like blue whales traveling over long distances. However, studies [67] show that the upper and the lower range of the frequency hearing capability is similar in most cetaceans, even though they primarily use different bands.

1.6.2 Biology, Neurobiology of Sound Production

Biology of the sound production in the cetaceans is best studied in the bottlenose dolphins (*Tursiops aduncus*), along with many other species like Atlantic dolphin (*Tursiops truncatus*), Pacific dolphins (*Tursiops gilli*), killer whale (*Orcinus orca*), beluga whale (*Delphinapterus leucas*), etc. Several studies [66,68,69] report the usage of different instruments like multi-hydrophone arrays and digital recorder and experimental methods to record the acoustics of the different cetacean species.

In the cetacean family, the odontocetes and mysticetes classes produce sound differently. Many odontocetes generally emit whistles, burst pulses, and echolocation clicks, usually of high-frequency. In comparison, mysticetes like baleen whale produces low-frequency sounds [66].

The larynx and the nasal sac system play a major role in the sound production in odontocetes. The bony structure and air sacs in a dolphin's head guide the acoustic signals through the dolphin's head in water, studied in details by Cranford et al. [70]. "Monkey lips" help to propagate sounds in the forward direction, while the vestibular sacs reflect upward-directed sound [66,70]. Wood [71] investigated about the role of the fatty melon in dolphin forehead to couple sounds from within its head into the water. The melon and jaw fat are composed of translucent lipid very rich in oil, helping to focus outgoing acoustic emissions [66] (see Figure 1.11 for the sound production mechanism). It is interesting to mention in this context that sonar transducers are often designed with rubber housing with low viscosity mineral oil, a low loss acoustic medium, to couple piezoelectric sensing elements in water [72].

Whales generate single pulse by forcing a burst of air through the tightly stretched lips at the anterior of the forehead. The pulse propagates through the "spermaceti organ," getting reflected and focused by the parabolic air sacs. Thus, the whales emit successively decaying echolocating clicks, attenuated between the two spermaceti air sacs [66,73]. The time interval between successive clicks depends on the distance between the two air sacs and thus can be used to estimate the size of the whale [74].

The ratio of the body mass to the size of the cerebral cortex (gray and white matter) is comparable to the primates, and a percentage of the total brain second to human beings. In dolphins, the olfactory nerve, bulb, tract are absent,

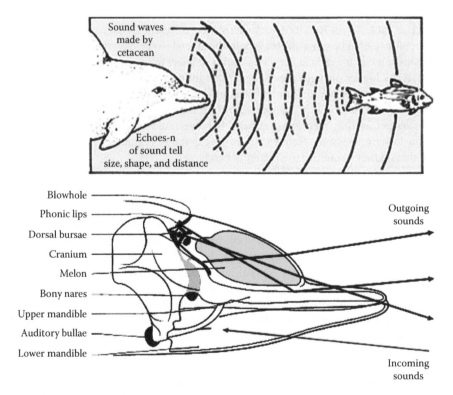

FIGURE 1.11 Sound production and echolation in cetaceans.

resulting in very little smell sensing capability [75]. In marine dolphin (*T. truncatus*), visual system is well developed, with good vision in air. However, river dolphins (*Platanista gangetica*) possess very poor vision. The brain stem of dolphins provides the "acousticomotor" circuitry for nasal phonation and body propulsion while the cerebellum hosts the maps needed for orientation in space and three-dimensional navigation [75].

Due to advanced neurobiology, cetaceans, especially bottlenose dolphins and whales (*O. orca*), have been successfully taught to repeat a wide range of actions [76]. *T. aduncus* can perform mimicry and flexibly access memories of recent actions [77].

Cetaceans show other advanced cognitive functions, often attributed to the humans. This include self-awareness, emotional responses, group living, cultural aspects, and possibly language [76].

1.6.3 UTILIZATION: ECHOLOCATION

One of the main purposes of the development of the advanced acoustic mechanism in the cetaceans is to assist in navigation and prey detection. This is mainly done by the echolocation capability, owing to the high-frequency clicks, like the bats discussed in Section 1.2. Figure 1.11 demonstrates the use of echolocation for prey hunting by cetaceans.

Studies like Ref. [69] show that the inter-pulse intervals of the cetacean sonar ranges from 10 to 200 ms. The inter-pulse interval is linearly correlated with target ranging [68].

With predictable patterns of attenuation, cetaceans, particularly whales, can perform long-distance ranging to estimate distance to a sound source. It is believed that humpback whales use their long-distance ranging capability to range distant singing whales, mating partners, and possible threats. Experiments with several hours of whale songs, frequency analysis, indicate that humpback whales are particularly able to perform long-distance ranging, in the order of even more than 100 km [78].

1.6.4 UTILIZATION: SOCIAL LIVING

Dolphins are known to have individually distinctive whistles, called the "signature whistles" [79]. Signature whistles, mostly recorded from isolated dolphins, are highly distinctive among individuals [79]. Experiments and studies [80] also reveal significantly different pattern in the frequency in calls between siblings, mother to siblings, etc. Studies like Ref. [81] show that dolphins can recognize one another using information encoded in the frequency modulation pattern, like human names.

This certainly points out the intense social living experienced by these animals. Signature whistles were increasingly noticed if mother–infant pairs of dolphins are separated. Field studies by Connor [82] point out several incidents like group of whales accompanying a dying group member and protecting against shark attacks.

Collaborative efforts are also seen in group hunting behaviors of the cetaceans. Family units of *O. ocra* are very stable, living together for more than 70 years often, including communal care of young and cooperative behaviors [66,76].

Bottlenose dolphins usually exhibit "fission–fusion" societies, in which many individual animals tend to come and go at will, even though not born in that group [66].

1.6.5 Utilization: Cultural Aspect and Emotions

Various emotions including stress and aggression have been noticed in cetaceans. Field observations showed two male orcas exhibiting grief after the body of an older female found dead, which was apparently their mother [76]. Herzing [83] reported examples of parental love in cetaceans, exhibited by orcas that suffered prolonged grief over the loss of a calf.

Herzing's study over 10 seasons for population of Atlantic dolphins (*Stenella frontalis*) in the Bahamas reports behaviors such as mother producing signature calls to departed calves which rejoined her after that, female and male dolphins repetitively broadcasting their own signature calls during courtship and mating, several instances of "joy," etc. [66,83].

1.6.6 Utilization: Whale Song

Whale songs are defined as "sequences of notes occurring in a regular sequence and patterned in time" [84]. Many of the aforesaid behavioral attributes of the cetaceans, especially whales, are manifested by these unique singing patterns. Among many species, songs of humpback whales (*Megaptera novangliae*) have been mostly studies and received most attention from 1971 till date [85].

Studies [66,85] reveal several facts about the whale songs as mentioned in the following:

* Songs differ in pattern with geographical difference, for example, Pacific, Atlantic, etc.
* Peak season of singing happens during winter months when the whales migrate to warmer waters; summer months are usually non-singing times.
* Singers are mostly males, who are lonely.
* Some instances of group singing are also observed.
* Singers tend to remain stationary during singing.
* Songs from consecutive years are very similar; but songs from distant time periods differ very significantly.

1.6.7 Utilization: Language

Cetaceans are certainly one of the most vocal animals. However, the existence of a language is highly debatable [76]. Besides "signature whistles," dolphins exhibit a wide range of communication, varying widely depending whether in wild or captivity. Wild cetaceans show breaching and tail-slapping with distinctive sound patterns.

Dialect is sound emission differences on a local scale among neighboring populations, while geographic difference is associated with widely separated populations. Different dialects have been observed in the cetaceans. The dialects of killer whales (*O. orca*) has been studied by Ford [86]. Different dialects in bottlenose dolphins are also reported [66].

Indo-pacific bottlenose dolphins (*T. aduncus*) show asymmetry of flipper and eye usage during social behavior. Studies [87] reported experiments and analysis of laterality of "flipper-to-body" rubbing, in which one dolphin ("rubber") rubs the body of another ("rubbee") with its flipper, in order to perceive the partner better, usually while swimming side by side.

1.6.8   UTILIZATION: OTHERS

Dolphin-assisted therapy has been employed for about 20 years to help mentally and physically challenged people [88]. Birch and Cole made a very speculative hypothesis that ultrasound from the echolocation clicks of dolphins may have a healing effect. The usage of ultrasound in medical treatments requires application of it with certain intensity and duration. This indicates that even if so, patients have to be exposed to dolphins' ultrasound for at least 2 min per session, for several of those [88]. However, the success or failure or technical understanding of the matter remains unclear to date.

1.6.9   INFLUENCE OF NOISE

Increasing concerns are raised by the biological and ecological communities, pointing toward noises caused by powerboats, low-flying aircrafts, etc., heavily affecting the lives of the cetaceans. Field observations showed whales (*Balaena mysticetus*) fleeing away from diesel-powered boats [66]. Lemon et al. [89] reported changes in the surface and acoustic behaviors of *T. aduncus* with powerboats [89]. Foote et al. [90] reported prolonged call duration by killer whales in the presence of boats, however, without altering the call rates.

1.6.10   ENGINEERING ADAPTATIONS

Extraordinary capabilities of the cetaceans has motivated the development of bio-inspired SONAR systems for small underwater vehicles (UAV). Like whales or dolphins, the reported developed sonar uses full signal bandwidths using sparse array configuration [91].

Whales particularly have the capability to detect type of targets using the low-frequency bio-sonar. This has motivated the development of signal processing for target recognition specially for underwater applications [92].

In robotics application, the usage of ultrasound sonar signals to automatically distinguish tress from metal poles has been demonstrated by Gao and Hinders [93]. Ratner and McKerrow [94] reported utilization of sonar sensors for outdoor mobile robot navigation along certain landmarks, identifying those.

1.7 OTHERS

1.7.1 SWIFTLETS: ECHOLOCATION

We have seen that bats and dolphins have the capability of echolocation. Among the birds, South American oilbird (*Steatornis caripensis*) [95] and many Swiftlet species (e.g., Collocaliini: *Collocalia troglodytes*) [96] show unique ability to echolocate. Unlike bats which primarily use echolocation to hunt prey, *S. caripensis* and Collocaliini use their echolocation capability to navigate in the darkness of their roosting and nesting grounds [96].

1.7.2 FISH: ELECTROLOCATION

The field generated by the electric organ of weakly electric fish varies with the electrical properties of nearby objects [97]. The differential current fluxes stimulate the receptors, organized over a large area of the skin, in the electric fishes differently. Based on this, both pulse-discharge fish (*Gnathonemus petersii*) and wave-discharge fish (*Sternopygus macrurus*) use their "electrolocation" [97,98] capability to use weak electric fields for the exploration of their surrounding environment for hunting. This is analogous to the "echolocation" in bats. Experiments show that both types of fishes, particularly *G. petersii*, can evaluate not only distance but also type and size of objects as well. These fishes can also recognize spatial patterns, memorize configurations in space, and even identify common spatial characteristics in different objects [97]. The fishes can differentiate between the "resistor" (e.g., other fish) and the "insulator" (e.g., stone) in the water as well as pick the difference between the "capacitances" of plants and insect larvae with respect to water or stone [97].

1.7.3 SHRIMP: LIGHT POLARIZATION

It is known that some animals, including humans, have a slight ability to perceive the linear polarization of light. Linear polarized light has an electric field oriented in only one direction or plane. The study [99] reports that mantis shrimp (*Stomatopod crustacean*) can distinguish between left- and right-rotated circularly polarized light. The eyes of the *Stomatopod crustacean*, often found in Australia's Great Barrier reef, can also convert linearly polarized light into circular polarization and vice versa.

Commercial products like CD, DVD players, digital cameras, and other optical devices like fiber-optic gyroscope can detect polarization of light. That principle of converting between linearly and circularly polarized light is commonly used in commercial devices, the polarization being manipulated only at a single wavelength. In comparison, the stomatopod can achieve this feat across the visible spectrum, from blue to red [99], a feat that is yet to be achieved by man-made engineering.

The eyes of mantis shrimp may make them capable of recognizing different types of coral, prey species (which are often transparent or semi-transparent), or predators, such as barracuda, which have shimmering scales. Alternatively, the manner in which mantis shrimp hunts (very rapid movements of the claws) may require very accurate ranging information, which would require accurate depth perception.

The species *Gonodactylus smithii* is the only organism known to simultaneously detect the four linear and two circular polarization components required for Stokes parameters, which yield a full description of polarization [100].

Mantis shrimp–inspired future CD or DVD players could read information at multiple wavelengths, or future digital cameras could have better circular polarizing filters to reduce image glare from water or the sky, improve color contrast, and create sharper images.

1.7.4 MARINE LIVES: BIOLUMINESCENCE

Bioluminescence [101] is the production and emission of light by a living organism. The vast majority of more than 700 genera of bioluminescent organisms come from the ocean [102], even though species like *Photinus pyralis* (popularly known as "firefly") or female *Lampyris noctiluca* (popularly known as "glowworm") also show bioluminescence. The creatures use bioluminescence for different purposes like finding food, attracting mates, evading predators, camouflaging for defense, etc.

Blue is the most common color found in bioluminescent species, most of which evolved in ocean, followed by green often found in shallow coastal species, while red, orange, violet, and yellow occur only rarely [102].

In bioluminescence, energy is released by a chemical reaction in the form of light emission, with the help of pigment and enzyme like "luciferin" and "luciferase." The "luciferin" reacts with oxygen to create light, while "luciferase" acts as a catalyst to speed up the reaction [103].

In clinical applications, green fluorescent protein (GFP), isolated and cloned from a bioluminescent jellyfish, has been extensively used as an in vivo fluorescent marker of gene expression, protein synthesis, and cell lineage [103].

1.7.5 INSECT KATYDID: ACOUSTIC MIMICRY

Insects like male cicadas are famous for singing choruses and rapid series of high-pitched clicks. The songs, which only the male cicadas can generate using their tymbal organ, are aimed at finding female mating partners who respond to it at intervals with loud clicks. The male cicada uses it as a cue to locate the partner.

However, studies show that predatory katydid, like *Chlorobalius leucoviridis*, befools the male cicadas by imitating the female response to their songs [104]. Mimicry is often used by insects or animals as a defense mechanism, usually in the form of smell and sight. However, the acoustic mimicry used by the katydid to catch prey that is lured by false promise of food or mating is rather unique [104].

Studies by Marshal and Hill [105] indicate that female cicadas respond in about 70 ms. Field recordings of about 30 min by Marshal and Hill [105] show that the katydids clicked after the correct song for about 22 out of 26 species, responding successfully in 18 cases, that is, more than 90% of the time.

1.7.6 MOSQUITOES: SEXUAL RECOGNITION

Mosquitoes hear with their antennae with the help of the "Johnston's organ" at the base of the antennae, discovered by Johnston in 1855. Speculated by Johnston that the auditory organ might influence mating behaviors of the mosquitoes, recent studies show that interactive auditory behavior does exist between male and female mosquitoes [106]. Both male and female mosquitoes like *Toxorhynchites brevipalpis* respond to pure tones by altering wing beat frequency.

Wing beat frequencies of male and female mosquitoes do not vary significantly, being in the range of 426 and 415 Hz, respectively [106]. However, the frequency convergence and startle responses of the mosquitoes differ significantly. When pairs of the same sex fly, the wing beat frequencies always diverge suddenly, providing frequency separation between the two flight tones. The frequency separation in same sex pairs is somewhat similar to the jamming avoidance in bat echolocation [107] and fish electrolocation [108] capabilities. In comparison, when opposite sexes fly together, their wing-beat frequencies usually closely match in frequencies [106].

The auditory interaction between two mosquitoes of same or opposite sex can be explained by negative feedback systems, the mosquito trying to minimize the frequency difference between their flight tones by altering the wing beat frequency [109]. The difference in the frequency between the flight tones could be imagined as the error signal in the negative feedback loop.

1.7.7 WASPS: FINDING HIDDEN INSECTS

Wood borers (longhorned beetle, buprestid, scolytid, weevil, cossid, sesiid, etc.) feed on xylem or phloem of trees, eventually destroying them. These borers are one of the most notorious insect pests in forest ecosystems, which severely affect the ecological, social, and economic benefits. As their highly concealed life history is rarely influenced by changes in outside conditions, wood borers are usually very difficult to be controlled, and even chemical insecticides can hardly contact the insect bodies, which still remains a big problem in the world as yet [110].

However, following coevolution, natural enemies like wasp, *Hyssopus pallidus*, can localize host borer like *Cydia pomonella* even inside fruit and under tree bark [110]. Wasps employ different mechanisms to locate concealed hosts.

Chemical signals are often used by parasitic wasps. These chemical signals often originate from the hosts, for example, in the form of volatiles or symbiotic microorganisms [111,112]. Many plants also emit volatile compounds when attacked by insects. This is also used by the wasps as cue for finding the hosts. For example, the wasp *Phymastichus coffea*, which is a primary parasitoid that attacks adults of the coffee berry borer, *Hypothenemus hampei*, are attracted to mechanically damaged and infested coffee berries, but not to uninfested ones.

Visual information is usually used in the discrimination of covers of exposed or concealed hosts. Many parasitoids like *Adalia bipunctata* and *Coccinella novemnotata* are attracted by host movement. It also attacks models made of colored paper, metal, and wood [110].

Detecting acoustic vibration is an effective exploratory technique employed by the insects [113]. Wasps like *Pimpla instigator* tap the substrate (wood, stem, soil, etc.) and detect the position of potential hosts through the returned echoes [114]. Parasitoid *Lariophagus distinguendus* finds host *Sitophilus granarius* inside grain via perceiving the larval feeding sound. The vibrational patterns released by moving larvae are different from those released by wriggling larvae or pupae in time and frequency domains [110].

1.7.8 BEETLES: GAS SENSING

Some jewel beetles of the genus *Melanophila* (Buprestidae) can detect IR radiation emitted from forest fires with IR receptors [115,116] from as far as 50 km.

The IR receptors of the *Melanophila* beetles are composed of 50–100 individual sensilla (each approximately 15 μm in diameter) and located in a tightly packed manner on the bottom of a 100 μm deep cavity located next to their mesothoracic legs [117]. The IR receptors of the beetle are mechanoreceptors that convert IR radiation into a mechanical stimulus, which results in a deformation of the sensory cell membrane [118].

As forest fires cause damages worth billions of dollars each year, it is very important to control such fire at an early stage, which makes detecting that remotely even more important. Thus, the principles employed by the jewel beetles could be of extreme interest for forest firewatchers [118].

Scientists at the Center of Advanced European Studies and Research (CAESAR) [118] aim to build IR sensor using silicon-bulk micro-machining technology, motivated by the biological sensors of the jewel beetles.

1.7.9 COCKROACHES: CLIMBING OBSTACLES

Cockroaches, *Blaberus discoidalis*, have unique leg structures and adaptively use different strategies as functions of the ground morphology.

Each cockroach leg is divided into several segments from the most proximal to the most distal segment. These are known as coxa, trochanter, femur, tibia, and at the end, into a series of foot joints collectively called tarsus [119]. Although front, middle, and rear legs have the same segments, they are different in lengths, yielding a ratio of front:middle:rear leg lengths of 1:1.2:1.7 [120]. According to the unique leg structure, the body of the cockroach is also divided into three articulated segments.

The experimental data by Watson et al. [121] show the kinematic changes when *B. discoidalis* is climbing. The experiments reveal that cockroaches do not deviate from normal running kinematic in surmounting obstacles whose height is smaller than one reached by front legs during swing trajectory. Once one or both front tarsi are naturally placed on top of the barrier, they push downward, changing its posture so that the subsequent movements of all legs drive the center of mass upward. In comparison, while climbing obstacles whose top is beyond the height of front legs during swing phase, cockroaches normally accomplish an anticipatory attitude change tilting the body upward. Cockroaches perform this postural adjustment, before front legs are placed on top of the barrier, principally by rotating the middle legs in order to bring them more perpendicular to the ground [121].

Motivated by the legs of *B. discoidalis*, robots are designed with single body segment and symmetrically connected leg pairs, each divided into three segments [122]. This type of bio-mimetic design is aimed at better climbing efficiency of robots.

1.7.10 SALAMANDER: LIMB REGROWING

Salamanders can regenerate lost limbs perfectly throughout their life irrespective of how many times they have been amputated. Also, frogs can rebuild limbs during tadpole stages [123].

After limb amputation, in salamander's body, the nearby blood vessels contract quickly, with a layer of skin covering quickly the wound. This skin epidermis then gradually transforms into a layer of signaling cells, the apical epithelial cap (AEC). Also, the fibroblasts from the connecting tissue network move to the center of the wound, where they form a blastema which is an assembly of stem-type cells. This eventually helps to regrow the limb [124].

Limb generation has always fascinated human beings because we can possibly replace the limbs lost accidentally. Natural human regeneration of amputated fingertips has been reported [125]. Motivated by the salamander's ability to regrow limbs, Muneoka et al., aim to regrow complete limbs [123]. The studies indicate that during regeneration, the fibroblasts migrate across the wound and crosstalk with cells to assess the extent of the wound and eventually start establishing the regeneration boundaries [123].

1.7.11 ANTS: PHEROMONE COMMUNICATION

Ants are one of the most amazing social insects in the *Formicidae* family. Ants form highly organized colonies which typically consist mostly of sterile wingless females forming castes of "workers," "soldiers," or other specialized groups, some fertile males called "drones" and one or more fertile females called "queens" [126].

Ants usually have poor vision. They work in groups with effective communication using the chemical called "pheromone" [126]. Pheromone is released in a wide range of structures from the ant's body like the Dufour's glands, poison glands and glands on the hindgut, pygidium, rectum, sternum, and hind tibia [126]. Ants can perceive smells using their paired antennae which provide information about direction and distance. Ants usually leave pheromone trails that other ants can follow. For example, if a forager ant finds food, it marks the path to the food from the anthill by the pheromone trail which is then followed by other ants [127]. When the food source gets exhausted, usually no new trails are marked by returning ants and the scent slowly dissipates. This shows the remarkable ability of the ants to deal adaptively in the changing environment. Also, it has been observed that when an established path to a food source is blocked by an obstacle, the foragers leave the path to explore new routes, preferably in the shortest way [128].

Ants use pheromones also for alarming purpose. For example, a dying ant emits an alarm pheromone that drives nearby ants into an attack frenzy and attracts more ants that are further from reach. Several ant species are known to have some "propaganda pheromones" that they use to confuse enemy ants and make them fight among themselves [129].

Inspired by the amazing foraging and social communication capabilities of the ants to solve complex real-life problems in harsh and changing environment, algorithms like ant colony optimization (ACO) [130,131] have been developed in computer science and operations research. ACO is a probabilistic technique to find an optimal path in a graph, with several applications in the field of optimization [132].

1.7.12 PADDLEFISH: STOCHASTIC RESONANCE

Stochastic resonance (SR) is a phenomenon that takes place in a nonlinear measurement system or device (e.g., man-made device or biological organs, etc.), where the measurement of information gets maximized in the presence of non-zero level of stochastic noise [133]. SR occurs if the signal-to-noise ratio (SNR) of a nonlinear system or device increases for moderate values of (added) noise intensity. It typically requires a weak coherent signal (e.g., periodic signal) in a bistable system, an inherent source of noise, and an energetic activation barrier (e.g., threshold). Then under SR, the system response might show resonance like behavior as a function of the noise level [133].

SR has found several applications in technology, chemistry, and biology. There are abundance of examples of SR happening in the biology, for example, sensory neurons firing with noise, hydrodynamically sensitive mechanoreceptor hair cells in crayfish, sensing environmental noise by cricket for predator evasion, foraging behavior of paddlefish, etc. [134].

Moss and coworkers [135] reported the significance of SR in the feeding behavior of paddlefish (*Polyodon spathula*). Their study and experiments revealed that paddlefish uses passive receptors to detect electric signals emitted by their main prey zooplankton. Experiments with artificial random electric noise showed active reactions of the paddlefish correlated very nicely at certain level of noise presence, not zero noise.

1.7.13 DRAGONFLY: AERODYNAMICS

Dragonfly has thin wings with tiny peaks on the surface. These peaks create a series of swirling vortices as air flows through dragonfly body. In turn, this helps the dragonfly to remain stable in flight even against the strongest winds. Early research on this shows that this type of design would potentially augment micro-wind turbine blades to avoid unwanted fast spinning when a storm hits [136].

REFERENCES

1. J. D. Altringham, *Bats: Biology and Behaviour*, Oxford University Press, New York, 1996.
2. C. Tudge, *The Variety of Life*, Oxford University Press, New York, 2000.
3. J. E. Hill, J. D. Smith, *Bats–A Natural History*, University of Texas Press, Austin, TX, 1984.
4. M. J. Novacek, Evidence for echolocation in the oldest known bats, *Nature*, 315: 140–141, 1985.
5. T. Nagel, What is it like to be a bat?, *Phil. Rev.*, 83: 435–450, 1974.
6. M. Kössl, M. Vater, Cochlear structure and function in bats, in *Hearing by Bats*, A. N. Popper, R. R. Fayvol (eds.), Springer, New York, 1995.
7. M. B. Fenton, Natural history and biosonar signals, in *Hearing by Bats*, A. N. Popper, R. R. Fayvol (eds.), Springer, New York, 1995.
8. C. F. Moss, S. R. Sinha, Neurobiology of echolocation in bats, *Curr. Opin. Neurobiol.*, 13: 751–758, 2003.
9. M. Denny, The physics of bat echolocation: Signal processing techniques, *Am. J. Phys.*, 72(12): 1465–1477, 2004.
10. J. A. Simmons, M. J. O'Farrell, Echolocation by the long-eared bat, *Plecotus phyllotis*, *J. Comp. Physiol. A*, 122: 201–214, 1977.
11. G. Jones, E. C. Teeling, The evolution of echolocation in bats, *Trends Ecol. Evol.*, 21(3): 149–156, 2006.
12. M. A. Wund, Learning and the development of habitat-specific bat echolocation, *Animal Behav.*, 70: 441–450, 2005.
13. T. Fenzl, G. Schuller, Dissimilarities in the vocal control over communication and echolocation calls in bats, *Behav. Brain Res.*, 182: 173–179, 2007.
14. U. Firzlaff, G. Schuller, Directionality of hearing in two CF/FM bats, *Pteronotus parnellii* and *Rhinolophus rouxi*, *Hear. Res.*, 197: 74–86, 2004.
15. R. S. Heffner, G. Koay, H. E. Heffner, Sound-localization acuity and its relation to vision in large and small fruit-eating bats: I. Echolocating species, *Phyllostomus hastatus* and *Carollia perspicillata*, *Hear. Res.*, 234: 1–9, 2007.
16. A. M. Glover, J. D. Altringham, Cave selection and use by swarming bat species, *Biol. Conserv.*, 141: 1493–1504, 2008.
17. R. A. Page, M. J. Ryan, Social transmission of novel foraging behavior in bats: Frog calls and their referents, *Curr. Biol.*, 16: 1201–1205, 2006.
18. R. A. Page, M. J. Ryan, The effect of signal complexity on localization performance in bats that localize frog calls, *Curr. Biol.*, 76: 761–769, 2008.
19. M. I. Skolnik, *Introduction to Radar Systems*, 3rd edn., Chapter 10, Radar transmitters, McGraw-Hill, New York, pp. 690–725, 2001.
20. D. R. Griffin, Echolocation in blind men, bats and radar, *Science*, 100: 589–590, 1944.
21. D. Kish, Echo vision: The man who sees with sound, *New Scientist*, 2703: 31–33, April 2009.
22. World Access for the blind, Documents, trainings. Available: www.worldaccessfortheblind.org, 2002–2011.
23. D. A. Waters, H. H. Abulula, Using bat-modelled sonar as a navigational tool in virtual environments, *Int. J. Human-Comput. Stud.*, 65: 873–886, 2007.
24. J. M. Carmena, J. C. T. Hallam, The use of Doppler in Sonar-based mobile robot navigation: Inspirations from biology, *Inf. Sci.*, 161: 71–94, 2004.
25. J. M. Carmena, J. C. T. Hallam, Narrowband target tracking using a biomimetic sonarhead, *Robot. Auton. Syst.*, 46: 247–259, 2004.
26. B. Barshan, R. Kuc, A bat-like sonar system for obstacle localization, *IEEE Trans. Syst. Man Cybern.*, 22(4): 636–646, 1992.

27. K. D. Roeder, *Nerve Cells and Insect Behavior*, Harvard University Press, Cambridge, MA, 1967.

28. L. A. Miller, A. Surlykke, How some insects detect and avoid being eaten by bats: Tactics and counter tactics of prey and predator, *BioScience*, 51: 570–581, 2001.

29. J. H. Fullard, J. A. Simmons, P. A. Saillant, Jamming bat echolocation: The dogbane tiger moth *Cycnia tenera* times its clicks to the terminal attack calls of the big brown bat *Eptesicus fuscus*, *J. Exp. Biol.*, 194: 285–298, 1994.

30. D. C. Dunning, K. D. Roeder, Moth sounds and the insect-catching behavior of bats, *Science*, 147: 173–174, 1965.

31. D. L. Bates, M. B. Fenton, Aposematism or startle? Predators learn their responses to the defences of prey, *Can. J. Zool.*, 68: 49–52, 1990.

32. J. H. Fullard, J. W. Dawson, D. S. Jacobs, Auditory encoding during the last moment of a moth's life, *J. Exp. Biol.*, 206: 281–294, 2003.

33. J. Frederick et al., Keeping up with bats: Dynamic auditory tuning in a moth, *Curr. Biol.*, 16: 2418–2423, 2006.

34. J. H. Fullard, The tuning of moth ears, *Experientia*, 44: 423–428, 1988.

35. J. E. Yack, J. H. Fullard, Ultrasonic hearing in nocturnal butterflies, *Nature*, 403: 265–266, 2000.

36. Wikipedia resources, F-117 Nighthawk.

37. T. H. Bullock, R. B. Cowles, Physiology of an infrared receptor: The facial pit of pit vipers, *Science*, 115: 541–543, 1952.

38. E. A. Newman, P. H. Hartline, Integration of visual and infrared information in bimodal neurons in rattlesnake optic tectum, *Science*, 213: 789–791, 1981.

39. J. Ebert, Infrared sense in snakes—Behavioural and anatomical examinations (*Crotalus atrox*, *Python regius*, *Corallus hortulanus*), Dr rer. nat. thesis, Rheinische Friedrich Wilhelms University, Bonn, Germany, 2007.

40. E. O. Gracheva et al., Molecular basis of infrared detection by snakes, *Nature*, 464: 1006–1011, 2010.

41. A. B. Sichert, P. Friedel, J. L. van Hemmen, Snake's perspective on heat: Reconstruction of input using an imperfect detection system, *Phys. Rev Lett.*, 97: 068105-1–068105-4, 2006.

42. Wikipedia resources, Infrared sensing in snakes.

43. A. B. Safer, M. S. Grace, G. J. Kemeny, Mid-infrared transmission and reflection microspectroscopy: Analysis of a novel biological imaging system: The snake infrared-imaging pit organ, Application notebook, Pike Technologies, pp. 16–18, September 2007.

44. G. Gaussorgues, *Infrared Thermography*, Chapman & Hall, London, U.K., 1994.

45. E. Y. K. Ng, R. U. Acharya, Remote-sensing infrared thermography, *IEEE Eng. Med. Biol. Mag.*, 28(1): 76–83, January/February 2009.

46. FLIR Infrared Cameras, FLIR -i, -P, -T series, 2009. Available: www.flir.com

47. Fluke Thermal Infrared Cameras, Fluke Ti- series, 2009. Available: www.fluke.com

48. K. von Frisch, *The Dance Language and Orientation of Bees*, The Belknap Press of Harvard University Press, Cambridge, MA, 1967.

49. K. von Frisch, Decoding the language of the bee, *Nobel Lecture*, 76–87, 1973.

50. Wikipedia resources, Waggle dance.

51. J. Riley, The flight paths of honeybees recruited by the waggle dance, *Nature*, 435: 205–207, 2005.

52. J. Tautz, K. Rohrseitz, D. C. Sandeman, One-strided waggle dance in bees, *Nature*, 382: 32, 1996.

53. A. M. Wenner, D. L. Jhonson, Honeybees: Do they use direction and distance information provided by their dancers? *Science*, 158: 1076–1077, 1967.

54. P. H. Wells, A. M. Wenner, Do honey bees have a language? *Nature*, 241: 171–175, 1973.

55. A. M. Wenner, P. H. Wells, *Anatomy of a Controversy: The Question of a "Language" Among Bees*, Columbia University Press, New York, 1990.

56. R. D. Marco, R. Menzel, Encoding spatial information in the waggle dance, *J. Exp. Biol.*, 208: 3885–3894, 2005.

57. C. Williams, Show me the honey, *New Scientist*, 2726: 40–41, September 2009.

58. C. Grüter, M. Balbuena, W. Farina, Informational conflicts created by the waggle dance, *Proc. Biol. Sci. R. Soc.*, 275: 1321–1327, 2008.

59. K. R. Abbott, R. Dukas, Honeybees consider flower danger in their waggle dance, *Anim. Behav.*, 78: 633–635, 2009.

60. B. A. Klein et al., Sleep deprivation impairs precision of waggle dance signaling in honey bees, *PNAS*, 107: 22705–22709, 2010.

61. S. Su et al., East learns from West: Asiatic honeybees can understand dance language of European honeybees, *PLoS ONE*, 3(6): e2365, 2008.

62. J. Kennedy, R. Eberhart, Particle swarm optimization, in *Proc. IEEE Int. Conf. Neural Netw.*, Perth, WA, Australia, pp. 1942–1948, 1995.

63. G. Crina, A. Abraham, Stigmergic optimization: Inspiration, technologies and perspectives, *Stud. Comput. Intell.*, 31: 1–24, 2006.

64. D. T. Pham et al., The bees algorithm, Technical Note, Manufacturing Engineering Centre, Cardiff University, Cardiff, U.K., 2005.

65. B. Basturk, D. Karaboga, An Artificial Bee Colony (ABC) algorithm for numeric function optimization, in *Proc. IEEE Swarm Intell. Symp.*, Indianapolis, IN, 2006. Available: http://mf.erciyes.edu.tr/abc/

66. W. W. L. Au, R. R. Fay (eds.), *Hearing by Whales and Dolphins*, Springer, New York, 2000.

67. J. D. Hall, C. S. Johnson, Auditory thresholds of a killer whale, *J. Acoust. Soc. Am.*, 51: 515–517, 1971.

68. P. T. Madsen, M. Wahlberg, Recording and quantification of ultrasonic echolocation clicks from free-ranging toothed whales, *Deep-Sea Res. I*, 54: 1421–1444, 2007.

69. T. Akamatsu et al., Comparison of echolocation behaviour between coastal and riverine porpoises, *Deep-Sea Res. II*, 54: 290–297, 2007.

70. T. W. Cranford et al., Visualizing dolphin sonar signal generation using high-speed video endoscopy, *J. Acoust. Soc. Am.*, 102: 3123, 1997.

71. F. G. Wood, *Evolution and Environment*, pp. 293–324, Yale University Press, New Haven, CT, 1968.

72. O. B. Wilson, *An Introduction to the Theory and Design of Sonar Transducers*, US Government Printing Office, Washington, DC, 1985.

73. B. Mohl, M. Amundin, Sperm whale clicks: Pulse interval in clicks from a 21 m specimen, in *Sound Production in Odontocetes with Emphasis on the Harbour Porpoise, Phocena phocena*, Doctoral dissertation, Stockholm University, Stockholm, Sweden, 1991.

74. J. C. Goold, Signal processing techniques for acoustic measurement of sperm whale body lengths, *Acoust. Soc. Am.*, 100: 3431–3441, 1996.

75. H. H. A. Oelschläger, The dolphin brain: A challenge for synthetic neurobiology, *Brain Res. Bull.*, 75: 450–459, 2008.

76. M. P. Simmonds, Into the brains of whales, *Appl. Anim. Behav. Sci.*, 100: 103–116, 2006.

77. E. Mercado et al., Memory for recent actions in the bottlenose dolphin (*Tursiops truncatus*): Repetition of arbitrary behaviours using an abstract rule, *Anim. Learn. Behav.*, 26: 210–218, 1998.

78. E. Mercado, S. R. Green, J. N. Schneider, Understanding auditory distance estimation by humpback whales: A computational approach, *Behav. Process.*, 77: 231–242, 2008.

79. M. C. Caldwell, D. K. Caldwell, Individualized whistle contours in bottlenosed dolphins (*Tursiops truncatus*), *Science*, 207: 434–435, 1965.

80. H. E. Harley, Whistle discrimination and categorization by the Atlantic bottlenose dolphin (*Tursiops truncatus*): A review of the signature whistle framework and a perceptual test, *Behav. Process.*, 77: 243–268, 2008.

81. R. A. Barton, Animal communication: Do dolphins have names? *Curr. Biol.*, 16(15): R598–R599, 2006.

82. R. C. Connor, Group living in whales and dolphins, in *Cetacean Societies: Field Studies of Dolphins and Whales*, pp. 199–218, The University of Chicago Press, Chicago, IL, 2000.

83. D. L. Herzing, A trail of grief, in *The Smile of the Dolphin*, Discovery Books, London, U.K., 2000.

84. C. W. Clark, Acoustic behavior in mysticete whales, in *Sensory Abilities of Cetaceans*, pp. 571–583, Plenum Press, New York, 1990.

85. D. A. Helweg et al., Humpback whale song: Our current understanding, in *Sensory Abilities of Aquatic Mammals*, pp. 459–483, Plenum Press, New York, 1992.

86. J. K. B. Ford, Call traditions and dialects of killer whales (*Ornicus orca*) in British Columbia, Doctoral dissertation, University of British Columbia, Vancouver, BC, Canada, 1984.

87. M. Sakai et al., Laterality of flipper rubbing behaviour in wild bottlenose dolphins (*Tursiops aduncus*) caused by asymmetry of eye use? *Behav. Brain Res.*, 170: 204–210, 2006.

88. K. Brensing, K. Linke, D. Todt, Can dolphins heal by ultrasound? *J. Theor. Biol.*, 225: 99–105, 2003.

89. M. Lemon et al., Response of travelling bottlenose dolphins (*Tursiops aduncus*) to experimental approaches by a powerboat in Jervis Bay, New South Wales, Australia, *Biol. Conserv.*, 127: 363–372, 2006.

90. A. D. Foote, R. W. Osborne, A. R. Hoelzel, Whale-call response to masking boat noise, *Nature*, 428: 910, 2004.

91. M. P. Olivieri, Bio-inspired broadband SONAR technology for small UUVs, in *Proc. IEEE OCEANS Conf.*, Biloxi, MS, pp. 2135–2144, 2002.

92. R. A. Altes, Signal processing for target recognition in biosonar, *Neural Netw.*, 8(7–8): 1275–1295, 1995.

93. W. Gao, M. K. Hinders, Mobile robot sonar interpretation algorithm for distinguishing trees from poles, *Robot. Auton. Syst.*, 53: 89–98, 2005.

94. D. Ratner, P. McKerrow, Navigating an outdoor robot along continuous landmarks with ultrasonic sensing, *Robot. Auton. Syst.*, 45: 73–82, 2003.

95. M. Thomassen, E. I. Knudsen, The oilbird: Hearing and echolocation, *Science*, 204: 425–427, 1979.

96. H. A. Thomassen et al., Do Swiftlets have an ear for echolocation? The functional morphology of Swiftlets' middle ears, *Hear. Res.*, 225: 25–37, 2007.

97. C. Graff et al., Fish perform spatial pattern recognition and abstraction by exclusive use of active electrolocation, *Curr. Biol.*, 14: 818–823, 2004.

98. H. Meyer et al., Behavioral responses of weakly electric fish to complex impedance, *J. Comp. Physiol. [A]*, 145: 459–470, 1982.

99. N. W. Roberts et al., A biological quarter-wave retarder with excellent achromaticity in the visible wavelength region, *Nat. Photon.*, 3: 641–644, 2009.

100. S. Kleinlogel, A. G. White, The secret world of shrimps: Polarisation vision at its best, *PLoS ONE*, 3(5): e2190, 2008.

101. J. W. Hastings, Biological diversity, chemical mechanisms, and the evolutionary origins of bioluminescent systems, *J. Mol. Evol.*, 19(5): 309–321, 1983.

102. E. A. Widder, Bioluminescence in the ocean: Origins of biological, chemical and eco-logical diversity, *Science*, 328: 704–708, 2010.
103. O. Shimomura, *Bioluminescence: Chemical Principles and Methods*, World Scientific, Singapore, 2006.
104. S. Pain, What the katy did next, *New Scientist*, 2727: 44–47, September 2009.
105. D. C. Marshall, K. B. R. Hill, Versatile aggressive mimicry of cicadas by an Australian predatory katydid, *PLoS ONE*, 4(1): e4185, 2009.
106. G. Gibson, I. Russell, Flying in tune: Sexual recognition in mosquitoes, *Curr. Biol.*, 16: 1311–1316, 2006.
107. N. Ulanovsky et al., Dynamics of jamming avoidance in echolocating bats, *Proc. R. Soc. Lond. B. Biol. Sci.*, 271: 1467–1475, 2004.
108. W. Heiligenberg, *Principles of Electrolocation and Jamming Avoidance in Electric Fish*, Springer, Berlin, Germany, 1977.
109. M. C. Göpfert, D. Robert, Active auditory mechanics in mosquitoes, *Proc. R. Soc. Lond. B. Biol. Sci.*, 268: 333–339, 2001.
110. W. Xiaoyi, Y. Zhongqi, Behavioral mechanisms of parasitic wasps for searching con-cealed insect hosts, *Acta Ecol. Sin.*, 28(3): 1257–1269, 2008.
111. S. B. Vinson, Chemical signals used by parasitoids, *Redia*, 74(3): 15–42, 1991.
112. J. T. Lill, R. J. Marquis, R. E. Ricklefs, Host plants influence parasitism of forest cater-pillars, *Nature*, 417: 170–173, 2002.
113. D. L. G. Quicke, *Parasitic Wasps*, Chapman & Hall, London, U.K., 1997.
114. G. R. Broad, D. L. Quicke, The adaptive significance of host location by vibrational sounding in parasitoid wasps, *Proc. R. Soc. Lond. B. Biol. Sci.*, 267: 2403–2409, 2000.
115. H. Schmitz, H. Bleckmann, M. Murtz, Infrared detection in a beetle, *Nature*, 386: 773–774, 1997.
116. S. Schütz et al., Insect antenna as a smoke detector, *Nature*, 398: 298–299, 1999.
117. A. Schmitz, A. Sehrbrock, H. Schmitz, The analysis of the mechanosensory origin of the infrared sensilla in *Melanophila acuminata* (Coleoptera; Buprestidae) adduces new insight into the transduction mechanism, *Anthr. Struct. Dev.*, 36: 291–303, 2007.
118. M. Lacher, S. Steltenkamp, Pyrophilous Jewel Beetle as model for a microtechnological infrared sensors, Report: Center of Advanced European Studies and Research, 2010. Available: www.caesar.de
119. R. D. Quinn, R. E. Ritzmann, Construction of a hexapod robot with cockroach kinemat-ics benefits both robotics and biology, *Connect. Sci.*, 10(3–4): 239–254, 1998.
120. F. Delcomyn, M. E. Nelson, Architectures for a biomimetic hexapod robot, *Robot. Auton. Syst.*, 30: 5–15, 2000.
121. J. T. Watson et al., Control of obstacle climbing in the cockroach, *Blaberus disoidalis*, I. Kinematics, *J. Comp. Physiol. A*, 188: 39–53, 2002.
122. P. Arena et al., Climbing obstacles via bio-inspired CNN-CPG and adaptive attitude control, *Proc. IEEE Symp. Circuits Syst.*, 5: 5214–5217, 2005.
123. K. Muneoka, M. Han, D. M. Gardiner, Regrowing human limbs, *Sci. Am.*, 36–43, April 2008.
124. M. Han et al., Limb regeneration in higher vertebrates: Developing a roadmap, *Anat. Rec. Part B: New Anat.*, 287B(1): 14–24, 2005.
125. P. A. Tsonis, *Limb Regeneration*, Cambridge University Press, New York, 1996.
126. B. Holldobler, E. O. Wilson, *The Ants*, Belknap Press of Harvard University Press, Cambridge, MA, 1990.
127. D. E. Jackson, F. L. Ratnieks, Communication in ants, *Curr. Biol.*, 16(15): R570–R574, 2006.
128. S. Goss et al., Self-organized shortcuts in the Argentine ant, *Naturwissenschaften*, 76(12): 579–581, 1989.

129. P. D'Ettorre, J. Heinze, Sociobiology of slave-making ants, *Acta Ethol.*, 3(2): 67–82, 2001.
130. M. Dorigo, Optimization, learning and natural algorithms, PhD thesis, Politecnico di Milano, Italy, 1992.
131. M. Dorigo, T. Stützle, *Ant Colony Optimization*, MIT Press, Cambridge, MA, 2004.
132. M. Dorigo, V. Maniezzo, A. Colorni, Ant system: Optimization by a colony of cooperating agents, *IEEE Trans. Syst., Man, Cybern. Part B*, 26(2): 29–41, 1996.
133. L. Gammaitoni et al., Stochastic resonance, *Rev. Mod. Phys.*, 70(1): 224–287, 1998.
134. P. Hänggi, Stochastic resonance in biology, *ChemPhysChem*, 3: 285–290, 2002.
135. D. F. Russel, L. A. Wilkens, F. Moss, Use of behavioural stochastic resonance by paddle fish for feeding, *Nature*, 402: 291–294, 1999.
136. W. Bird, Dragonfly wings hold the key to solving windy problem, *NewScientist*, 19, January 2011.

2 Physics and Modeling of DNA-Derivative Architectures for Long-Wavelength Bio-Sensing

Alexei Bykhovski and Dwight Woolard

CONTENTS

2.1 INTRODUCTION

The physics and modeling of biological molecular components is discussed in the context of defining DNA-based biological molecule switches (BMSs) that will be useful for incorporating into larger DNA-based nanoscaffolds for the purposes of defining a novel class of smart materials for terahertz (THz) and/or very far-infrared (far-IR)-based biological sensing. Here, theoretical research will be discussed that is applied to define molecular-level functionality that will be useful for realizing "THz/IR-sensitive" materials. Synthetic DNA-derivative architectures with sensitivity to THz and/or far-IR signals are of interest because many biological (and chemical) molecules possess unique spectral fingerprints in the far-IR and THz frequency

39

(~3.0–0.03 mm) regimes. However, practical problems (e.g., weak signatures, limited number of discernable features, and sensitivity to environmental factors) often reduce the effectiveness of THz/IR spectroscopy in practice, and this motivates the development of novel smart material paradigms that can be used to extract nanoscale information (e.g., composition, dynamics, and conformation) through electronic/ photonic transformations to the macroscale. Hence, this research seeks to define DNA-derivative architectures with new spectral-based sensing modalities that will be useful for long-wavelength bio-sensing applications [1]. In addition, since these same smart material paradigms can be used to define antibody (i.e., structures that capture) and receptors (i.e., structures that capture and report) mimics, they also have important relevance to recognition-based detection and future medical applications (e.g., in developing synthetic vaccines) [2].

2.2 BIOLOGICAL-SEMICONDUCTING INTERFACES

The exploration of bio-organic device functionality and sensing in the future will require interfacing with traditional electronic materials and/or structures. An example of one such interface was recently considered in the context of the resonant sensing of bio-molecules [3]. Resonant far-IR spectroscopy is a common technique for the characterization of biological (bio) molecules. The lower portion of the THz spectrum of DNA, RNA, and proteins is also being actively studied using both experimental and computational methods. To date, some progress has been made in the detection and identification of bio-materials and interest is rapidly increasing across the scientific and technology communities. Most experimental and theoretical works have considered bulk quantities of bio-molecules in solution or as dehydrated powders or thin films. However, reliable and repeatable collection of the spectral signatures of DNA/RNA targets by conventional optical and quasi-optical techniques is hampered by effects (e.g., statistical, geometrical, environmental, etc.), which often alter and/or degrade the available sequence (i.e., structural) information. Hence, this issue motivates the search for alternative methods for extracting long-wavelength spectral information from single (or few) numbers of bio-molecules.

One potential alternative for the detection of ultra-low concentrations of bio-material was recently suggested [4]. In this sensor concept, a known single strand of DNA is to be bonded to Silicon nano-probes and active THz illumination will be applied to induce phonon vibrations that are sensed by direct electron current measurements. In order to fully realize the potential of any type of ultra-sensitive DNA detection that involves contact probing, a realistic analysis of the DNA bonded to nanoscale substrate features is required. Currently, the bonding of DNA to semiconductor substrates is less understood than the dynamics and spectral characteristics of DNA in solution. In order to fill this gap, a first principle study of deOxyGuanosine (dG) residues chemically bonded via $(CH)_2$ linkers to the surface of hydrogen-terminated Silicon(111) nano-sized clusters was recently performed [3]. These investigations utilize differing types of Silicon surface structure models, and QM molecular descriptions were generated to accommodate one or two dG residues. In particular, a model for two different dG residues that form a single strand of DNA in the lateral direction (along the surface of a nano-dot) was developed. The number of Silicon atoms included in the four-layer nano-sized Silicon quantum dot (q-dot) structure that were considered varied from 49 (smallest) to 82 (largest). First principle simulations with valence electron basis and effective core potentials (ECPs) were conducted. Here, for each molecular system defined by a chosen DNA chain bonded to a particular Silicon nano-dot, all-atom geometric optimizations without constraints were performed to determine the final conformations, and normal mode analyses were applied to derive the spectral absorption information. More specifically, stable dG conformations on Silicon were obtained for varying types of DNA chain length and nano-dot size/shape. These simulation studies provided new insights into the collective dynamics of the combined DNA chain and Silicon nano-dot system and the resulting optically active absorption modes within the THz and far-IR regimes. Furthermore, these studies demonstrated that it is possible to identify THz spectral signatures of DNA chains chemical bonded to nano-dots that are essentially insensitive to the specific geometrical characteristics of the nano-dots. For example, Figure 2.1 provides the simulation results for two dG residues (dGG) bonded to Silicon q-dots that contain 82 (Si82) and 49 (Si49) atoms, respectively. Figure 2.1a illustrates snapshots of the conformation results (side and above views) for the two compound structures, with the larger dGG/Si82 on the left and the smaller dGG/Si49 on the right. Figure 2.1b compares the spectral absorption arising within the dGG/Si82 and dGG/Si49 structures below 6 THz. Here, one can easily identify absorption lines that are uniquely defined by the size of the q-dot (e.g., dGG/Si49 exhibits unique lines at 80 and 89 cm^{-1}) but there are also lines that persist independent of q-dot size (e.g., at 176 cm^{-1}). These results are particularly important in the context of the DNA-derivative architectures under discussion here for application to long-wavelength-based biological sensing. Specifically, these results demonstrate the general feasibility of using highly complex DNA nanoscaffolds for the purpose of organizing target DNA derivatives to enable an enhanced and extended capability for collecting the THz signatures associated with the target DNA molecules. A novel concept for using biological architectures to enhance and multiply the available THz spectral fingerprints in DNA molecules will be discussed in the next section.

2.3 NOVEL BIOLOGICAL ARCHITECTURAL CONCEPTS BASED UPON LIGHT-INDUCED SWITCHING

While long-wavelength spectroscopic analysis has been shown to have a potential for the detection and characterization of biological materials and agents, the practical application of THz fingerprinting to macroscopic bio-samples is very difficult because the already very-weak spectral signatures can be masked and/or altered by structural or geometrical effects and by external environmental influences, and there are a relatively limited number of available spectral signatures (~ <100) associated with any particular specimen in its natural (e.g., ground) state [5] These facts have motivated a research into new types of biological architectures that can be used to actively control the conformational properties of target bio-molecules at the nanoscale and to precisely extract information regarding their THz-frequency spectral absorption properties [7] Here, the first goal is to define man-engineered nanoscaffolds that allow for the use of external stimulus (e.g., light-based excitation) to excite the target

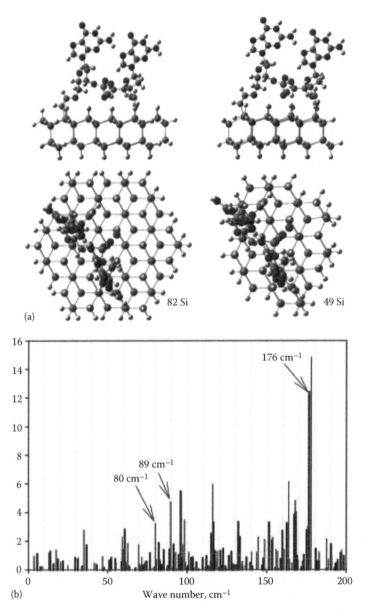

FIGURE 2.1   Simulation results for two compound structures dGG/Si82 and dGG/Si49. (a) Snapshots of the conformation for dGG/Si82 (left) and dGG/Si49 (right). (b) Spectral absorption for dGG/Si82 (black lines) and dGG/Si49 (grey lines) below 200 cm⁻¹ (6 THz).

molecules into as many alternative metastable states as possible (i.e., where they will exhibit multiple sets of THz fingerprints) as this will enable a type of multistate spectral sensing (i.e., with increased amount of information for identification and characterization). The second goal is to use these nanoscaffolds to define highly organized systems (e.g., where the target molecules might be aligned as in a crystal) such that the absorption associated with the target molecules is significantly enhanced, and/or a novel electro-optical-based transduction can be defined that allows for efficient extraction of the THz properties (e.g., dielectric response) of the target molecules. Here, it is also desired that these bio-architectures initially offer measurement apertures of sufficient size (~order of the spectral wavelength) amenable for traditional spectroscopic characterization, although, if novel interfacing with the nanoscale could be defined sometime in the future, the sensitivity and discrimination capability would be increased even for very small nanoscaffold-based systems.

One novel bio-based architectural concept of the general type discussed earlier currently under consideration by our research group [1] involves the molecular engineering of THz/IR smart materials that can be used for biological (and chemical) threat agent sensing. Here the specific goal is to define switchable molecular components that when incorporated into larger DNA-based nanoscaffolds lead to THz and/or IR regime electronic and/or photonic material properties that are dictated in a predictable manner by novel functionality paradigms. This research is considering both organic molecular switches (OMSs) and biological molecular switches (BMSs) that can be incorporated into DNA-based nanoscaffolds to define novel transduction of the spectral information related to the particular target molecules that are under consideration. Note that these smart material type paradigms are envisioned both for use in sensing an exposure event to some external molecular agent (i.e., either by binding or bonding interactions) and for use in extracting nanoscale information (e.g., composition, dynamics, and conformation) related to select molecules that were used in the construction of the bio-architecture itself (e.g., in the case where the BMS was made up of genetic material).

These smart materials may be constructed through the use of designer DNA origami structures [6,7] and/or general motif DNA nanoscaffolds [8] that will allow for the incorporation of organized molecular functionality. Figure 2.2 provides an illustration of how DNA origami architectures could be applied to realize THz/IR smart material systems. First, Figure 2.2a illustrates the construction of a hypothetical DNA-based unit cell that spans three-dimensional (3D) coordinate space. Here, each face of this DNA unit cell is made up of a DNA origami cassette that has been designed with a window for incorporating a switchable molecule. Next, Figure 2.2b illustrates how these DNA unit cells might be combined (e.g., through predefined bonding points on the origami structures or through the use of additional DNA motif structures) to arrive at large scale two-dimensional (2D) structures with a overall dielectric tensor that is strongly influenced by the switchable molecules that were incorporated into the windows. It is relevant to note that though these structures might be novel, they are definitely not a fantasy, as active self-assembly research is being pursued at this time [6–8] that has already produced the real-world constructs of this type. Here, the DNA cassettes shown in Figure 2.2c and the chain of DNA cassettes shown in Figure 2.2d represent the first nanoscaffold prototypes for these architectural concepts. These structural designs are of relevance to the discussion presented here because if one were to choose DNA derivative as the BMSs, then it should be possible to extract information on the spectral absorption of the DNA based upon how this property modifies the electro-optical transmission (amplitude and phase) through the material. Furthermore, if DNA derivatives are used that

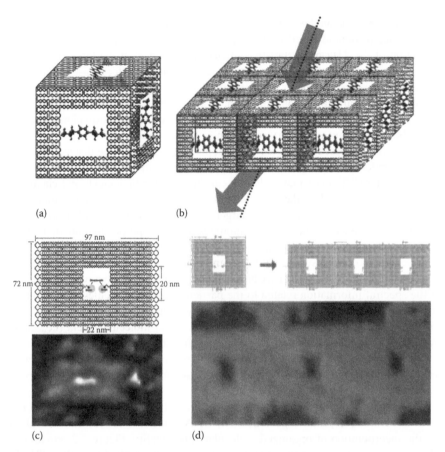

FIGURE 2.2 (a) DNA-based unit cell that spans 3D coordinate space. (b) Smart material using the unit cell from (a) with measurable absorption/refraction. (c) An actual 97×72 nm DNA origami cassette design with biotin/streptavidin adapter molecules (upper) and AFM image of assembled structure (lower). (d) Linear array design (upper) using DNA cassette from (c) and AFM image of assembled array (lower).

allow for being light-induced into many different metastable states, then the large amount of resulting spectral signature information would allow for full identification and characterization of the associated genetic sequence information. The following sections will address the theoretical problem of defining and analyzing DNA-based BMSs (e.g., stilbene-DNA complexes) useful for these types of sensing applications.

2.4 HYBRID AB-INITIO/EMPERICAL MODELING OF STILBENE-DERIVATIVE/DNA COMPLEXES

2.4.1 BACKGROUND

Both molecular mechanical (MM) and first principle (or ab-initio) QM methodologies are actively used in nanoscale phenomenology and nano-structures-related research that involve organic and biological (bio) molecules. In particular, simulations on molecular conformation, dynamics, and associated interactions (i.e., including far-IR and THz spectral absorption and emission) usually require the application of ab-initio models. However, research into new nanotechnology concepts that seeks to identify novel functionality derived from nanoscale-dictated electronic, photonic, and/or phonon processes usually involves modeling and simulation of very large and complex bio-organic molecular systems [3]. Even a single bio-molecule can consist of hundreds of thousands of atoms. Moreover, it is often necessary to consider the effects of local atomic interactions and chemical reactions. Essentially all QM methods (i.e., including parameterized methods such as tight-binding) are usually not feasible for such complex problems since too many atoms are involved in the simulations and this same fact makes all first principle approaches prohibitively expensive. On the other hand, empirical molecular mechanics is unable to solve many important problems including those involving changes in electronic states, chemical reactions, and breaking or formation of chemical bonds. Furthermore, it is not possible to adequately model the structure and dynamics of some bio-organic nano-structures with the required accuracies using predefined MM force fields which only depend on the local atoms that are involved and which have effective charges that do not depend on molecular conformations.

Therefore, serious difficulties arise in the context of molecular-scale modeling and simulation when one seeks to explore new types of nanoscale devices and architectures that incorporate organic- and/or bio-structures with complex geometries and/or functionality. Specifically, very high quality (accuracy, fidelity, etc.) in physical modeling and simulation is required to achieve a basic understanding of the fundamental mechanisms and to successfully investigate specific bio-electronic applications that rely upon organic-/bio-based molecular structure and/or functionality [5]. Furthermore, since these investigations will inherently require the interfacing of traditional electronic materials or devices with novel organic- and/or bio-based systems, the progression to practical sensors and sensor systems will present computational problems of ever-increasing size and complexity. Hybrid approaches that combine different levels of theory such as QM and MM methodologies may help to bridge the gap in understanding electronic and atomic structure and light-induced interactions in complex bio-organic nano-structures. Therefore, identification of novel functionality derived from nanoscale-dictated electronic, photonic, and phonon processes requires simulation tools that execute a successful merger of the MM and first principle QM techniques. To this end, the feasibility of using hybrid theoretical approaches for a proper description of bio-organic nanoscale device structures needs to be explored.

Stilbene-functionalized bio-molecular structures have a potential to become future sensing elements in both single devices and integrated arrays. Previous studies have demonstrated that stilbene derivatives have potential uses in various applications and fields, in particular biological nanoscale sensors. For example, a chemically reactive blue-fluorescent optical sensor based on stilbene derivatives was demonstrated that exhibits strong fluorescence in specific protein environments [9]. Another example can be found in synthetic DNA structures where the fluorescence quenching from stilbene derivatives was observed in the presence of G bases and modified Z bases [10]. In Ref. [11], hybrid QM/MM modeling and simulation techniques were explored for their utility in describing complex covalently bonded bio-molecular structures such as a double-stranded (ds) DNA fragment capped with trimethoxystilbene carboxamide (TMS), which is a well-known stilbene derivative. Also, various single-stranded (ss) DNA fragments capped with TMS were studied. This broad study was conducted because both ss- and ds-based structures play important roles in bio-sensing. Since the stilbene-functionalized DNA molecules are viewed as prime candidates for integration into new types of nanoscale sensing architectures, the dependence of spectral absorption (i.e., in the THz, infrared [IR], and ultraviolet [UV] regimes) on conformation and DNA sequence types were investigated to provide information on the potential use of active modalities (i.e., light-induced transitions) in detection and identification applications. In Ref. [11], for each molecular system defined by a chosen DNA chain bonded to TMS, all-atom geometric optimizations without constraints were performed to determine the final conformations, and normal mode analyses were applied to derive the spectral absorption information. Time-dependent (TD) Hartree–Fock (TDHF) and density functional theory (TDDFT) were applied to derive electronic states and light-induced transitions for studied hybrid systems. For TMS-ss-DNA, hybrid semiempirical QM/first principle simulations were performed using the very well-known AM1 [12] method (which determines various molecular energy terms from semiempirical expressions with parameters specifically tailored for the atomic components of DNA—i.e., H, C, O, P, and N) to describe the DNA part of the molecule and the HF approximation to model the TMS section of the molecule bonded to one of the nucleotides. The simulation studies of Ref. [11] provided a detailed insight into the electronic structure and dynamics of the combined DNA chain and TMS system and the resulting electronic transitions and optically active absorption modes within the THz and far-IR regimes. These studies demonstrated that it is possible to identify THz spectral signatures of specific DNA chains capped with TMS. This work also suggested that light-induced transitions could be a useful tool for the identification of the DNA sequence type. Furthermore, since the predicted two conformations of the TGCGCA duplex capped with TMS agree well with nuclear magnetic resonance (NMR*) data [13], this research demonstrated that hybrid QM/MM methodologies can be a valuable theoretical tool for the study and design of novel sensor functionality based upon electronic, photonic, and phonon processes in bio-organic structures.

2.4.2 MOLECULAR-MECHANICAL STUDIES

Though MM modeling techniques have their obvious limitations (e.g., not able to provide excited-state information), they are very useful for estimating the basic molecular conformation and for providing approximate insights into the static and dynamic phenomenology associated with both small and large molecular systems. Indeed, molecular mechanics is widely used in simulations of multi-atom/molecular

---

* **NMR** is an effect whereby magnetic nuclei (with non-zero spin) in a magnetic field absorb and re-emit *electromagnetic* (EM) energy. This energy is at a specific *resonance* frequency which depends on the strength of the magnetic field and atom surroundings.

systems that sometimes include hundreds of thousands of atoms. At the same time, it is often a prerequisite for generating input parameters needed in hybrid QM/MM simulations. MM simulations rely on MM force fields that effectively describe covalent/van-der-Waals (VDW)/Coulomb interactions that are known to occur between individual-bonded and non-bonded atom pairs. Here, specific force characteristics are assigned to the atoms that depend on the expected types of chemical bonds. Individual force types such as long-range electrostatic, VDW forces, covalent bond stretch, bond angles, and torsions are possible contributors to the overall effective force field. The force field models offered by the Amber simulation package are proven to perform well in describing biological systems that include nucleic acids and proteins [14–18]. Here, it is important to note that the accurate specification of the effective MM force is highly dependent on the overall molecular system. For example, the use of Amber to model general organic molecules (i.e., other than carbohydrates) requires the user to specify additional force field parameters to obtain accurate results. Also excluded from the Amber library are many "non-standard" nucleic acid structures, such as nucleotides with the phosphate group protonated. Despite these general limitations, the Amber force field model is highly applicable to the TMS-DNA molecules under discussion here, and Amber-generated results established an important physical foundation for the hybrid QM/MM studies of the next section [11].

2.4.3 Hybrid QM/MM Studies

Results from prior research [11] have clearly illustrated the strong influence of the TMS molecule on the resulting conformations, spectral absorption, and stability behavior of compound TMS-DNA type structures. Indeed, the previous MM-based studies demonstrated that when stilbene derivatives are bound to ss-DNA or ds-DNA that there are identifiable structural characteristics of the stilbene derivatives that dictate many properties of the entire molecule. Hence, when one views these molecular mechanisms from a classical point of view where certain semi-isolated components (or segments) of the compound structure are known to determine the overall properties, it naturally defines a hybridization of the physical modeling problem. More specifically, this type of situation suggests that when one is interested in predicting the conformation or dynamics of the compound molecule that it should be physically acceptable to invoke a structural partitioning of the problem. In particular, one can apply a hybrid QM/MM modeling approach where the "passive" part (i.e., primarily receiving the influence) is characterized by MM techniques that are incapable of predicting electronic effects, and the "active" part (i.e., primarily dictating the response) is treated by a first principle QM-based approach. Such hybrid modeling approach allows one to study both the electronic structure and atomic motions of TMS-DNA conjugates.

A very simple example of a hybrid partitioning for TMS that is relevant to the problem under discussion is shown in Figure 2.3. Here, consider the task of modeling two conformations of the TMS molecule where the only change is produced by a 180° flip (see Figure 2.3a) of the top carbon ring relative to the bottom carbon ring. More specifically, in this example a double-bond originally pointing to the left

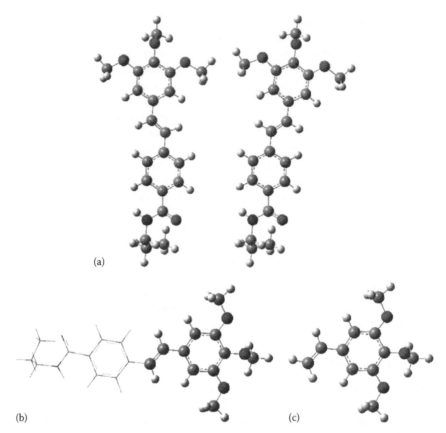

FIGURE 2.3 An example partitioning of the TMS molecule. (a) Two conformations of TMS that differ only by a 180° flip (i.e., compare left conformation to the right conformation) of the upper carbon ring relative to lower carbon ring in the chain. (b) A partitioning about the rotation bond in (a) where the upper portion (now shown on right) is treated primarily by a QM approach, and the lower portion (now shown on left) is treated primarily by an MM technique. (c) QM molecular partition (i.e., upper part from (a) and right part from (b)) where the interface bond has been saturated by hydrogen to enable both MM and QM simulations on this portion of the TMS molecule.

(see left-side of Figure 2.3a) flips, or rotates about the next lower bond, such that it ends up pointing to the right (see right side of Figure 2.3a). In this type of situation, it may be useful to execute a partitioning (see Figure 2.3b) where the portion of the molecule above the rotating bond (i.e., now shown on the right side of Figure 2.3b) is to be treated primarily by a QM approach, and the portion of the molecule below the rotating bond (i.e., now shown on the left side of Figure 2.3b) is to be treated primarily by an MM technique. Note that the best choices for the QM and MM components will depend on which part of the molecule is primarily responsible for influencing the conformational change, and how to make this choice (or guess) will become more apparent when the entire TMS-DNA problem is discussed in the following. Also, since the hybridized simulation method will require an independent assessment for how the QM component (i.e., represented by spheres in Figure 2.3b) contributes to the total molecular potential energy (see the following discussion), it will be necessary to calculate both the MM and QM solutions for this molecular partition where the broken bond on the left end is saturated by a hydrogen atom (see Figure 2.3c).

In the hybrid QM/MM computer simulations that are utilized for these studies, the active site (i.e., the TMS molecule combined with some number of bases) which represents a small portion of the entire interacting molecule region is considered on a QM level of theory using HF or DFT method. The excited states are obtained using TD HF or DFT. The remainder of the system (i.e., a fragment of ss- or ds-DNA) is described using MM force fields or semiempirical quantum mechanics. In studying these TMS-DNA conjugate structures, a partitioning method (i.e., called ONIOM) is utilized that followed the procedure described in [18]. A Gaussian-based approach is then used to calculate the optimized geometries and associated energies. This procedure involves calculating the energy of the combined structure at a lower level of theory and modeling of the active site using both higher and lower levels of theory. Here, the difference of the two active site calculations is combined with the results from the entire structure to determine a consistent overall energy balance (i.e., MM result for the active site is subtracted from the MM result for the combined structure and then added to the QM result for the active site). Hence, the recognized active TMS component is characterized by the more accurate QM theory and the entire structure (which includes the passive part) is first treated by the more computationally efficient MM theory and then double-counting of the active part is removed by subtracting the results from the MM treatment of the TMS component. Note that a partitioning across covalent bonds is required, so link-atoms must be used to saturate the bonds that were broken to define the smaller active site molecule (e.g., analogous to what was done in Figure 2.3c). Hydrogen was used as a link atom in all simulations. In the studies presented here, it is reasonable to assume that only valence electrons participate in the forming of all chemical bonds. The Stevens/Basch/Krauss ECP split-valence basis [19,20] was used in both the HF and DFT simulations. In addition, to account for electron–electron interactions a long-range and VDW-corrected hybrid DFT functional was applied [21,22].

Hybrid DFT/MM studies on the symmetric $(TMS-TGCGCA)_2$, where each identical DNA strand is bonded to TMS, may be used to demonstrate that at least two stable conformations should exist (see Figure 2.4). The initial simulations utilized a partitioning where the active site includes only TMS and the first base (i.e., the active site molecule is TMS-T) because TMS has a structure similar to the molecule discussed in Figure 2.3 and therefore can be expected to exhibit a similar ring flip that should be somewhat independent of the base pair. The common features of the two conformations include the proximity between the two $-OCH_3$ groups (methoxy groups) of the stilbene derivative and the $-CH_2-$ group (methylene group) of the deoxynucleoside of the complementary DNA chain (see Figure 2.2a and b). As was expected, the two predicted conformations are distinguished primarily by 180° TMS ring-structure flips (see Figure 2.4c and d). Here, the conformation with the left-handed ring flip (i.e., Figure 2.4c) has a slightly lower energy. Previously, this TGCGCA duplex capped with TMS was fabricated and studied experimentally using NMR [13]. The authors concluded that the two sets of signals they observed

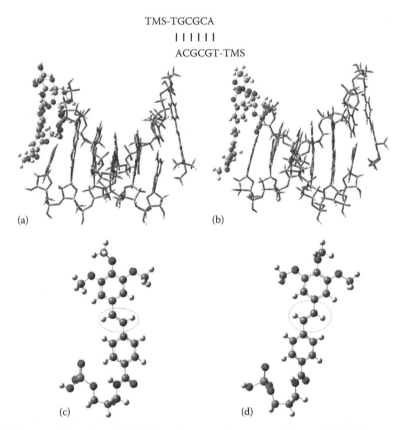

TMS-TGCGCA

| | | | | |

ACGCGT-TMS

FIGURE 2.4   (Top) Two DFT/MM-predicted conformations of the ds-TMS-DNA structure (TMS-TGCGCA)$_2$. (a) Conformation I and (b) conformation II. For each conformation, color-coded spheres represent the portion of a structure simulated with the DFT model and color-coded tubes represent the portion of a structure described only by the MM model. (Bottom) Isolated views of the TMS components from the two conformations of (TMS-TGCGCA)$_2$ given in (a) and (b). Here the two TMS substructures differ by a 180° flip of the end carbon ring (i.e., the upper ring in the illustration) relative to the next carbon ring in the chain (i.e., the lower ring in the illustration) where (c) illustrates the ring flip to the left associated with the molecule from (a), and (d) illustrates the ring flip to the right associated with the molecule from (b).

in ^{1}H NMR spectrum of the structure result from the two different orientations of the stilbene relative to the neighboring nucleobases. Structural details deduced by the authors of Ref. [13] included a 180° flip of the stilbene ring system between the major and minor conformations. Therefore, the hybrid DFT/MM modeling results presented in [11] agree well with NMR-based deductions. The switching between the two conformations was explored by freezing the torsion that is responsible for the ring flip and then by optimizing the structure over the rest of internal coordinates. A transient structure that corresponds to a 90° (approximately halfway) turn in the torsion (within QM region) is plotted in Figure 2.5a. Notice that the proximity

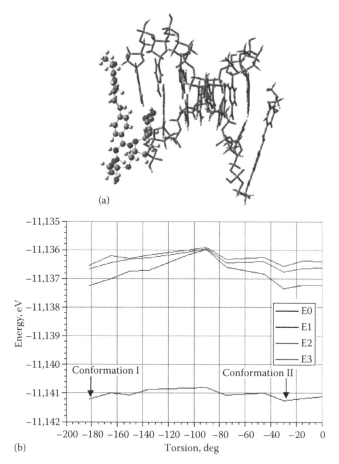

(a)

(b)

FIGURE 2.5   (a) Transient DFT/MM-predicted structure that corresponds to a 90° (approximately halfway) turn in the torsion angle (within QM region) that is responsible for switching between conformations I and II of Figure 2.2. The proximity between tail oxygen groups of TMS and the second (non-bonded to TMS) DNA strand is retained. Also, the T base of the first DNA strand is pushed back to allow for the transition. (b) Optimized energy versus torsion (within QM region) for TDDFT/MM-predicted structures of Figure 2.2. Minima corresponding to conformations I and II are highlighted with black arrows. Ground level and first three excited levels are plotted for the QM part of the system.

between the tail methoxy groups of TMS and the complementary DNA strand is retained, and the T base of the first DNA strand is pushed back to allow for the transition. The optimized energy versus torsion (within the QM region) for TDDFT/MM-predicted structures of Figure 2.4 is plotted in Figure 2.5b. The minima corresponding to conformations I and II are highlighted with black arrows. Only the ground energy level and first three excited states are plotted for the QM part of the system. The predicted barrier for flipping is low, so the transitions can be activated by thermal effects. In Figure 2.6, the predicted vibrational absorption spectra of the electronic ground state of the symmetric (TMS-TGCGCA)$_2$ structure are plotted for

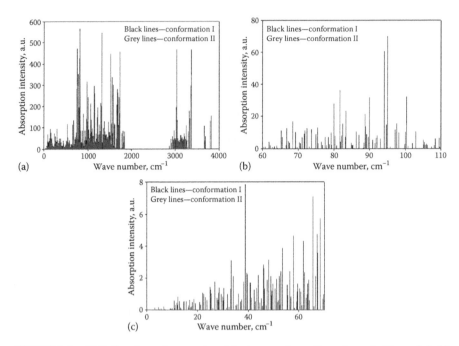

FIGURE 2.6   (a) Entire absorption spectra of the two conformations shown in Figure 2.4. (b) Spectra for frequency range of 2.0–3.5 THz. (c) Spectra for frequency range below 2.1 THz.

the two conformations. The entire spectrum as given in Figure 2.6a demonstrates many similarities for both conformations. However, when one carefully considers subregions of the THz regime, distinctly unique spectral lines can be observed near ~67, 82, 90, and 100 cm^{-1} (see Figure 2.6b and c) and also near 280 cm^{-1} as discussed in Ref. [11]. These absorption features can be attributed to vibrations in torsions and bond angles. Finally, spectral signatures that are unique to one of the two conformations can be identified near 10, 12, 27, 33, 39, 54, 58, 65, and 68 cm^{-1} (see Figure 2.6b) with all of these residing in the low THz (collective motion) regime. In addition, the results for the six lowest electronic transitions (i.e., with state labeling 0, 1, ..., 6) that are associated with the two conformations (I, II) along with the corresponding oscillator strengths were given in Ref. [11]. The predicted 0–6 transition is 0.13 eV higher and substantially stronger for conformation II as compared to conformation I. The predicted 0–1 transition is also 19% stronger for II type conformation. Hence, both the spectral results and transition characteristics show a potential for discriminating between the two conformations. Both conformations depicted in Figure 2.4 correspond to trans TMS. However, a *cis–trans* conformation of $(TMS-TGCGCA)_2$ was also predicted by a DFT/MM analysis. For this case one TMS is in *cis* conformation (i.e., top TMS molecule perpendicular to DNA strand) and another one is in *trans* (i.e., bottom TMS molecule is aligned to DNA strand). There also exists an all-*cis* conformation for this molecule with a slightly lower energy. Additional studies were also performed to investigate single TMS capping of ds-DNA, that is, $TMS-(TGCGCA)_2$. Since no experimental data were available

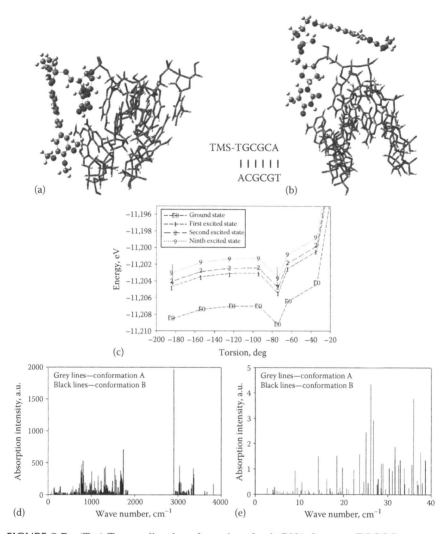

FIGURE 2.7 (Top) Two predicted conformations for ds-DNA fragment TGCGCA capped on one end with TMS obtained from hybrid HF/MM. (a) TMS chain is aligned along the end base-pair T–A and (b) TMS chain is aligned along the ds-DNA backbone. (Middle) Energy-scan profiles (from hybrid DFT/MM model) for the zeroth, first, second, and ninth states of the TMS-TGCGCA molecule versus torsion angle. (c) The two stable conformations are highlighted with arrows where (a) is on left and (b) is on right. (Bottom) The DFT/MM-predicted absorption spectra for each conformation is also shown for (d) the THz-to-IR regime and (e) THz regime.

to infer an approximate stable conformation as in the last case considered, initial simulations were performed using the less physically accurate but much more computationally efficient hybrid HF/MM model to obtain the two conformations given in Figure 2.7. Here, the active site for these simulations consists of the TMS molecule bound to a single T base, or TMS-T. The results in Figure 2.7a and b show that the major difference in the two conformations is related to the orientation of the TMS molecule with the ds-DNA, that is, the result in Figure 2.7a has TMS aligned with the end base-pair T–A and the result in Figure 2.7b has TMS aligned with the ds-DNA backbone. The accuracy of both results was confirmed by the DFT/MM model using the same active-site partition and an expanded active-site partition that included the first three bases adjacent to the binding site of the TMS (i.e., TMS-TGC). These results demonstrate conformations that maintain the same basic qualitative orientation between the TMS and ds-DNA molecules. Simulations that sampled a number of the available coordinates belonging to the TMS molecule revealed that changes to the torsion coordinate associated with the chemical bond between the TMS molecule and the DNA might be useful in defining an energy-space pathway between the two refined minimum energy conformations that possess the general geometries as illustrated in Figure 2.7a and b. The energy diagram in Figure 2.7c depicts this energy-space pathway for the case of the ground-state and a few excited-state conformations that were obtained from hybrid DFT-/MM-based simulations using an active-site partition of TMS-T. In particular, the energy profile for the zeroth, first, second, and ninth electronic states are given as a function of the torsion angle that connects TMS to the DNA. As is clearly shown, all the energy profiles along the single torsion coordinate variation between the two stable conformations contain a significant barrier. This means that changes in other coordinate spaces will need to be accessed in order to realize light-induced switching between the positions of the two stable conformations (i.e., around $-75°$ and $-185°$) which have been highlighted with arrows in Figure 2.7c. Note that determining energy-space pathways for even moderately complicated molecules can be a significant challenge, and active work by our group on developing computational methods for addressing this problem will be discussed in Section 2.4. While a light-based switching methodology is lacking at this point, TMS-(TGCGCA)$_2$ is an attractive BMS candidate because the spectral absorption results for these two stable conformations for the IR-to-THz regime (Figure 2.7d) and for the THz regime (Figure 2.7e) both show substantial differences in their individual spectral fingerprints [11]. Note that line broadening at THz frequencies is typically on the order of $0.5\,cm^{-1}$ [21]; therefore, most of these THz absorption lines can be detected. Hence, there appears to be a significant opportunity for utilizing the spectral absorption signatures associated with the multiple conformational states of stilbene-DNA conjugates to infer the base-pair sequences of DNA fragments.

These collective results demonstrate that first principle and DFT hybrid QM/MM methodologies can bridge a gap in understanding electronic and atomic structure and light-induced interactions in complex bio-organic systems. In particular, these simulation techniques were used to successfully derive two distinct low-energy conformations of (TMS-TGCGCA)$_2$ that differ in the orientation of the stilbene ring system relative to the terminal base pair, and which are in agreement with experimental NMR data. In addition, the approach was able to predict the all-*cis* and *cis–trans* conformations, along with two distinct conformations for the asymmetric structure with only one TMS bonded to the TGCGCA duplex, that is, TMS-(TGCGCA)$_2$. Both base-pair composition and conformation differences are predicted to impact vibrational absorption spectra especially in the THz regime and to manifest themselves in conformation-dependent and sequence-dependent shifts in electronic transitions.

Hence, this work suggests that stilbene derivatives may be useful for synthesizing DNA conjugates that facilitate the spectral characterization of the associated base-pair sequences. More specifically, this research suggests that stilbene-DNA conjugates offer the type of switchable spectral characteristics that will be useful for THz-based and possibly IR-based detection and identification purposes. These facts are particularly important when one considers the development of advanced biologically based architectures that use light-induced interactions and the resulting THz-frequency spectral signatures to extract information about the molecular structure and dynamics as discussed in [5]. The next section will discuss how these same modeling techniques can be used to facilitate the independent experimental analysis of component molecules (e.g., biotin) that will be required for the successful construction of novel biological architectures of the type discussed in Section 2.3. Section 2.6 will then conclude with a discussion of novel algorithm development for deriving light-induced energy-space pathways.

2.5 CHARACTERIZATION OF ACTIVE MOLECULES IN POLYETHYLENE MATRICES

As discussed earlier, the self-assembly of novel biological architectures for use in long-wavelength spectral sensing application will require the incorporation of active molecules into nanoscaffolds (e.g., the DNA cassettes shown in Figure 2.2) which will in turn be used to create the larger molecular matrices and supramolecular assemblies. These installation procedures for incorporating the functional polymeric components (e.g., stilbene-DNA BMSs) will require the use of other linker molecules (e.g., biotin/streptavidin) to facilitate the molecular construction. Hence, it will be very important to understand the dynamics of these individual molecular components and their subsequent influence on the larger nanostructures [3]. Biotin is an example of an important active molecule that is expected to play a key role in the molecular assembly of DNA-based nanostructures. The characterization of biotin is especially important since it is known to be optically active that therefore could strongly influence the functionality of the molecular switches used in DNA-based nanostructures, for example, DNA-Origami. Therefore, it is very important to understand the fundamental dynamics of isolated biotin, as well as molecular compounds that contain biotin. Fortunately, materials such as polyethylene (PE) form a non-bonded (hydrogen-bonded) material matrix that is transparent to long-wavelength radiation and as such they are an important experimental tool for understanding the molecular dynamics of quasi-isolated molecules that are embedded into these matrices if one is able to accurately predict the interactions between the embedded molecules and the PE superstructure. Therefore, it will be very useful to explore the impact of the PE matrix embedment on the THz spectra of these types of active biological molecules.

In a very recent work, PE was chosen as a matrix for the investigation of active biological or organic molecules. This material is weakly absorbing at THz frequencies and is widely used in spectroscopy [23]. It consists of (C_2H_4) polymer chains that are arranged in a crystal lattice [24]. The quality of PE depends on processing. This study focuses on high-density polyethylene (HDPE) that has little branching. In the case of branching, main chains have minor chains bonded to them in places where a normal CH_2 group is replaced by a CH, thus creating one extra valence. Therefore, the HDPE structure is close to an ideal crystal lattice of the PE. Several atomic structures of PE slabs suitable for incorporating in hybrid ab-initio/MM structural models were simulated. In all structural models, orthorhombic PE lattice symmetry was used to generate a starting atomic arrangement. PE II was modeled which means that all polymer chains were terminated with methyl end groups ($-CH_3$ or –Me). Also, no lattice defects were taken into account with the exception of defects that were related to an active molecule embedding. Previously, the vibrations and light absorption in PE lattice were obtained theoretically using specially developed MM force fields [23,25]. In this work, all-purpose force fields such as universal force field (UFF) and Amber's force field were employed in the MM portions of simulations. The THz-to-IR absorption spectra of PE that were obtained with DFT/MM approach for the structural models of PE are in a good agreement with observed spectral lines that were reported in [23].

Biotin is a flexible biologically active molecule with a number of distinct low-THz spectral features [26]. It is a vitamin coenzyme and it functions as a bio-reactions catalyst, and the bio-activity of biotin is utilized for bio-chemical sensing [27]. HF ab-initio and DFT simulations of biotin were performed both in harmonic approximation and with corrections for anharmonicity. Specifically, third- and fourth-order anharmonicity corrections in atomic displacements to harmonic vibrational states were calculated numerically for biotin within the first principle theoretical framework by the perturbation theory [28]. In Figure 2.8a, results for integrated line intensities that were obtained with HF/6-311G(2d,d,p) are plotted for low-THz together with observed low-temperature (4.2 K) absorption peak positions from Ref. [26]. Predicted lines in the harmonic approximation (solid lines in Figure 2.8a) are in a good agreement with observed ones, in particular, a strongest line is close to $43 \, cm^{-1}$ (1.3 THz), second strongest is at $\sim 94 \, cm^{-1}$ (2.8 THz) followed by lines near $33 \, cm^{-1}$ (1 THz) and near $20 \, cm^{-1}$ (0.53 THz). But not all observed lines can be explained, since there are more of them (see black dots in Figure 2.8a), than the theory predicts. The inclusion of anharmonicity corrections (dashed lines in Figure 2.8a) does not introduce significant changes in line positions; as with or without anharmonicity, more lines are observed than predicted. To explore this issue further, first principle molecular dynamical (MD) simulations of biotin were performed using the atom centered density matrix propagation (ADMP) molecular dynamics model that provides good energy conservation [29]. The Stevens/Basch/Krauss ECP split-valence basis [21] was used in these MD simulations. In addition, to account for electron–electron interactions, a long-range and VDW-corrected hybrid DFT functional was applied [22]. In the ADMP approach, the basis functions may be considered as "traveling" along with the classical nuclear coordinates. As a result of the ADMP simulations, the atomic trajectories, velocities, and a low-temperature (~ 4 K) dipole moment time history of biotin were obtained for about 15 ps with a 1 fs time step. At each step, the dipole moment was computed quantum-mechanically using the wave functions that depend on atomic positions, so both the electrical and mechanical anharmonicities were taken into account. The energy was well conserved during the course of simulations with the relative energy change on the order of 2×10^{-7}.

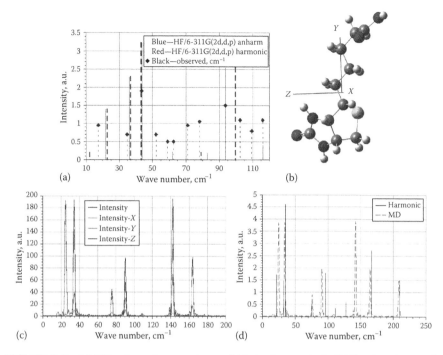

FIGURE 2.8 (a) THz absorption spectrum of biotin: harmonic (solid lines); anharmonic (dashed lines); and experimental observations at 4.2 K reported in Ref. [23] (black dots). Theory used HF/6-311G(2d,d,p) [25]. (b) A snapshot of biotin from the MD simulations. (c) Low-THz absorption spectra of biotin obtained from the DFT-based MD simulation. The entire spectrum (black line) and contributions from Cartesian components of dipole moment (lighter grey to darker grey lines, respectively). (d) Low-THz absorption spectra of biotin obtained from the DFT-based MD simulation (dashed line) and from the DFT-based energy minimization in harmonic approximation with the same density functional and basis (black lines).

The absorption spectrum was computed within the framework of linear response theory using the relationship between the equilibrium dipole moment fluctuations and the light absorption cross section [30]. In this approach, the total absorption intensity is proportional to a power spectrum that is represented by a sum of squared Fourier transforms of Cartesian components of dipole moment history. In Figure 2.8b, the simulated biotin molecule is shown with Cartesian axes. In Figure 2.8c, the total intensity (black lines) is plotted along with its Cartesian contributions (green, red, and blue) for the lower THz portion of the spectrum (0–6 THz or 0–200 cm^{-1}). Notice that the most significant contributions come from the Y-component of the dipole moment (red) for major lines around 30–35 and 140 cm^{-1}, which corresponds to a direction along the carbon chain (see Figure 2.8b). In Figure 2.8d, the low THz spectrum of total intensity (red dash) from Figure 2.8c is plotted against the absorption lines (black) that were calculated for the optimized biotin structure in harmonic approximation using the same DFT potential and basis as in the ADMP approach. This comparison clearly demonstrates that the harmonic spectrum yields an overall reasonable THz fingerprint of biotin. Most major lines are well reproduced with a notable exception of the ADMP result at 143 cm^{-1}. Here, harmonic approximation yields a line that is shifted with respect to the ADMP result to lower frequencies at 128 cm^{-1}. Therefore, the ADMP simulation confirms the harmonic result and also predicts that there are only five absorption lines below 150 cm^{-1}. Notice that the measured biotin was embedded in HDPE [26]. In order to study theoretically the effect of embedding, the QM/MM approach was used. Specifically, *first principle* QM methods and DFT were combined with MM empirical descriptions in order to perform large-scale atomistic spectral modeling of biotin in PE in a broad frequency range from THz to UV. The most important part of the system (i.e., biotin in a cavity) is described using first principle QM or DFT. In this case, PE strands were either in an MM-only region or in a QM region without the introduction of any link atoms between MM and QM parts of the system [28]. Using this approach, the energy was calculated and geometry optimizations were performed for developed structural models to minimize the energy. Then atomic vibrations and absorption spectra were obtained in harmonic approximation. Excited electronic states were studied using TD DFT. The two-step optimization process consisted of an MM step that optimized biotin's placement inside a PE matrix and a hybrid optimization step. In the second step, partitioning of a structure in QM and MM-only regions (layers) was performed and equilibrium coordinates for the structure were obtained. Atomic vibrations were then analyzed in the harmonic approximation for the equilibrium structure and line intensities were calculated. In Figure 2.9, an example of a studied structure consisting of a biotin molecule (spheres) embedded in a PE matrix (spheres and tubes) is shown. Here, tubes represent an MM-only region. In Figure 2.9a, the QM region is extended to include surrounding PE strands. In this particular case, the MM region was immobilized so that only ab-initio atoms were allowed to move during a geometry optimization. In Figure 2.9b, a model with biotin-only QM region is shown. In Figure 2.9c, an optimization result for PE with cavity without a biotin molecule is presented.

The resulting low-THz absorption spectrum for the entire structure is shown in Figure 2.9d. Here, dark grey lines represent the theory, b3lyp/ UFF, for model of Figure 2.9a, light grey lines represent the model of Figure 2.9b, and black dots represent centers of observed lines at 4.2 K as reported in [26]. The Stevens/Basch/Krauss ECP split-valence basis was used in this simulation [21]. Additional low-THz lines involve atomic vibrations in biotin and PE both and reflect THz coupling between biotin and its PE cavity. While line widths were not calculated, the theoretical spectrum of biotin in HDPE obtained from the extended QM model agreed well with those observed experimentally at 4.2 K within 15–120 cm^{-1} range.

In Figure 2.10a, predicted QM/MM THz absorption spectra below 7 THz are plotted for the case of biotin in PE that is associated with the model of Figure 2.9a (see dotted lines) and for the case of PE with a cavity (i.e., no biotin) that is associated with the model of Figure 2.9c (see black lines). Here, one can observe a major contribution of biotin to the overall THz absorption. In Figure 2.10b, the predicted broad spectrum THz-to-IR resonance absorption of biotin in PE (dotted lines) and an isolated biotin molecule (dashed lines) is plotted. Notice that major lines in the spectra are almost the same with the exception for C–H stretch modes of PE (see black arrow) that are visible around 3000 cm^{-1}. Table 2.1 presents the predictions

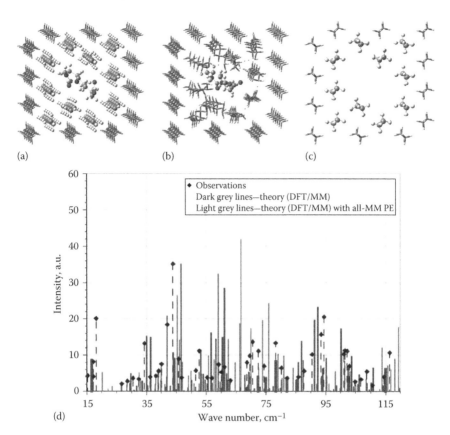

(a) (b) (c)

(d)

FIGURE 2.9 Simulation results for biotin in a cavity inside a PE matrix test cases include (a) extended QM region, (b) biotin-only QM region, and (c) PE with cavity. Spheres represent atoms in QM region and tubes represent atoms in MM region. (d) Low-THz absorption spectrum of the biotin-PE structures. Theory for model of (a) (dark grey lines), theory for model of (b) (light grey lines); and experimental observations at 4.2 K reported in [23] (black dots).

for the lowest-five non-forbidden ($i = 1$, 2, 3, 4, and 7) electronic excited states (and transition intensities) for biotin and biotin in PE that correspond to the models of Figure 2.9a and b, to biotin taken from the conformation of model of Figure 2.9a, and to an isolated biotin molecule. Predicted changes in transition energies suggest conformational differences between isolated biotin and biotin in PE. In particular, the presence of PE is predicted to cause a line shift to higher energy by 37 meV with simultaneous decrease in transition intensity by 42%. Other differences in the transition energies suggest a significant effect of PE cavity on the spectrum of biotin.

The molecule azobenzene, as well as its derivatives, is also of interest because its molecular properties (e.g., conformation and spectral absorption) are known to be dependent on excitation by light. More specifically, it can be switched by light between *trans* and *cis* forms [31]. Hence, hybrid QM/MM simulations were performed to study azobenzene embedded in PE matrices. In Figure 2.11a, an optimization result

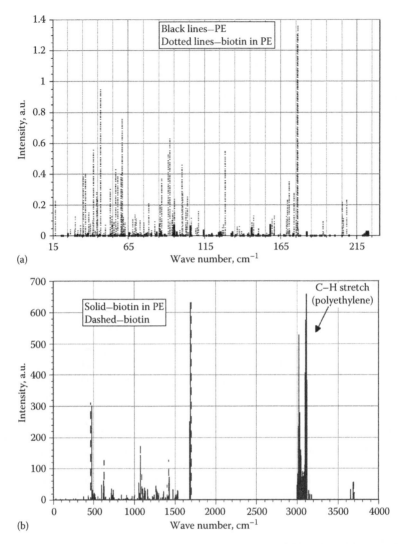

FIGURE 2.10 (a) Predicted THz absorption spectra below 7 THz. Biotin in PE as shown in Figure 2.9a (solid lines) and PE with a cavity as shown in Figure 2.9c (black lines). (b) Predicted resonance absorption spectra of biotin in PE (solid lines) and single molecule biotin (dashed lines). The spectral features are very similar with the exception for the PE C–H stretch modes occurring near $3000 \, cm^{-1}$.

for *trans*-azobenzene in a PE matrix is shown. In Figure 2.11b, a detail of the simulated structure with azobenzene is plotted. No substantial strand rearrangement to fit *trans*-azobenzene in PE is required. The deformation region is localized within 10–12 A around azobenzene (in the center of Figure 2.11a). However, azobenzene is constrained inside PE. On the other hand, substantial strand rearrangement is required to fit *cis*-azobenzene in PE according to the preliminary QM/MM results. This could affect the relaxation mechanism in excited *trans*-azobenzene molecules

TABLE 2.1

Predicted Electronic Excited States (and Transition Intensities) for Biotin and Biotin in PE

E (i)	Model of Figure 2.9a eV (cm⁻¹)	Model of Figure 2.9b eV (cm⁻¹)	Biotin Only from Figure 2.9a eV (cm⁻¹)	Isolated Biotin eV (cm⁻¹)
E(1)	4.9646 (0)	5.1343 (0)	5.0558 (0)	5.1120 (0)
E(2)	5.1145 (0.0004)	5.2460 (0.0001)	5.1282 (0)	5.1377 (0)
E(3)	5.3657 (0.0008)	5.3918 (0.0005)	5.3575 (0.0005)	5.3867 (0.0005)
E(4)	5.6819 (0)	5.9151 (0.0005)	5.7490 (0.0001)	5.7200 (0.0003)
E(7)	6.1983 (0.0165)	6.2480 (0.0268)	6.2158 (0.0240)	6.1616 (0.0234)

and therefore have an impact on their *trans–cis* photo-switching and sensing properties. These types of studies on azobenzene, and their conjugates with DNA, will be continued in the future in order to define novel BMSs that will be useful for functionalizing DNA-derivative architectures.

ab-initio, DFT, and hybrid QM/MM methodologies have been shown here to be effective tools for assessing the impact of polymer matrices on the THz spectra of biological molecules such as biotin. While PE is almost transparent at THz, additional low-THz lines are predicted in the biotin/PE system, which reflects a dynamic interaction between biotin and a surrounding PE cavity. The obtained results agree well with the available experimental data. Also, in accordance with the preliminary QM/MM results, *trans*-azobenzene can be trapped in a PE matrix which could alter its light switching and sensing characteristics. *Trans–cis* photoisomerization of azobenzene in PE matrices is a subject of an ongoing research.

2.6   LIGHT-INDUCED TRANSITIONS IN MULTIDIMENSIONAL MOLECULAR COORDINATE SPACES

Once DNA-based BMSs are identified that exhibit multiple stable conformation states with differing THz and/or far-IR spectral signatures, it will be necessary to determine energy-space pathways that can be used to make transitions between these conformational states by the application of light-based excitations. While a number of relatively small molecules are known to exhibit light-induced conformational transformations with simply defined energy-space pathways (e.g., the vision process in mammals is initiated by light-induced isomerizations of 11-*cis* retinal to all-*trans* retinal [32]), when one considers even moderately complex molecules, the desired transformation trajectory can become dependent on multiple molecular coordinates and/or it may involve multiple excited-state transitions. In such cases, the computational burden for determining these energy-space trajectories and the required light excitation(s) can become excessive, or even practically intractable.

The basic issues associated with these types of energy-space transition problems are nicely illustrated by 2-butene, which is known to undergo light-induced transformations between *trans*-to-*cis* conformations [6]. Figure 2.12a illustrates the

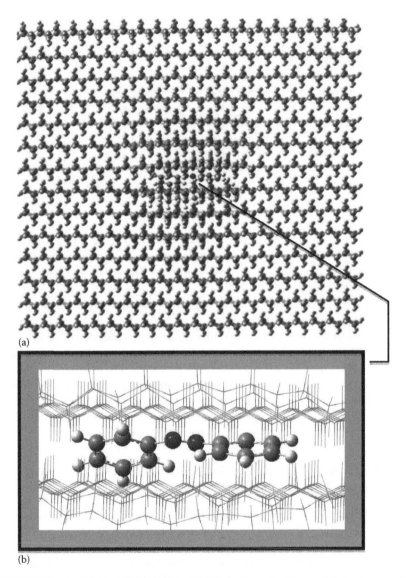

FIGURE 2.11   Azobenzene in PE matrix. (a) Illustration of entire model structure, and (b) detail showing the hybrid QM/MM model for the azobenzene (spheres) and PE (chains). Note that for these simulations, azobenzene molecule was treated QM and all PE was MM.

trans-2-butene conformation and the *cis*-2-butene conformation where the transition between the two has been defined as a rotation about the single molecular coordinate labeled as *D*5 (i.e., torsion rotation about the carbon-to-carbon double-bond). It is useful to note that the *trans* conformation refers to the fact that the two CH_3 functional groups reside on opposite sides of the double-bonded carbon atoms, whereas in the *cis*-conformation the two CH_3 functional groups reside on the same side of the double-bonded carbon atoms. While 2-butene is amenable to multiple coordinate

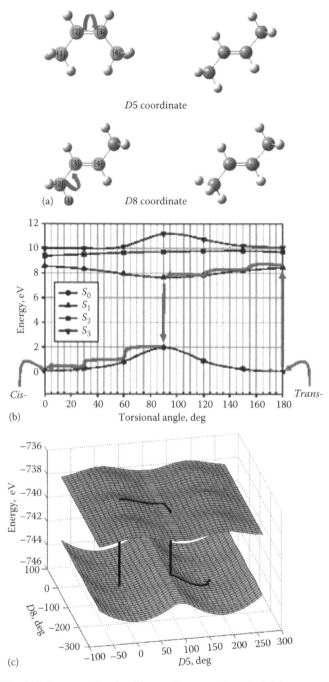

FIGURE 2.12 (a) 2-Butene molecule with coordinate rotations for $D5$ (upper) and $D8$ (lower) labeled. (b) One-dimensional analysis (i.e. $D5$ coordinate only) that successfully predicts the *trans*-to-*cis* transition of 2-butene. (c) Actual 2D transition path associated with conformational transformation defined by (b) that shows transversals (back and forth) in $D8$ space.

transformations when excited by light (e.g., $D5$ and $D8$ as shown in Figure 2.12a) it is possible to describe the *trans*-to-*cis* transformation most simply in terms of the $D5$ coordinate as shown in Figure 2.12b. As is illustrated symbolically by the green-line, a light-induced transition from the ground state (S_0) to the first excited state (S_1) modifies the charge distribution of the *trans*-2-butene molecule such that the lowest-energy of the first excited state lies directly above the potential energy barrier in the ground-state. Hence, the molecule will first undergo a non-radiative conformation change ($D5$ changes from 180° to 90°) that is followed by a radiative decay (i.e., from S_1 to S_0) where 50% of the excited molecules are placed into a position where they can naturally evolve (i.e., $D5$ changes from 90° to 0°) into the *cis*-2-butene con-formation. While this single-coordinate-based physical picture provides an accurate light-based methodology for inducing the *trans*-to-*cis* conformation change, it may ignore dynamics that might be useful for identifying alternative coordinate-space pathways. Indeed, a more robust analysis [33] can be used to more accurately char-acterize this same transition in terms of two (i.e., $D5$ and $D8$) dynamical coordinates as illustrated in Figure 2.12c. As shown, when the steepest descent evolution is deter-mined in terms of multiple coordinates, the molecule is also shown to exhibit varia-tions in the $D8$ coordinate (which is also mirrored in the corresponding coordinate of opposing CH_3) where a complete cycle is executed (i.e., angular increase in $D8$ during the S_1 state evolution is matched exactly by angular decrease in $D8$ during the S_0 state evolution) such that the initial and final values of $D8$ end up being equal. While the $D8$ coordinate can be argued as irrelevant in this particular transforma-tion of 2-butene, it illustrates the basic point that multiple-coordinate energy-space analyses will be of critical importance for defining the required BMSs.

At this time active research is being conducted to develop new simulation tools that will be able to utilize the richness of the higher-dimensional coordinate-space transfor-mations in combination with multiple frequency- and/or time-domain light-pumping to access complicated switching-trajectory pathways between known (and unknown) stable states of the target BMSs. As this is a very challenging quantum chemistry problem, it is expected that this will require new simulation tools that combine hybrid QM/MM modeling with efficient optimization algorithms and proper interfaces for human-guided analysis of light-induced transitions occurring within complex mol-ecules. Both the computational burden and the potential opportunity for discovery can be illustrated by the results from recent exploratory studies on the relatively simple molecule stilbene, which has already been discussed as an important switch-inducing component for use in defining DNA-based BMSs. Note that these particular studies sought to begin at a known ground-state conformation of stilbene (i.e., *cis*-stilbene) and then to use permutation sets of applied light excitations and assumed radiative decays (i.e., an example might be a light-induced excitation from S_0 to S_2, followed by the application of downward decays when local minima are encountered at S_2 and S_1) in an attempt to find an alternative metastable state within the ground level.

Consider first the simulation of energy-space trajectories of the stilbene molecule where two coordinates were considered at a time, and steepest descent was used to perform the optimization over smoothly defined energy surfaces that were populated by fully parallel calls to the program Gaussian (see Ref. [34] for an example of this mathematical approach). Here, it is useful to note that the numerical-optimization trajectory calculations required negligible simulation time as compared to defining the energy surface. While these two-coordinate at-a-time numerical searches failed to identify any alternative light-induced molecular conformations for stilbene, they are useful for illustrating the significant computational burden (~1 h for each $\{S_0, S_2, S_1, S_0\}$ permutation) even when each point of the potential energy mapping was calculated on a separate processor. These significant computational requirements and the realization that multiple coordinated interactions are important for light-induced stilbene transformations subsequently motivated the development and application of a more efficient optimization algorithm for testing.

Consider now the simulation of the energy-space trajectories for the stilbene molecule using this new algorithm where small incremental patches were dynamically defined to significantly reduce the computational burden of defining the required energy surfaces, and that enabled optimization over much larger coordinate spaces. A highly innovative aspect of this algorithm was the strategic use of Smolyak interpolation that can be made exact for second-order systems using only 61 grid points per hypercube (see Ref. [35] for an example of this mathematical approach). This simulation methodology was able to execute numerical searches that considered the evolution of five molecular coordinates at a time, and when those coordinates were those defined as shown in Figure 2.13a and used with the excitation or decay series as defined in Figure 2.13b, a light-induced pathway to a *newly discovered* alternative conformation was determined in only 27 h. This is a very efficient simulation as compared to the prior algorithm, which would have used 20 weeks of simulation time to perform the required grid-space mapping. A study of the time evolution of the five molecular coordinates (see Figure 2.13c) reveals that the motions of the coordinate pairs (D_2, D_2') and (D_3, D_3') track each other exactly (which is why this labeling was chosen) and that the two CH_3 functional groups simultaneously undergo (i.e., in stereo) an identical circular-flapping type motion that occurs relative to the D_1 coordinate, during a period of time where D_1 first rotates by almost $180°$ and then returns to its initial positions.

The evolution of the stilbene molecule into this very subtle difference in conformation (see Figure 2.14a) is clearly enabled by the unique multi-coordinate motions (see Figure 2.14b). More specifically, the two CH_3 functional groups have assumed final positions where they are spread more laterally apart from each other by virtue of the spiraling motions of D_1, D_2 (D_2'), and D_3 (D_3') that occur most strongly when the molecule is decaying through the S_1 and S_0 energy spaces. It should be noted that the use of the incremental patch method prevents an immediate assessment of the energy barrier heights between this new metastable state and the natural ground state, so additional studies are now underway to determine the thermal stability of the metastable state. In any event, these results demonstrate the power and potential of this new optimization algorithm for use in searching for new energy-space pathways. Hence, work is now performed to develop a useful graphical interface that will facilitate human-guided simulations in the future, after which the simulation tool will be used to design and optimize stilbene-derivative/DNA complexes for use as BMSs in sensing architectures.

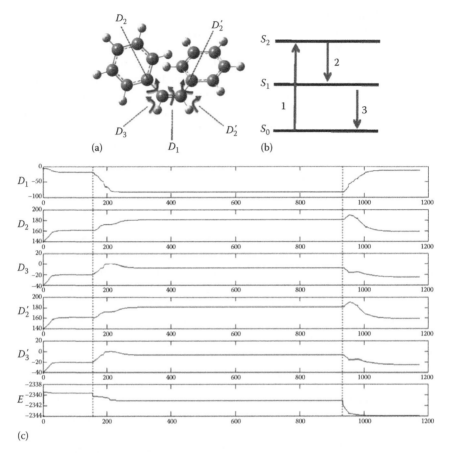

FIGURE 2.13 Five coordinate energy-space trajectory simulation information. (a) Illustration of the five coordinate angles D_1, D_2, D_2', D_3, and D_3'. (b) Illustration of the excitation/decay series $\{S_0, S_1, S_2\}$. (c) Time evolution of the five coordinate angles (in angular degrees) and molecular energy (in eV). Note that the vertical dotted lines divide the three $\{S_2, S_1, S_0\}$ regions.

2.7 CONCLUSIONS

An overview has been provided on the physics and modeling of DNA-based molecular components that have a potential for use in realizing DNA-derivative architectures with a capability for executing long-wavelength-based bio-sensing. More specifically, research work was presented that defines DNA-based BMSs that will be useful for incorporating into larger DNA-based nanoscaffolds for the purposes of defining a novel class of smart materials for THz and/or very far-IR-based biological sensing. Recent progress was discussed on the theoretical design and optimization of a novel class of DNA conjugate molecules that are amenable light-induced switching of their conformation states. These types of smart sensing

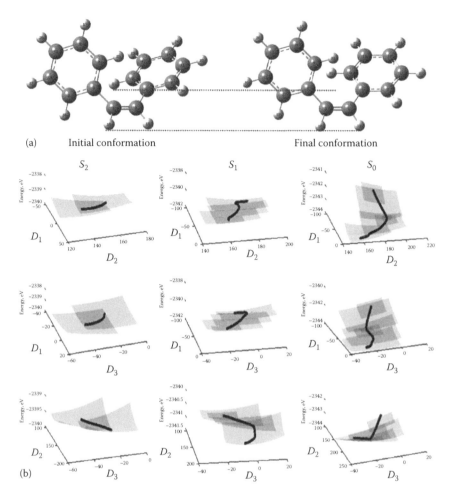

FIGURE 2.14 (a) Initial and final conformations of 2-butene from five coordinate energy-space trajectory simulation. Both views are oriented relative to the left CH_3 group and the horizontal dotted lines are included for perspective. (b) 2D trajectories (black lines) and energy-space patches (in grey) for pathways through each of the three energy-state spaces $\{S_2, S_1, S_0\}$. Note that a strong downward spiral within any 2D plot (and separation of the trajectory line from the patch) indicates a strong influence of one or more other coordinate dimensions.

materials are of interest because they may provide for the extraction of nanoscale information (e.g., composition, dynamics, and conformation) through electronic or photonic transformations to the macroscale. Therefore, this research has relevance to long-wavelength bio-sensing applications, and for use in the characterization and study of analogous biological systems such as antibody and receptor mimics, which could define new recognition detection methodologies and synthetic medical treatments.

ACKNOWLEDGMENT

The authors wish to recognize Professor Michael Norton, of Marshall University, and Professor Carl T. Kelley and Dr. David Mokrauer, of North Carolina State University for helpful discussions and information.

REFERENCES

1. D. Woolard, G. Recine, A. Bykhovski, and W. Zhang, Molecular-level engineering of THz/IR-sensitive materials for future biological sensing applications, *SPIE Proceedings*, **7763** (2010).
2. D. L. Woolard and J. O. Jensen, Functionalized DNA materials for sensing and medical applications, *SPIE Proceedings*, DSS11 SPIE Defense, Security, and Sensing (2011).
3. A. Bykhovski, P. Zhao, and D. Woolard, First principle study of the terahertz and far-infrared spectral signatures in DNA bonded to silicon nano dots, *IEEE Sensors*, **10** (3), 585–598 (2010).
4. P. Zhao and B. Woolard, Influence of base pair interaction on vibrational spectrum of a poly-dG molecule bonded to Si substrates, *IEEE Sensors Journal*, **8** (6), 998–1003, 2008.
5. D. L. Woolard, E. R. Brown, and M. Pepper, Terahertz frequency sensing and imaging: A time of reckoning future applications? *Proceedings of the IEEE*, **93** (10), 1722–1743 (2005); Y. Luo, B. Gelmont, and D. Woolard, Bio-molecular devices for terahertz frequency sensing, in *Theoretical and Computational Chemistry Volume 17, Molecular and Nano Electronics: Analysis, Design and Simulation*, J. M. Seminario (ed.), Elsevier, New York, pp. 55–81 (2007).
6. M. Rahman and M. L. Norton, Two-dimensional materials as substrates for the development of origami-based bionanosensors, *IEEE Transactions on Nanotechnology*, **9**, 539–542 (2010).
7. W. Shen, H. Zhong, D. Neff, and M. Norton, NTA directed protein nanopatterning on DNA origami nanoconstructs, *Journal of American Chemical Society*, **131**, 6660–6661 (2009).
8. W. Liu, H. Zhong, R. Wang, and N. Seeman, Crystalline two-dimensional DNA-origami arrays, *Angewandte Chemie*, **123** (1), 278–281 (2011).
9. A. Simeonov, M. Matsushita, E. A. Juban, E. H. Z. Thompson, T. Z. Hoffman, A. E. Beuscher IV, M. J. Taylor, P. Wirsching, R. Rettig, J. K. McCusker, R. C. Stevens, D. P. Millar, P. G. Schultz, R. A. Lerner, and K. D. Janda, Blue-fluorescent antibodies, *Science*, **290**, 307–313 (2000).
10. F. D. Lewis, J. Liu, X. Liu, X. Zuo, R. T. Hayes, and M. R. Wasielewski, Dynamics and energetics of hole trapping in DNA by 7-deazaguanine, *Angewandte Chemie International Edition*, **41** (6), 1026–1028 (2002).
11. A. Bykhovski and D. Woolard, Hybrid ab-initio/empirical modeling of the conformations and light-induced transitions in stilbene-derivatives bonded to DNA, *IEEE Transactions on Nanotechnology*, **9** (5), 565–574 (2010).
12. M. J. S. Dewar, E. G. Zoebisch, E. F. Healy, and J. J. P. Stewart, Development and use of quantum mechanical molecular models. 76. AM1: A new general purpose quantum mechanical molecular model, *Journal of American Chemical Society*, **107** (13), 3902–3909 (1985).
13. J. Tuma, R. Paulini, J. A. R. Stutz, and C. Richert, How much *pi*-stacking do DNA termini seek? Solution structure of a self-complementary DNA hexamer with trimethoxystilbenes capping the terminal base pairs, *Biochemistry*, **43**, 15680–15687 (2004).

14. D. A. Case, D. A. Pearlman, J. W. Caldwell, T. E. Cheatham, J. Wang, W. S. Ross, C. L. Simmerling, T.A. Darden, K.M. Merz, R.V. Stanton, A.L. Cheng, J.J. Vincent, M. Crowley, V. Tsui, H. Gohlke, R.J. Radmer, Y. Duan, J. Pitera, I. Massova, G.L. Seibel, U.C. Singh, P.K. Weiner, and P.A. Kollman, AMBER 8, University of California, San Francisco, CA, http://amber.scripps.edu/, p. 263, (2004).

15. W. D. Cornell, P. Cieplak, C. I. Bayly, I. R. Gould, K. M. Merz Jr., D. M. Ferguson, D. C. Spellmeyer, T. Fox, J. W. Caldwell, and P. A. Kollman, A second generation force field for the simulation of proteins and nucleic acids, *Journal of American Chemical Society*, **117**, 5179–5197 (1995).

16. C. I. Bayly, P. Cieplak, W. D. Cornell, and P. A. Kollman, A well-behaved electrostatic potential based method using charge restraints for determining atom-centered charges: The RESP model, *Journal of Physical Chemistry*, **97**, 10269–10280 (1993).

17. A. Bykhovski, T. Globus, T. Khromova, B. Gelmont, and D. Woolard, An analysis of the THz frequency signatures in the cellular components of biological agents, *International Journal of High Speed Electronics and Systems*, **17** (2), 225–237 (2007).

18. A. Bykhovski, T. Globus, T. Khromova, B. Gelmont, and D. Woolard, Resonant terahertz spectroscopy of bacterial thioredoxin in water: Simulation and experiment, in selected topics in electronics and systems, *World Scientific*, **48**, 367–375 (2008).

19. M. Svensson, S. Humbel, and K. Morokuma, Energetics using the single point IMOMO (integrated molecular orbital + molecular orbital) calculations: Choices of computational levels and model system, *Journal of Chemical Physics*, **105**, 3654–3661 (1996).

20. M. J. Frisch and A. Frisch, Gaussian 09 user's reference and IOps reference, http://www.gaussian.com (2009).

21. W. Stevens, H. Basch, and J. Krauss, Compact effective potentials and efficient shared-exponent basis sets for the first- and second-row atoms, *Journal of Chemical Physics*, **81**, 6026–6033 (1984).

22. J.-D. Chai and M. Head-Gordon, Long-range corrected hybrid density functionals with damped atom–atom dispersion corrections, *Physical Chemistry*, **10**, 6615–6620 (2008).

23. C. Painter, M. M. Coleman, and J. L. Koenig, *The Theory of Vibrational Spectroscopy and its Application to Polymeric Materials*, John Wiley & Sons, New York, p. 323 (1982).

24. G. Avitabile, R. Napolitano, B. Pirozzi, K. D. Rouse, M. W. Thomas, and B. T. M. Willis, Low temperature crystal structure of polyethylene: Results from a neutron diffraction study and from potential energy calculations, *Journal of Polymer Science: Polymer Letters Edition*, **13**, 351–355 (1975).

25. M. Tasumi and T. Shimanouchi, Crystal vibrations and interatomic forces in polymethylene crystals, *Journal of Chemical Physics*, **43** (4), 1245–1258 (1965).

26. T. M. Korter and D. F. Plusquellic, Continuous-wave terahertz spectroscopy of biotin: Vibrational anharmonicity in the far-infrared, *Chemical Physics Letters*, **385**, 45–51 (2004).

27. D. C. Schriemer and L. Li, Combining avidin-biotin chemistry with matrix-assisted laser desorption/ionization mass spectrometry, *Analytical Chemistry*, **68**, 3382–3387 (1996).

28. A. Bykhovski and D. Woolard, Spectra of biological and organic molecules in polymer matrices: A hybrid ab-initio/empirical study, in *2010 International Symposium on Spectral Sensing Research (2010 ISSSR)*, June 20–24, Springfield, MO (2010).

29. S. S. Iyengar, H. B. Schlegel, J. M. Millam, G. A. Voth, G. E. Scuseria, and M. J. Frisch, Ab initio molecular dynamics: Propagating the density matrix with Gaussian orbitals. II. Generalizations based on mass-weighting, idempotency, energy conservation and choice of initial conditions, *Journal of Chemical Physics*, **115**, 10291–10302 (2001).

30. P. H. Berens and K. R. Wilson, Molecular dynamics and spectra. I. Diatomic rotation and vibration, *Journal of Chemical Physics*, **74** (9), 4872–4882 (1981).
31. K. G. Yager and C. J. Barrett, Azobenzene polymers for photonic applications, in *Smart Light-Responsive Materials*, Y. Zhao and T. Ikeda (eds.), John Wiley & Sons, Inc., New York (2009).
32. D. L. Woolard, Y. Luo, B. L. Gelmont, T. Globus, and J. O. Jensen, Bio-molecular inspired electronic architectures for enhanced sensing of THz-frequency bio-signatures, *International Journal of High Speed Electronics Systems*, **16** (2), 609–637 (2006).
33. D. Mokrauer, C. T. Kelley, and A. Bykhovski, Efficient parallel computation of molecular potential energy surfaces for the study of light-induced transition dynamics in multiple coordinates, *IEEE Transactions on Nanotechnology*, **10** (5), 70–74 (2010).
34. D. Mokrauer, C. T. Kelley, and A. Bykhovski, Parallel computation of surrogate models for potential energy surfaces, DCABES, pp. 1–4, in *Ninth International Symposium on Distributed Computing and Applications to Business, Engineering and Science*, August 10–12, Hong Kong, China (2010).
35. D. Mokrauer, C. T. Kelley, and A. Bykhovski, Simulations of light-induced molecular transformations in multiple dimensions with incremental sparse surrogates, accepted for publication, *Journal of Algorithms and Computational Technology* (2011).

3 Label-Free Biosensors for Biomedical Applications
The Potential of Integrated Optical Biosensors and Silicon Photonics

Jeffrey W. Chamberlain and Daniel M. Ratner

CONTENTS

3.1   INTRODUCTION

Sensors, in their most general form, measure signals from the environment for inter-pretation and analysis—with the objective of providing actionable information to the user. Such is the case for biosensors, which have found widespread applications in medical research, healthcare, environmental monitoring, chem-bio defense, and food safety. In biomedicine, biosensors are used for discovering new drugs, elucidat-ing biological pathways, and studying biomolecular interactions. Within the health-care setting, biosensors can be implemented as biometric assays and can serve as diagnostic tools, potentially predicting disease or suggesting disease susceptibility via genetic screening. Biosensors are also used for environmental monitoring, such as detecting the presence of specific allergens or environmental contaminants, and for defense threat agents such as anthrax, ricin, or other toxins. While this chapter focuses on biosensor design considerations in medicine (research, diagnostics, etc.), nearly all of the devices discussed have far-reaching applications beyond the exclu-sive domain of biomedical research. For instance, a biosensor for detecting malaria could also be implemented to detect pathogenic organisms in food, or as a research tool to screen for inhibitors to prevent disease.

In Chapter 4, we address the requirements and desired features of biosensors for biomedical applications, emphasizing the goal of realizing fully integrated and dis-tributable lab-on-a-chip (LOC) devices. We highlight silicon photonics as an advan-tageous technology for such LOC applications, particularly for label-free optical biosensing. In addition, we address other promising label-free electrochemical and mechanical biosensors, with a brief survey of current research in the field. Finally, we provide a perspective on the remaining challenges that need to be addressed for biosensors to inform researchers and clinicians with the ultimate objective of improved healthcare outcomes.

3.1.1   DESIRED BIOSENSOR CHARACTERISTICS

The ideal biosensor, within the realm of biomedicine, should rapidly detect any ana-lyte of choice from a low-volume, unprocessed sample; be disposable or contain reusable low-cost components; require minimal training; integrate the source, trans-ducer, and detector into a single portable device; allow long-term storage in ambient conditions; and produce a clear and quantitative readout of the amount of target analyte in the sample. While such an ideal biosensor has not been realized, it is use-ful to keep the ultimate goal in mind when discussing the desired characteristics of biosensors. It is also important to note that all of these qualities are not required for a technology to have value, but they are the characteristics needed to achieve true point-of-care (POC) biosensing. Incremental steps toward such a system could still have important applications. For example, simplifying the operator requirements and allowing use with unprocessed samples would significantly improve the utility of diagnostics in a clinical setting.

3.1.1.1 Sensitivity and Selectivity

A biosensor must, first and foremost, be both sensitive and selective for the target agent or agents it is designed to detect. With the hope of using unprocessed samples such as saliva, blood, urine, or other bodily fluids, the analyte to be detected will usually be at low relative concentrations within the complex biological milieu of cells, proteins, lipids, and salts found within a typical biological sample. A comprehensive study of potential cancer biomarkers reported that 88% of cancer biomarkers found in plasma (out of 211 surveyed) are below 10 µg/mL, and 49% are below 10 ng/mL [1]. As a comparison, common plasma proteins such as albumin and fibrinogen exist at mg/mL levels, and the salinity is roughly 150 mM. Even when sample processing is an option in a properly equipped laboratory, it can be laborious and time consuming, and the motivation for using unprocessed or minimally processed samples reemerges. Rapid testing and analysis is of the utmost importance in a clinical setting, where diagnostics can be used to guide prophylactic or therapeutic intervention, significantly influence patient outcome, and dramatically reduce the time and cost associated with patient care.

3.1.1.2 Label Free

Along these lines, it is desirable to detect target analyte without the need for a label. In biosensors and diagnostic assays, the label is usually a chromophore, fluorophore, or enzyme that is either directly attached to one of the interacting molecules or attached to a secondary reporter molecule in order to amplify the binding event signal (Figure 3.1A). While assays that employ labels continue to find widespread use in research and medicine—for instance, the ubiquitous enzyme-linked immunosorbant assay (ELISA)—there are many reasons why it is desirable to eliminate the indicator (Figure 3.1B). Labeling not only increases the time and cost of an assay but also inherently alters binding interactions [2] and it obscures quantification [3]. Labels

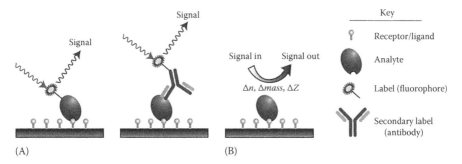

FIGURE 3.1 Label-based and label-free biosensing differ by the means with which the analyte is detected. (A) Example of label-based sensing via a fluorescently labeled analyte or secondary probe (antibody). (B) Illustration of label-free biosensing, where an inherent property of the analyte, such as refractive index (n), mass, or impedance (Z), alters the input signal such that detection can occur.

are also, in general, unstable molecules that require careful storage; this makes assay standardization difficult and limits their use in a POC setting. Finally, when the target analyte is unknown, such as high-throughput screening of molecular libraries, labeling is simply not an option. Label-free biosensors can decrease cost and assay time, provide quantitative information of unaltered binding interactions in real time, and potentially enable portable biosensing of unprocessed samples. The lack of a label, however, increases the requirements for sensitivity and necessitates on-chip controls to ensure that signals being analyzed are due to the analyte and not nonspecific interactions with the sensor.

3.1.1.3 Multiplexing

The need for on-device controls is complementary with the desire for biosensors to be multiplexible, where multiplexing describes a biosensor's ability to run multiple experiments simultaneously (Figure 3.2). For screening applications and disease diagnostics, the ability to test for multiple binding interactions is imperative. Disease states are most often described by multiple biomarkers, and the ability to provide a conclusive diagnosis is reliant on taking a systems approach of measuring the concentrations of more than one biomolecule [4–7]. A good example of the dangers of relying on a single biomarker for disease can be found in prostate-specific antigen (PSA), an FDA-approved cancer biomarker, which was once widely thought to be a direct indicator for prostate cancer. As a result, researchers developed a number of diagnostic assays and recommendations for diagnosing cancer based on the detected concentration of PSA. However, it became apparent that PSA levels varied widely between individuals and it served as a poor indicator for the presence or prognosis of cancer [4,8]. Investigators have found similar poor diagnostic values for the remaining eight FDA-approved cancer biomarkers, highlighting a shortcoming of single biomarker–based diagnostics [1]. In contrast, a systems approach operates on the hypothesis that disease introduces genetic or environmental perturbations that alter biological networks; resulting changes that are widespread being reflected in the levels of multiple proteins and other biomarkers present in the body. Considering the complexity of a physiologic system, it is clear why multiple biomarkers may be needed to diagnose disease.

In addition to accurately diagnosing single diseases, multiplexible biosensors could be used to screen for multiple diseases or pathogens at once, which would be especially useful for POC applications where resources are limited and visits to the clinic are rare. In the case of pathogen detection, multiple tests are often needed to positively identify a pathogenic organism. Currently, diagnosticians employ various culture and biochemical tests, requiring days to reach a positive conclusion. The most accurate tests rely on polymerase chain reaction (PCR), but this process is prohibitively expensive and unavailable in areas of the world where such tests are most needed. As an alternative to pathogen detection based on nucleic acid assays, many pathogens can be uniquely characterized by their ligand-binding affinities and antigenic profiles [9–14]; a quantitative biosensor containing a panel of known pathogen-binding ligands and antibodies could detect the presence of these pathogens from patient samples.

In biosensor design, multiplexibility requires multiple sensors or multiple sensing regions that can be differentially functionalized and interpreted. Microarrays using DNA [15,16], proteins [17,18], and carbohydrates [19–22] provide an excellent

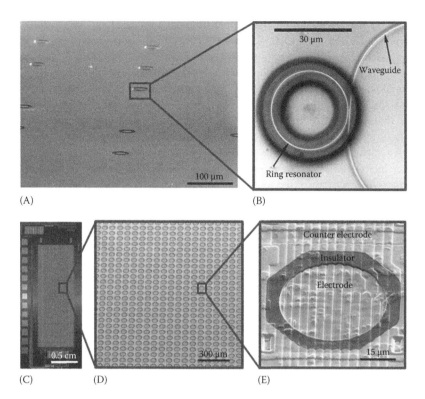

(A) (B)

(C) (D) (E)

FIGURE 3.2 Two examples of multiplexed biosensor platforms. (A) Each donut-shaped depression seen in the scanning electron micrograph represents an individual microring resonator. Device manufactured by Genalyte, Inc. (San Diego, California), (Photo courtesy of Nanophotonics Laboratory, University of Washington, Seattle, WA.) (B) Exposed through lithographic etching of a polymer coating. Microrings are interrogated with a bus waveguide that comes within 100 nm of the resonator. (Photo courtesy of Tate Owen.) (C–E) Illustration of a commercial microelectrode array manufactured by CustomArray, Inc. The array contains 12,544 individually addressable microelectrodes. Each electrode is 44 μm across and is separated from the reference electrode by an insulating layer (E). (C, Photo courtesy of CustomArray, Inc., Mukilteo, WA; D and E, Photo courtesy of Authors.)

example for the potential of high-throughput screening, yet they almost always rely on a fluorescent readout. Nonfluorescent microarrays are increasingly being employed with surface plasmon resonance imaging (SPRi) [23,24]. In addition to being high throughput and label free, SPRi allows real-time analysis of binding interactions.

3.1.2 GENERAL LABEL-FREE BIOSENSOR SETUP AND OPERATION

A biosensor is any device that converts a biological event—typically, but not limited to, binding between complementary biomolecules—to an output signal that can be analyzed to describe the sensed event. When no label is used, detecting the binding event relies on a transducer, the mechanism of which can be electrochemical, mechanical, or optical. The transducer must be functionalized with a bioactive surface that captures the target biomolecules or changes in response to the target, such that these changes on the surface of the transducer elicit a detectable signal (Figure 3.1B). The bioactive surface is generally composed of one or multiple biorecognition molecules such as oligonucleotides, peptides, antibodies, aptamers, phages, and carbohydrates. These molecules can be attached to the transducers using nonspecific adsorption or by specific covalent attachment via reactive groups that are either native to the biomolecules or introduced through molecular biology or synthetic techniques. Regardless of the mechanism of attachment, the bioactive surface must be resilient to degradation and prevent nonspecific binding to other biomolecules or contaminants. In addition to the transducer element of a biosensor, its operation relies on a variety of support instrumentation and additional components. First, a biosensor requires some way to deliver the sample to the sensing region. With the exception of gas-based sensors, this usually requires a fluidic handling system (pumps, tubing, channels, etc.) that can effectively deliver aqueous samples while simultaneously minimizing reagent consumption. Next, the signal generated by the transducer must be analyzed and correlated with the sample, whether quantifying analyte concentration, binding kinetics, or simply the presence of the target. These additional aspects of the biosensor, beyond initial signal transduction, require instrumentation that actuates (e.g., signal generator and light source), detects (e.g., oscilloscope and photodetector), and processes (e.g., microprocessor) the signal. In most cases, all of these components require power. Thus, considerations for constructing and implementing integrated POC biosensors extend beyond the sensor itself.

3.2   TOWARD FULLY INTEGRATED BIOSENSORS

Significant effort has been dedicated to miniaturization and integration of biosensing components to construct LOC devices. The motivation is obvious: fully integrated chip-based biosensors would allow biosensor applications to expand beyond research laboratories into clinics, households, and the POC. In addition to the aforementioned desired characteristics of a biosensor, the devices need to be cheap, robust, reliable, and easy to use and consume low amounts of power. While many technologies and disciplines are converging to achieve this goal, two stand out as especially enabling for the future of biosensors. The first is the fabrication of miniaturized and integrated electronic devices, an effort spearheaded by the microelectronics industry, whose techniques have had far-reaching applications in micro- and nanotechnologies. The other is microfluidics (a beneficiary of microelectronic fabrication processes), which enables the handling and manipulation of small volumes of fluidic samples and is particularly amenable to device integration. In order to realize an integrated LOC biosensor, researchers will, without a doubt, need to leverage both of these technologies. It follows that these devices would greatly benefit from having planar, chip-based components in order to integrate microfluidics while capitalizing on semiconductor fabrication technologies [25].

Silicon photonics has become a focal point of many parallel efforts to achieve fully integrated biosensors complete with on-chip light sources, detectors, and data processors [26–29]; the high-throughput, cost-effective, and scalable manufacturing techniques developed by the microelectronics industry and the sensitivity, efficiency, and pervasiveness of optical biosensing are united by their ability to be implemented on silicon. Jokerst et al. point out the components necessary for a fully integrated planar photonic biosensor [25]. Importantly, all of the components—a light source (e.g., thin-film III–V edge-emitting laser) [30], a sensor (e.g., microring resonators) [31], and a photodetector (e.g., InGaAs metal–semiconductor–metal photodetector) [32]—have been fabricated in planar formats and tested independently by different groups, so all that remains is piecing everything together. This is no trivial task, especially considering the other hurdles such as microfluidic integration, sensor functionalization, and device characterization, yet the technologies exist and efforts are under way to make fully integrated and distributable biosensors. Steps toward this goal have been made. Microresonators, an optical biosensing device, have been integrated with chip-based photodetectors that were able to monitor the resonant condition of the microresonators [33,34]. Chip-based light sources (e.g., thin-film edge-emitting lasers [35] and dye lasers [36]) have also been integrated with an interferometric coupler [35], as well as a waveguide and a photodetector in series [30]. A variety of traditional optical components such as microlenses [37], mirrors [38], filters [39,40], laser diodes, and photodiodes [39], as well as sensing components like interferometers and microresonators have been integrated into microfluidic devices [25–27,41]. Microfluidic devices have also incorporated valves [42,43], pumps [44–46], and sample-processing capabilities such as mixers [47,48], target concentrators [49,50], and target separators [51–53].

3.2.1 SILICON PHOTONICS FOR DEVICE INTEGRATION

3.2.1.1 Note on Classification

As with many broad classification systems, defining the scope of silicon photonics is a difficult task, and the silicon photonic community remains divided on means of classification. A decent approach is classifying based on the materials used in the system; some in the field argue that silicon photonics is strictly limited to devices composed only out of silicon waveguides, while others include materials such as silicon oxides, silicon nitride (Si_3N_4), and SiON. We take the latter approach of using a more broad definition of silicon photonics, and place an emphasis on planar devices that leverage CMOS-based microfabrication techniques and thus have potential for component integration. Some of the devices used as examples do not use silicon-based waveguides, while others do not use waveguides at all. Nevertheless, it is useful to introduce some of the basics of silicon photonics at this point in the discussion.

3.2.1.2 Why Silicon?

Traditional optical components are bulky and expensive; they require exotic materials such as indium phosphide (InP), gallium arsenide (GaAs), and lithium niobate ($LiNbO_3$). In addition, compound optical devices are usually assembled by hand [54], such that the difficulty of assembly increases exponentially with the number of components. This makes the large-scale production of complex free space optical systems onerous. Silicon, on the other hand, has a history of automated processing and the microelectronics industry has developed extensive manufacturing infrastructure. This industry has invested hundreds of billions of dollars to the processing and implementation of silicon in microelectronic devices, driven by silicon's advantageous electronic properties and its low cost. More recently, much attention has been directed to determining ways to implement silicon as an optical material to address the cost and manufacturing limitations of current optical devices and to advance the microelectronics industry [55,56]. Importantly, silicon has properties that also make it attractive for photonics, most notably its transparency to wavelengths of light greater than 1100 nm and its high refractive index ($RI = 3.5$). These properties, coupled with CMOS-processing techniques, allow optical devices to be defined in silicon substrates such that fabricating thousands of silicon photonic components has become a trivial task. Device alignment, typically the most critical and time-consuming step of assembling traditional compound optical systems, becomes a fully passive process because the photolithographic masks define the device layout.

An important feature of silicon-based devices is the facile modification by oxidation, doping, and metallization. The incorporation of oxygen into silicon (forming SiO_2) is the most common modification, and it is ubiquitously used as the insulator in silicon-on-insulator (SOI) microelectronic devices. As applied to silicon photonic structures, the lower RI of SiO_2 (1.46) compared to silicon allows it to serve as a cladding material and confine the light modes within features, known as waveguides, defined in the silicon. Doping is necessary to impart electrical functionality, such as diodes, into silicon. A simple example that demonstrates this feature is a p-i-n diode, where group III and group V ions are implanted in discrete regions on the silicon to introduce p-type and n-type behavior, respectively [54]. Finally, the metallization of silicon (i.e., deposition of metallic structures on silicon) allows for the inclusion of device interconnects and contact pads for external manipulation and interrogation. These common processes utilized by the microelectronics industry can also be incorporated into the fabrication of silicon photonic devices to act as, for example, optical switches and modulators [57].

Waveguides can shuttle and manipulate light on the devices in a way that is analogous to how metallic wiring guides electrical signals, and since using light in place of electrons significantly increases potential data transfer rates and decreases loss, silicon photonics is of particular interest to the communications industry and assures continued investment in this area. Fiber-optic cables are already used to rapidly transmit data over long distances with little loss, and the transition to replace on-chip electronics with optical components has begun [58]. Further, because of silicon's excellent electrical properties, the extensive tools available for modifying silicon (e.g., silicon doping and metallization) [54], and the demonstrations of on-chip optical components discussed earlier, it is clear that fully integrated optoelectronic chips are achievable in the near future.

3.2.1.3   Waveguides: Fabrication and Basic Principles

Waveguides are defined on silicon substrates using standard lithographic processes that are widely implemented in CMOS fabrication techniques. A variety of lithographic techniques exist for creating silicon structures, including wet and dry chemical etching approaches and e-beam lithography. Dry etching is accepted as the more precise and reproducible technique for large-scale fabrication, achieving critical dimensions of 10 nm. This is smaller than what is required for many basic waveguide structures (with the exception of slot waveguides), but the capability remains, and very high densities of waveguides can be attained. E-beam lithography, while not amenable to high-throughput fabrication, allows rapid prototyping of device structures which enables research in silicon photonics. Both planar and two-dimensional waveguides have been used in biosensors, but in most practical applications that extend beyond biosensing, light must be confined in two dimensions so as to manipulate and direct light in a controlled manner while minimizing loss [54]. One of the most basic 2D waveguide structures, a ridge waveguide, is shown in Figure 3.3.

Light is coupled into and confined within the waveguide core on account of its higher RI relative to the cladding layers that surround it. The lower cladding layer is almost always silicon dioxide, while the upper cladding can be air ($RI = 1$), water ($RI = 1.33$), more silicon dioxide, or just about any material with an RI below that of the waveguide core. Rectangular rib waveguides can support multiple light propagation modes, the number of which is determined by the dimensions of the waveguide. In general, the larger the waveguide cross section, the more modes it will support. However, given the correct waveguide geometry, only the fundamental mode will propagate because higher-order modes will leak out over very short distances [59]. A portion of the light, the amount of which depends on the waveguide structure and the cladding material, propagates outside of the core. The light outside of the waveguide is known as the evanescent field; the intensity of this field decays exponentially away from the core and it is influenced by the RI of the cladding (Figure 3.3B). Thus, the propagation of light through the waveguide is sensitive to changes outside of the core within the evanescent field, a property which is exploited for applications in biosensing. The distance which the evanescent field extends into the dielectric can be tuned by altering the dimensions of the waveguide, the wavelength of light, and the dielectric itself, but it is typically on the order of tens to a few hundred nanometers [60,61]. Like surface plasmon resonance (SPR) biosensors, this gives waveguide-based biosensors the ability to sense localized biomolecular interactions at the waveguide surface while being largely insensitive to changes in the bulk fluid.

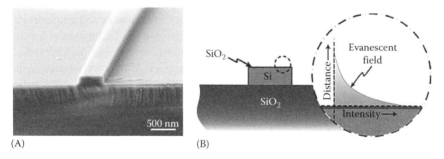

(A)   (B)

FIGURE 3.3 Basic ridge waveguide. (A) An SEM showing the cross section of a silicon ridge waveguide on top of a silicon oxide substrate. (Photo courtesy of Nanophotonics Laboratory, University of Washington, Seattle, WA.) (B) A schematic illustrating the evanescent field associated with the light mode traveling through a ridge waveguide.

3.3 LABEL-FREE BIOSENSORS

As previously discussed, biosensors that do not require labeling of the target mole-cules significantly increase the potential applications and the amount of information that can be obtained. This facilitates quantitative real-time binding analysis of native molecules for extracting kinetic binding parameters and makes it possible to use unprocessed samples, thus enabling high information content POC and distributed devices. Instead of using a label, these biosensors rely on inherent properties of the target such as impedance, mass, or RI to measure binding. Although the focus of this chapter is silicon-based optical biosensors, we would be remiss not to acknowledge the role of electrochemical and mechanical biosensors in label-free sensing. These devices have shown potential for realizing fully integrated devices and stand to bene-fit from some of the same fabrication and integration capabilities provided by silicon.

3.3.1 ELECTROCHEMICAL BIOSENSORS

Generally speaking, electrochemical biosensors operate by detecting the change in the resistance or capacitance on an electrical sensing component in response to the formation of binding complexes or to environmental perturbations. The most preva-lent and well-known example of electrochemical biosensors are those used in the majority of glucose monitors [62]. However, all of these devices, along with most electrochemical biosensors for other applications, possess limited sensitivity and require the use of an electroactive indicator to generate a detectable signal [62,63]. Nonetheless, electrochemical biosensors remain attractive because they can be mass produced at low costs, they have low power requirements, and they can be scaled down to allow miniaturization and multiplexing [64]. Given these benefits, there are significant ongoing efforts to improve label-free electrochemical biosensors, through the construction of more sensitive nanoscale devices based on nanowires, nanotubes, and nanofibers [65–67].

3.3.1.1 Electrical Impedance Spectroscopy with Microelectrodes

Electrical impedance spectroscopy (EIS), in the case of affinity biosensors, mea-sures the change in the impedance of an electrical circuit due to the binding of ana-lyte to a functionalized electrode. EIS is favored over voltammetry and amperometry because the measurement technique is less damaging to the biofunctional capture layer [64]. The electrodes used for EIS can be miniaturized and multiplexed, as indeed they have been [68,69]. A significant advantage of microelectrodes is that by applying a current or voltage at the electrodes, one can create localized reaction environments—this opens up the possibility of on-chip synthesis and functional-ization in a highly multiplexed fashion. A microelectrode microarray developed by CombiMatrix Corp., now CustomArray, Inc., Mukilteo, WA, enabled electrochemi-cally controlled synthesis of unique peptides and sequences of DNA on individual electrodes [70]. The initial devices containing 1,024 electrodes were then scaled up to contain 12,544 electrodes (Figure 3.2C through E), a device that the com-pany has used to develop a completely automated, high-throughput assay to screen for and identify subtypes of influenza A by way of genotyping [69]. In addition to directly synthesizing DNA probes on the electrodes, they could be used to polymerize pyrrole, which could then be used to immobilize capture agents such as DNA and antibodies, the latter of which was used to develop an assay for staphylococcal enterotoxin B [71,72]. Despite these examples, with EIS sensing regimes, device performance is tied to the properties of the electrodes, and it remains unclear how binding can be quantified and how miniaturization may alter the equivalent circuits used to describe binding [64]. Further, due to the lower sensitivity of label-free EIS, there exists a limit to how small the electrodes can be before the surface area available for biomolecule functionalization is too small to generate a detectable signal upon target binding.

3.3.1.2 Nano Field Effect Transistors

Nanoelectrochemical sensors primarily act as field effect transistors (FETs), a sensing technique that is free from the size limitations discussed for the electrodes used for EIS. In a standard transistor setup, a semiconducting material attaches a source and drain electrode; a third electrode, known as a gate electrode, separated from the semiconductor by a thin dielectric, controls the conductance of the semiconductor by applying positive and negative voltages. In this manner, the gate electrode acts as a switch for the current flowing from the source to the drain. In a FET-based biosensor, biomolecules take the place of the gate electrode, whereby the conductance of the semiconductor is altered by the binding of biomolecules (Figure 3.4). Thus, changes in current between the source and the drain are correlated with binding events. Researchers have investigated nanowires [73–77], nanofibers [78], and carbon nanotubes [74,79,80] as semiconducting materials to use in nanoelectrochemical biosensors. A primary motivation is the potential sensitivity of these materials to biomolecular binding events—these one-dimensional materials have similar sizes to biomolecules, such that extremely small amounts of bound analyte will significantly affect the electrical properties of the transistor. Investigators have demonstrated detection of proteins [74,75,79], single viruses [73], and DNA with concentrations reported down into the picomolar range [80] and even down to 10 fM [76,77]. The limited size of these sensors, coupled with the preestablished microelectronics infrastructure, also make them good candidates for multiplexing, and devices containing up to 200 sensors have been reported [73,81]. Some hurdles remain before FET-based

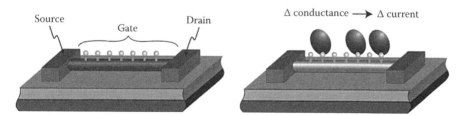

FIGURE 3.4 A nano field effect transistor biosensor is shown schematically. The intrinsic electrical properties of biomolecules that bind to the functionalized semiconducting material (e.g., nanowire and carbon nanotube) changes its conductance and therefore influences the amount of current that flows from the source to the drain.

nanoelectrochemical biosensors can be used as reliable research instruments, let alone as biosensors in a healthcare setting. Despite the apparent potential to achieve highly multiplexed devices, there are no good solutions for high-throughput fabrication. Carbon nanotubes must be synthesized off-chip and placed in position; nanowires, to their credit, can be patterned lithographically [77,82] or grown on-chip [75,76], but they still suffer from fabrication inconsistencies. Given their size and high sensitivities, heterogeneity between sensors can result in altered performance and poor reproducibility. FET sensors are also ion sensitive, and ions in solution will act as a gating mechanism and dramatically reduce sensitivity. This, along with the structural fragility of the FETs, limits the potential samples and experimental setups that can be used. Lastly, the mechanisms by which biomolecules influence electrical properties are not well understood, and different biomolecules alter the sensor response in ways that do not correlate predictably with the size of the biomolecule or its concentration in solution.

3.3.2 MECHANICAL BIOSENSORS

Mechanical biosensors directly detect the change in mass on the sensor surface due to the binding of biomolecules, viruses, or cells. Mechanical sensors represent the most sensitive of the sensing techniques, with noise floors and mass resolutions reported as low as 20 and 7 zg, respectively [83]. Given their potential sensitivity, mechanical biosensor research has largely been directed toward reaching very low detection limits for applications such as rare analyte sensing and weighing individual viruses and cells. Surface acoustic wave sensors, including the quartz crystal microbalance (QCM) [84–86], utilize the sensitivity of piezoelectric crystal resonance to perturbations in the surrounding environment. In QCM, the quartz surface is usually coated with an anchoring layer to which biological receptors are immobilized. Electrodes attached to the quartz apply an alternating voltage which elicits a resonant mechanical oscillation. Tracking changes in the oscillatory frequency in response to binding at the sensor surface produces the signal, with reported limits of detection (LOD) down to 10 pg/mm^2 [61]. QCM biosensors have been used to detect binding interactions of proteins [87], oligonucleotides [88,89], carbohydrates [90–92], lipids [93,94], viruses [95,96], and cells [97,98]. A distinct advantage of QCM over both electrical and optical biosensors is the wide range of materials that can be deposited on top of the quartz. Since the sensing mechanism does not rely on the transmission of an optical signal or the propagation of light, QCM supports the study of interfacial interactions using a wide variety of materials [84,99]. However, QCM is not without its limitations. While QCM is amenable to performing binding experiments in a liquid environment, sensitivity is reduced and it can be difficult to separate the effects of mass, density, and viscosity in the QCM signal [100]. It is also difficult to fabricate dense arrays of acoustic wave devices, although it has been demonstrated [101].

3.3.2.1 Micro- and Nanoelectromechanical Systems

Researchers have investigated micro- and nanobiosensors using mechanical transduction mechanisms in an effort to increase sensitivity and allow multiplexing. These devices employ standard photolithography techniques and are usually made out of silicon or silicon nitride, allowing high densities of devices and the possibility of integration with electronic components and flow cells. The majority of these devices are based on analyte binding to functionalized cantilevers, which either changes the deflection of the cantilever (static devices) or its resonant frequency of oscillation (dynamic devices). Static devices (Figure 3.5A) have the advantage of being able to operate in both gas and liquid environments, but they have decreased sensitivity because it requires the attachment of a near monolayer of analyte to deflect the cantilever [67]. Nevertheless, static cantilever devices have been shown to detect single base-pair mismatches of 12-mer DNA strands and picomolar LOD of oligonucleotides [102] and nanomolar concentrations for proteins [102,103]. An impressive study also showed detection limits of PSA down to 0.2 ng/mL (6 nM) in a background of both 1 mg/mL BSA and HSA, which matches detection limits of ELISA for PSA and is physiologically relevant [104]. Dynamic devices (Figure 3.5B) have significantly higher potential sensitivities, and researchers have shown the ability to detect viruses [105] down to the single virus [106–108], single cells [109,110], DNA with a single 1587-mer strand [111], and PSA down to 10 pg/mL [112]. The Bashir group has also used their devices for weighing single viruses [107] and cells [110]. As with QCM, an important drawback of most dynamic mechanical biosensors is that their sensitivity is limited by the dampening effects of liquid, so detection must be done in vacuum or air. However, several recent papers have implemented nanofluidic channels fully confined within the cantilevers (Figure 3.5C) [113,114]. In this configuration, the cantilevers are maintained in a vacuum and the channel within

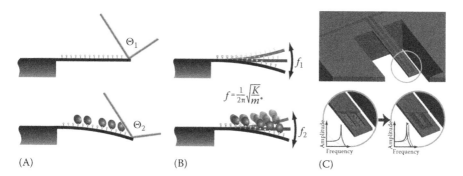

(A) (B) (C)

FIGURE 3.5 The two different modes of operation for microcantilever-based mechanical biosensors are demonstrated schematically. In both cases, the cantilevers are functionalized with the appropriate receptor or capture molecule. (A) Static microcantilever biosensors correlate the deflection of the cantilever arm upon the binding of biomolecules. (B) Dynamic microcantilever biosensors sense binding of biomolecules through changes in the oscillatory frequency. The mass of bound material can be calculated because the deflection or the change in the frequency (f) of oscillation can be related to the spring constant (K) and the effective mass (m^*) of the cantilever. (C) Since viscous damping of liquids leads to a dramatic decrease in sensitivity for dynamic microcantilever biosensors, Burg et al. have fabricated devices that contain a nanofluidic channel inside of the cantilever arm. (From Burg, T.P. et al., *Nature*, 446(7139), 1066, 2007.)

the cantilevers allows biological interactions to occur in solution—not only are the interactions measured in a physiologic (aqueous) environment, but they are detected in real time. Studies using these devices have measured the changing masses of individual cells during growth [115,116], detected cancer biomarkers in serum down to 10 ng/mL [117], and IgG protein below nM concentrations [113].

3.3.3 OPTICAL BIOSENSORS

Optical biosensors are the most widely used label-free biosensing platforms to study biomolecular interactions because of their relative ease of use, high sensitivity, and the high information content of the data they generate. In 2008 alone, there were over 1400 articles published on optical biosensors [118]. Compared to many electrochemical and mechanical platforms, optical biosensors are more flexible and easy to use from an operational standpoint. Additionally, the sensitivity of optical biosensors is not drastically reduced by physiological salinity or viscosity in the analyte buffer, making them amenable to a wide range of samples. Label-free detection methods using optical biosensors include RI detection, optical absorbance detection, and Raman spectroscopic detection [119], with the most common form being RI detection. RI detection is based on the sensitivity of light to changes in RI; biomolecules have a higher RI than buffer solutions (e.g., 1.45 for proteins vs. 1.33 for water), allowing their detection by monitoring the properties of the interacting light. A number of RI-based optical biosensors exist, including SPR, optical fibers, planar waveguides, interferometers, photonic crystals, and resonant cavities.

3.3.3.1 SPR and SPRi

First reported in 1983 [120], biosensors based on SPR are among the most widely used optical biosensor. As of 2008, a total of 24 manufacturers offered commercial platforms, including instruments made by GE, Bio-Rad, Biosensing Instruments, and Reichert [118]. SPR detection relies on the sensitivity of evanescent fields to changes in the local RI of the dielectric. In most SPR instruments, the evanescent field is associated with surface plasmon modes that are created from the coupling of light with a metallic film, usually gold, via total internal reflection (TIR) within a prism (Figure 3.6). The conditions of TIR (the wavelength of light and the incident angle of light that couples with the metallic film) vary with the RI of the dielectric above the metal film. A flow cell delivers biomolecules to the surface of the metallic film, where binding of analyte to immobilized receptors causes changes in the local RI. The instrument tracks these changes in real time and reports them as a shift in resonant wavelength (angular SPR) or as changes in the intensity of reflected light (SPRi). Traditional angular SPR has superior LOD than SPRi—the RI detection limit for SPR typically ranges from 10^{-6} to 10^{-8} refractive index units (RIU), whereas SPRi is usually in the range of 10^{-5}–10^{-6} RIU [119,121]—but SPR is only able to monitor binding in a single region at once for each light source. SPRi, on the other hand, uses a CCD array to detect the intensity of reflected light from the entire chip surface, allowing arrayed and simultaneous sensing of multiple binding interactions. The number of interactions that can be monitored simultaneously is limited only by the spatial resolution of the instrument ($\sim$4 μm) [24] and the arraying

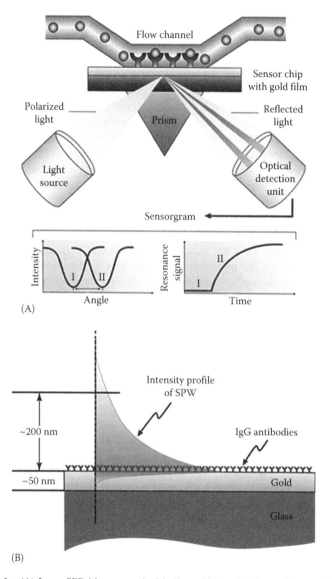

FIGURE 3.6 (A) In an SPR biosensor, the binding of biomolecules to immobilized recep-tors changes the coupling properties (in this case, the coupling angle) of light reflected off of a metallic film. These coupling angle changes, documented over a given time period, generate a sensorgram that describes the binding events. (From Cooper, M.A., *Nature Reviews Drug Discovery*, 1(7), 515, 2002. With permission.) (B) The coupling properties of light are sensi-tive to changes in refractive index within the surface plasmon wave (SPW) which extends ~200 nm (intensity = 1/*e*) from the metal–dielectric interface. The schematic includes IgG antibodies drawn to scale for perspective.

density of functionalization, the latter of which has been widely addressed by the microarray community. Realizing 100 interaction regions on a single SPRi chip is common, with reports of systems that allow up to 10,000 spots [122]. Given its high-throughput potential, SPRi appears to have significant potential as a multiplexed POC device. Most studies using SPRi have employed it for multiplexed interaction screening and characterization rather than detection [123], favoring angular SPR for the higher sensitivity measurements. Several commercial instruments exist, including the FlexChip from BIA-CORE [124,125], GWC's SPRimager®II [126,127], and Texas Instruments' Spreeta system [128,129]. In addition to the Spreeta instrument, other groups have developed portable SPRi systems that make important inroads to realizing POC applications [130–132]. Unlike the aforementioned multiplexed devices, SPRi benefits from the fact that different sensing areas can be defined by the user based on the locations of the functionalized regions, and the metal surface (usually gold) of SPRi sensor chips is relatively robust. This greatly reduces the alignment difficulties that are encountered when attempting to functionalize specific devices on multiplexed electrochemical and mechanical devices. As a biosensing platform, SPR benefits from the extensive literature on the biofunctionalization of gold surfaces—one of the best understood surfaces for functionalization and the standard for biological surface analysis [133]. Among the important properties are gold's biocompatibility and its ability to bind strongly to thiol groups (at near covalent strength), which allows for facile tethering of biomolecules and non-fouling self-assembled monolayers (SAMs) [134,135]. However, despite these advantages, SPRi has yet to realize widespread use in the clinic or in POC settings.

As mentioned, the popularity of SPR is partly a result of the detailed binding information that the technique can produce. Properly designed experiments can yield both qualitative and quantitative information, such as the selectivity, the strength, the kinetic binding parameters, and the thermodynamic parameters of a binding interaction, as well as identify the active concentration of the target molecule. However, emphasis must be placed on the careful design, execution, and analysis of optical biosensing experiments if meaningful information is to be extracted. An unfortunate reality of the widespread use of SPR, and optical biosensors more generally, is that many investigators make incorrect conclusions from their data as a result of poorly run experiments or faulty analyses. Rich and Myszka reviewed the optical biosensor literature every year from 1998 to 2008 and found that a large majority contain major flaws in some aspect [118,136,137]. These reviews are excellent sources for understanding the proper utilization of optical biosensing technologies and they also communicate the wide range of applications of these instruments.

3.3.3.2 Grating-Based Sensors

Optical fibers and planar waveguides can both be incorporated into surface plasmon wave (SPW) biosensors, and they operate similarly to SPR. In these cases, an optical fiber or a waveguide acts in place of the prism to couple light with a metallic layer, which generates the SPW and the corresponding evanescent wave that is used for sensing RI changes in the dielectric. Alternatively, in non-SPW biosensing conformations, optical fibers and planar waveguides often rely on coupling light with a grating structure. A grating consists of a periodic physical perturbation on the surface of the sensor; light couples into the grating at a specific angle and wavelength that are determined by the effective RI (n_{eff}) of the fiber or waveguide and the grating period. Binding of biomolecules changes n_{eff}, enabling real-time detection. In the case of fibers, the gratings are etched into the optical core or into the cladding immediately surrounding the core, such that a biofunctionalized grating provides the sensing region (Figure 3.7A). While these devices are more widely used for sensing load, strain, temperature, and vibration [138], examples of their application for biosensing include the detection of DNA 20 base pairs in length down to 0.7 µg/mL using a device with an RI DL of 7×10^{-6} RIU [139], real-time monitoring of antibody binding with a dynamic range from 2 to 100 µg/mL and antigen detection from crude *Escherichia coli* lysate [140], and the detection of hemoglobin in sugar solutions with an inferred sensitivity to a change in hemoglobin concentration of 0.005% [141]. While fiber gratings are inexpensive and straightforward to manufacture, they suffer from relatively poor sensitivities [119].

Grating-coupled planar waveguides are also easy and cheap to fabricate, as they consist of a thin-film waveguide deposited on a glass support into which a grating can be etched using photolithography or imprinting [142]. Optical waveguide light-mode spectroscopy (OWLS) is one well-known implementation of this sensing modality; these devices measure the change in the coupling angle due to changes in the RI on the grating (Figure 3.7B). They have been used for biosensing applications including antibody capture of the herbicide trifluralin down to 100 ng/mL [143] and the detection of mycotoxins down to 0.5 ng/mL [144]. OWLS has been more widely applied to studying biomolecular adsorption kinetics and conformation on a variety of material surfaces [145–147]. OWLS does not permit multiplexing capabilities, but a very similar technique employing planar waveguide gratings known as wavelength-interrogated optical sensors (WIOS) addresses this issue, and a device with 24 different sensing sites has been used to simultaneously monitor 4 different classes

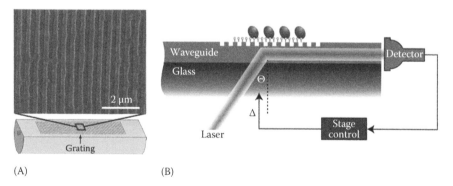

(A) (B)

FIGURE 3.7 Grating-based biosensors have been shown in a variety of conformations, two of which are depicted as follows: (A) A Bragg grating is etched into the cladding of a D-shaped fiber. (From Smith, K.H., et al., *Applied Optics*, 45(8), 1669, 2006. With permission.) (B) An OWLS biosensor changes the angle of the stage in response to biomolecule binding to maintain coupling. These changes are recorded over time to generate the sensorgram.

of veterinary antibiotics in milk with a DL ranging from 0.5 to 34 ng/mL, depending on the class of antibiotic [148].

3.3.3.3 Interferometric Sensors

3.3.3.3.1 Mach–Zehnder Interferometers

In a Mach–Zehnder interferometer (MZI), a single-frequency, coherent polarized light source is split into two paths. The sample is placed in one of these paths where light interactions with the sample cause a shift in the phase of the light, and the other path acts as the reference. The light is then recombined and the phase shift caused by the sample in the sensing arm leads to interference which can be detected by a change in the light's intensity. Although traditionally done in free space, MZIs can be fabricated in a planar structure using waveguides to split and recombine the light; these are called integrated MZIs. In such a setup, the sensing arm is functionalized and binding of the sample changes the RI within the evanescent field of the waveguide, thus modulating the phase of the propagating light and leading to interference upon recombining with light from the reference arm (Figure 3.8). The first biosensing demonstration of integrated MZIs detected human chorionic gonadotropin (hCG) down to 50 pM using immobilized capture antibodies [149]. This device had an RI LOD of 5×10^{-6} RIU, but improvements in MZI fabrication and analysis have led to demonstrations of LODs down to 10^{-7} RIU [150], which is on par with most SPR instruments. Other demonstrations of MZI biosensing include detection of IgG down to 1 ng/mL [151], and the ability to distinguish the wild-type DNA (58-mer) from a mutated sequence down to 10 pM concentrations [152]. Very few reports of biosensing with MZIs have emerged following the initial interest in the 1990s. This could be due to difficulties in multiplexing and the requirement of a relatively long sensing region to generate a detectable signal. Long sensing regions not only require a larger footprint on the device, but they also work against sensitivity because of increased loss. A more recent publication addressed both of these issues by demonstrating a multiplexed device that used coiled waveguides as sensors [153]. The device had six sensors, four of which were functionalized with two different

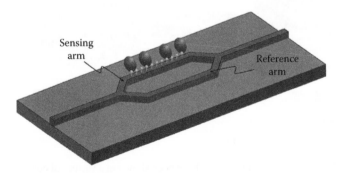

FIGURE 3.8 Schematic representation of a chip-based MZI biosensor. Changes in the RI surrounding the sensing arm induce a phase change, resulting in interference upon recombination with the reference arm.

antibodies (two sensors for each antibody) and the remaining two were used as reference sensors. The sensor response corresponded to a surface coverage of just 0.3 pg/mm². However, it remains to be seen if integrated MZIs for biosensing applications will have a significant impact in biomedical research.

3.3.3.3.2 *Young's Interferometers*

The Young's interferometer (YI) can be integrated onto a chip surface in much the same way as MZIs and can be used for biosensing. Similar to MZIs, YIs split light from a single waveguide into multiple arms including one reference arm. However, instead of recombining the light back into a single waveguide, a CCD is used to record the interference fringes that result from the optical output (Figure 3.9), permitting multiplexed sensing with just one reference. An integrated YI for sensing was first demonstrated in 1994 [154], and the technique has an established RI LOD of 10^{-7} RIU [155]. YIs have been used subsequently in a number of proofs of concept applications. For instance, a multiplexed device containing three sample arms and one reference arm enabled biosensing of herpes simplex virus type 1 (HSV-1) [156]. The authors were able to detect as few as 10^5 HSV-1 particles in serum and 10^3 particles in buffer, highlighting the potential for the device to be miniaturized and integrated for POC applications [157]. Hoffman et al. developed a planar waveguide YI that they used to determine the binding kinetics for protein G capture of immunoglobulin G (IgG) and demonstrated its compatibility with biotin–streptavidin functionalization techniques [158]. The authors reported an RI LOD of 10^{-9} RIU which

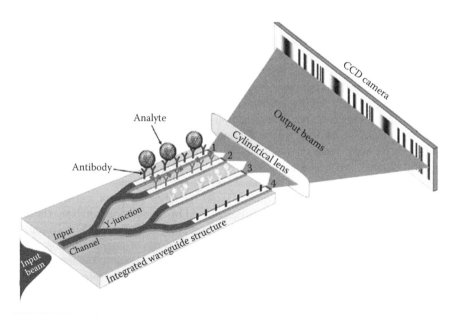

FIGURE 3.9 Schematic representation of a multiplexed integrated YI. Antibodies with different specificities are functionalized on the waveguide arms such that specific recognition and binding of analyte alters the local RI, leading to changes in the interference pattern detected by the CCD camera. (From Ymeti, A. et al., *Nano Letters*, 7(2), 394, 2007. With permission.)

corresponds to a surface coverage of just 13 fg/mm^2, one of the lowest reported values of any optical biosensor. However, it should be noted that while there have been a number of reports demonstrating MZI and YI multiplexed sensing, interferometric biosensing has not proven to be readily amenable to high-throughput multiplexing due to the large sensing regions required and because the complexity of analysis increases significantly with each additional sensing arm.

3.3.3.4 Resonant Cavity Sensors

Resonant cavities represent one of the most rapidly expanding and promising label-free optical biosensing techniques due largely to their high sensitivity and their potential to be integrated into multiplexed chip-based devices. In resonant cavity sensors, which include microspheres (Figure 3.10A), microtoroids (Figure 3.10B), microrings (Figure 3.2A and B), and microcapillaries, light is coupled into an optical cavity where certain wavelengths are confined; this confinement generates a narrow dip in the transmission spectrum. The wavelengths of light that travel around the outside of the cavity and return in phase are the resonant wavelengths and can be described by the equation $\lambda = 2\pi r n_{eff}/m$, where λ is the wavelength of light, r is the radius of the cavity, n_{eff} is the effective RI of the waveguide mode, and m is an integer. A fiber or integrated bus waveguide in close proximity to the resonant cavity delivers light to the cavity for coupling and away from the cavity so that the transmission spectrum can be tracked. Resonant wavelengths appear as dips in the transmission spectrum because the resonant condition extracts power from the light in the fiber/waveguide that reaches the detector [159,160]. The dependence of the resonant wavelength on n_{eff} is due to the evanescent field that extends and decays exponentially away from the surface of the cavity, and, like the other optical biosensors discussed in this chapter, it is this relationship that creates the sensing mechanism. By changing n_{eff}, biomolecules binding to the resonant cavity shift the resonant wavelengths supported by the structure. In contrast to the other evanescent sensing techniques already described (SPR, grating-coupled devices, and interferometers), where each photon only interacts with the biomolecules one time, a photon coupled into a resonant cavity interacts with biomolecules each time it travels around the cavity, which can reach into the thousands for some resonant cavities [160]. This feature bestows high sensitivities to small devices, which is not possible with other optical biosensors (e.g., interferometric sensors). The number of revolutions a photon makes around a resonant cavity before dissipating is related to the quality factor (Q) of the resonator and determines the sensitivity of the device [160]. Q is determined by the full-width half maximum ($\delta\lambda_r$) of the resonant dip at the resonant wavelength (λ_r), according to the equation $Q = \lambda_r/\delta\lambda_r$. Thus, a higher Q corresponds to a more narrow dip in the transmission spectrum which facilitates sensitive tracking of the resonant wavelength. Demonstrations of optical cavities implemented as biosensors include microspheres, microcapillaries, and microfabricated chip-based structures such as micro-toroids and microrings.

3.3.3.4.1 Microspheres and Micro-Toroids

Microspheres exhibit Q-factors over 10^6 and they have had RI LODs reported as low as 10^{-7} RIU [161,162]. Resonant cavity microspheres are generally constructed by melting the tip of an optical fiber or a glass rod [119], which must then be brought into

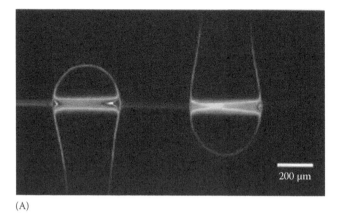

(A)

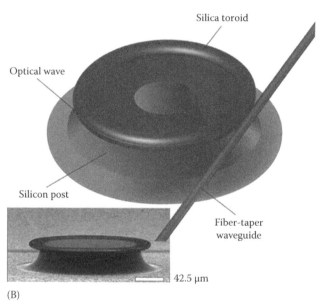

Silica toroid

Optical wave

Silicon post

Fiber-taper
waveguide

42.5 μm

(B)

FIGURE 3.10 Resonant cavity biosensors confine the wavelengths of light which, after circumnavigating the cavity, constructively interfere with itself. Biosensing is possible because the resonant frequency is sensitive to perturbations in the surrounding RI. Examples include (A) silica microspheres (From Vollmer, F., et al., *Biophysical Journal*, 2003. With permission.) and (B) silica microtoroids fabricated on a silicon support. (From Vahala, K.J., *Nature*, 424(6950), 2003. With permission.)

close proximity to and aligned with a tapered fiber. Demonstrations of resonant microsphere biosensors include the detection of protease activity with an LOD of trypsin at 10^{-4} units/mL [163], detection and mass determination of single influenza A virus particles [160], and the detection of single nucleotide mismatch of DNA with an LOD of 6 pg/mm^2 [164]. The device used for DNA detection used two microspheres of different sizes brought into proximity of a single tapered fiber. Because of their different size, each microsphere had a unique resonant wavelength and they could be interrogated simultaneously. Despite this proof-of-concept multiplexed device, microsphere-based resonant cavity biosensors are resistant to large-scale multiplexing because of the sensitive alignment required between the microspheres and the tapered fiber and because they are incompatible with planar fabrication techniques [25]. Armani et al. developed micro-toroid resonant cavities with extremely high Q ($>10^8$), which were fabricated using planar lithography. The authors report remarkable single-molecule label-free detection of interleukin-2 (IL-2) via capture by immobilized IL-2 antibody in 10-fold diluted fetal bovine serum [165]. This, while impressive, also raises a number of questions related to the reported sensing mechanism [166] and the observed mass transport [167], suggesting that we still have much to learn about ultrasensitive optical biosensing. Armani's results have set a high standard for optical sensing, but they do not address our need for high-throughput multiplexed sensing. While the micro-toroids were fabricated on-chip using photolithography, the technique requires alignment of a tapered optical fiber waveguide for coupling. Further, the inherent fragility of both microsphere and micro-toroid systems make them sensitive to flow, particularly with viscous fluids, such as blood plasma.

3.3.3.4.2 Microrings

Planar microrings have arguably become the most popular form of resonant cavity biosensors, owing to their small size, high sensitivity, ease of manufacture, and multiplexing potential. Ring size can vary, but nearly all are on the order of tens of microns in diameter, which is favorable as compared to interferometric devices which require sensing lengths on the order of a centimeter [61]. Microring resonators do not have a decreased sensitivity on account of their small size because of the increased light interaction imparted by the resonance, as previously discussed. They do have lower Q ($10^4 - 4 \times 10^4$) and slightly higher (worse) reported RI LODs (10^5–10^7 RIU) [31,168,169] than microspheres and microtoroids, but their simple and scalable fabrication, multiplexing capability, and potential for integration with other components make them attractive for biosensing applications. Microring resonators can be fabricated using standard silicon wafer processes, enabling passive alignment of multiple microrings with on-chip bus waveguides, which is a significant advantage over the microsphere and micro-toroid devices. While they are almost universally fabricated on a silicon substrate, the waveguides and rings can be made out of polymers [169,170], silicon oxide [171,172], silicon nitride [168,173,174], and SOI [175–177]. Sensitive multiplexed detection and binding assays using microring resonators are demonstrative of the advantages of this biosensing platform. Using a device containing five independent microrings, Ramachandran et al. showed specific binding of E. coli O157:H7 to microrings functionalized with antibodies, detection of complementary DNA probes, and quantitative detection of IgG [172]. Although not unique to this device, three disadvantages become apparent: (1) relatively low acquisition rates, (2) the lack of integrated fluidics, and (3) a paucity of high-throughput functionalization techniques. In the device reported by Ramachandran et al., the scan rate was 15 s per microring, limiting measurement frequency to 75 s per device, if all rings were interrogated [172]. Faster scan rates are required to extract binding kinetics and for truly high-throughput multiplexed measurements. An instrument containing integrated fluidics and peripheral instrumentation for using disposable chips was reported by Carlborg et al. [168] Device characterization showed an RI LOD of 5×10^{-6} RIU and a mass density detection limit of $0.9 \, pg/mm^2$. The authors have since published on characterization of the temperature sensitivity of this device [178], but they have yet to report on its implementation in a biosensing experiment. Another instrument used extensively by ourselves and the Bailey group at the University of Illinois Urbana Champagne directly addresses the issues of scanning speed and fluidic integration, and both of our groups have devised improved techniques for differential functionalization of the microrings [179,180]. The platform has a detachable microfluidic chamber and uses high-speed scanning instrumentation which interrogates all 32 rings on the device in fewer than 10 s [31]. Bailey's group has reported on detection of carcinoembryonic antigen (CEA) in undiluted serum down to 2 ng/mL [177], detection of Jurkat T lymphocyte secretions of IL-2 and IL-8 [181], detection of multiple micro RNAs with the ability to distinguish between single nucleotide polymorphisms [182], and quantitative detection of five protein biomarkers in mixed samples [183]. This group also did a thorough theoretical and empirical analysis to characterize the mass sensitivity and the evanescent sensing field of the microrings, finding a mass sensitivity of $1.5 \, pg/mm^2$ and a $1/e$ evanescent decay distance of 63 nm [60]. Such characterization is rare within the field of biosensors, yet this information is critical for experimental design and interpretation of results. For multiplexing functionalization, Bailey has employed a six-channel microfluidic device to differentially functionalize groups of microrings [183]. Using the same microring-based biosensor, our group has implemented a piezoelectric spotter to differentially functionalize microrings on multiple chips in a single run, thus demonstrating a rapid and scalable approach [180]. An alternative method for addressing the issue of scan speed mentioned earlier was demonstrated by Xu et al. [175]. Instead of increasing the scan speed, the investigators used a single waveguide to interrogate five rings with different radii. Since rings of varying diameter will support resonances of different wavelengths, Xu was able to distinguish shifts in the resonant frequency due to binding of species-specific IgG capture on each microring. In addition to demonstrating specific and simultaneous detection of two different IgG antibodies, the researchers deduced an impressive mass density sensitivity of $0.3 \, pg/mm^2$. It is clear that devices implementing microring resonators have made significant advances toward realizing applications beyond the lab bench and into the clinic. The combination of high sensitivities, ease of manufacture, multiplexibility, and potential for integration has positioned the microring resonator-based device as one of the most promising optical sensing technologies to emerge from the biosensing community.

3.4 OUTLOOK AND CONCLUSIONS

Frustratingly, biosensor technology remains largely confined to the research setting, and very few technologies have made it to the clinic, to the general public, or to the POC setting—where the need is great. In a survey of the biosensing literature, and even within this chapter, it becomes apparent that this is not for want of new sensing techniques or increased sensitivity. Instead, the biosensing community continues to produce new devices with new or improved approaches for accomplishing similar goals. All too often, promising new technologies are falling short of the goal of making an impact in healthcare, drug discovery, environmental monitoring, defense, etc. Clearly, increased attention needs to be directed toward realizing impactful applications of the technology.

Fully integrated devices open up many possibilities for real-world applications, but in order to gain traction and establish biosensors as an effective tool, research must focus on a few strategic areas where biosensors can make the most immediate and meaningful impact. More focused, application-driven, and collaborative research and development efforts would increase the likelihood of overcoming the hurdles that are currently preventing biosensors from being implemented in POC settings. For instance, targeting specific applications that demonstrate the most need will attract the funding that will be needed to fully develop the biosensor and get it through clinical trials. Simply put, technology is no longer the limiting factor to more fully incorporating biosensors into healthcare—increasingly it has become a problem of systems integration and design of application-centric biosensors.

Over the past several decades, significant effort has been invested with the aim of developing sensing technologies that will impact the practice of biomedical research and healthcare. This investment has yielded a plethora of sensing technologies built upon a host of sensing modalities (i.e., electrochemical, mechanical, and optical). Ultimately, there is no one-size-fits-all solution for biosensing, and in this chapter, we have argued for a few important design considerations for developing application-based sensors. (1) A biosensor should be *sensitive* and *selective* for the intended analyte(s) within complex samples, such as saliva, blood, or urine. (2) *Label-free* detection can decrease assay time, costs, and complexity, and is generally more flexible than its label-based counterpart. (3) *Multiplexing* confers enhanced reliability by allowing in-line controls and increased assay information density, thereby reducing costs associated with multiple tests. Finally, (4) a *fully integrated* platform, including peripheral instrumentation (e.g., a light source, a detector, and a microprocessor) and sample handling capabilities (e.g., pumps and microfluidic channels) in addition to the sensor, is essential for these devices to expand beyond the lab and to the POC.

Silicon photonic optical biosensors are the most promising candidate technology with the potential to integrate all of these design features. As an optical biosensing technique, these devices are label-free because they rely on the inherent refractive indices of the analyte to generate the signal. Their limited size and high sensitivity will enable massively parallelized multiplexed sensing using wafer-scale processing, dramatically reducing the cost and complexity of fabricating thousands of devices onto a single chip. In addition, by leveraging microelectronic fabrication techniques, silicon photonic biosensors can be integrated with planar on-chip light sources, detectors, and microprocessors. Microfluidics, including pumps, sample preparation strategies, and optical components, can be readily incorporated onto these planar features. Ultimately, the barriers to achieving a fully integrated biosensor using silicon photonics appear to be lower than they are for other sensing modalities.

ACKNOWLEDGMENTS

This work was supported by NSF CBET (award no. 0930411) and the Washington Research Foundation. JWC wishes to thank the NSF graduate research fellowship program. The authors would also like to thank Jim Kirk and Mike Gould for their edits and valuable discussions.

REFERENCES

1. Polanski, M. and N.L. Anderson, A list of candidate cancer biomarkers for targeted proteomics, *Biomarker Insights*, 2007, **1**: 1–48.
2. Sun, Y.S. et al., Effect of fluorescently labeling protein probes on kinetics of protein–ligand reactions, *Langmuir*, 2008, **24**(23): 13399–13405.
3. Kodadek, T., Protein microarrays: Prospects and problems, *Chemistry & Biology*, 2001, **8**(2): 105–115.
4. Hood, L. et al., Systems biology and new technologies enable predictive and preventative medicine, *Science*, 2004, **306**(5296): 640–643.
5. Soper, S.A. et al., Point-of-care biosensor systems for cancer diagnostics/prognostics, *Biosensors & Bioelectronics*, 2006, **21**(10): 1932–1942.
6. Sidransky, D., Emerging molecular markers of cancer, *Nature Reviews Cancer*, 2002, **2**(3): 210–219.
7. Wulfkuhle, J.D., L.A. Liotta, and E.F. Petricoin, Proteomic applications for the early detection of cancer, *Nature Reviews Cancer*, 2003, **3**(4): 267–275.
8. Hernandez, J. and I.M. Thompson, Prostate-specific antigen: A review of the validation of the most commonly used cancer biomarker, *Cancer*, 2004, **101**(5): 894–904.
9. Karlsson, K.A., Bacterium-host protein–carbohydrate interactions and pathogenicity, *Biochemical Society Transactions*, 1999, **27**(4): 471–474.
10. Nagahori, N. et al., Inhibition of adhesion of type 1 fimbriated *Escherichia coli* to highly mannosylated ligands, *Chembiochem*, 2002, **3**(9): 836–844.
11. Autar, R. et al., Adhesion inhibition of F1C-fimbriated *Escherichia coli* and *Pseudomonas aeruginosa* PAK and PAO by multivalent carbohydrate ligands, *Chembiochem*, 2003, **4**(12): 1317–1325.
12. Disney, M.D. et al., Detection of bacteria with carbohydrate-functionalized fluorescent polymers, *Journal of the American Chemical Society*, 2004, **126**(41): 13343–13346.
13. Disney, M.D. and P.H. Seeberger, The use of carbohydrate microarrays to study carbohydrate–cell interactions and to detect pathogens, *Chemistry & Biology*, 2004, **11**(12): 1701–1707.
14. Smith, A.E. and A. Helenius, How viruses enter animal cells, *Science*, 2004, **304**(5668): 237–242.
15. Eisen, M.B. and P.O. Brown, DNA arrays for analysis of gene expression, in c*DNA Preparation and Characterization*, pp. 179–205, Academic press, San Diego, CA, 1999.
16. Heller, M.J., DNA microarray technology: Devices, systems, and applications, *Annual Review of Biomedical Engineering*, 2002, **4**: 129–153.
17. Templin, M.F. et al., Protein microarray technology, *Trends in Biotechnology*, 2002, **20**(4): 160–166.
18. Zhu, H. and M. Snyder, Protein chip technology, *Current Opinion in Chemical Biology*, 2003, **7**(1): 55–63.
19. Ratner, D.M. et al., Probing protein–carbohydrate interactions with microarrays of synthetic oligosaccharides, *Chembiochem*, 2004, **5**(3): 379–382.

20. Wang, D.N. et al., Carbohydrate microarrays for the recognition of cross-reactive molecular markers of microbes and host cells, *Nature Biotechnology*, 2002, **20**(3): 275–281.
21. Feizi, T. et al., Carbohydrate microarrays—A new set of technologies at the frontiers of glycomics, *Current Opinion in Structural Biology*, 2003, **13**(5): 637–645.
22. Blixt, O. et al., Printed covalent glycan array for ligand profiling of diverse glycan binding proteins, *Proceedings of the National Academy of Sciences of the United States of America*, 2004, **101**(49): 17033–17038.
23. Bally, M. et al., Optical microarray biosensing techniques, *Surface and Interface Analysis*, 2006, **38**(11): 1442–1458.
24. Campbell, C.T. and G. Kim, SPR microscopy and its applications to high-throughput analyses of biomolecular binding events and their kinetics, *Biomaterials*, 2007, **28**(15): 2380–2392.
25. Jokerst, N. et al., Chip scale integrated microresonator sensing systems, *Journal of Biophotonics*, 2009, **2**(4): 212–226.
26. Myers, F.B. and L.P. Lee, Innovations in optical microfluidic technologies for point-of-care diagnostics, *Lab Chip*, 2008, **8**(12): 2015–2031.
27. Monat, C., P. Domachuk, and B.J. Eggleton, Integrated optofluidics: A new river of light, *Nature Photonics*, 2007, **1**(2): 106–114.
28. Momeni, B. et al., Silicon nanophotonic devices for integrated sensing, *Journal of Nanophotonics*, 2009, **3**: 031001.
29. Balslev, S. et al., Lab-on-a-chip with integrated optical transducers, *Lab Chip*, 2006, **6**(2): 213–217.
30. Seo, S.W., S.Y. Cho, and N.M. Jokerst, A thin-film laser, polymer waveguide, and thin-film photodetector cointegrated onto a silicon substrate, *IEEE Photonics Technology Letters*, 2005, **17**(10): 2197–2199.
31. Iqbal, M. et al., Label-free biosensor arrays based on silicon ring resonators and high-speed optical scanning instrumentation, *IEEE Journal of Selected Topics in Quantum Electronics*, 2010, **16**(3): 654–661.
32. Seo, S.W. et al., High-speed large-area inverted InGaAs thin-film metal–semiconductor–metal photodetectors, *IEEE Journal of Selected Topics in Quantum Electronics*, 2004, **10**(4): 686–693.
33. Cho, S.Y. and N.M. Jokerst, Integrated thin film photodetectors with vertically coupled microring resonators for chip scale spectral analysis, *Applied Physics Letters*, 2007, **90**(10): 101105.
34. Cho, S.Y. and N.M. Jokerst, A polymer microdisk photonic sensor integrated onto silicon, *IEEE Photonics Technology Letters*, 2006, **18**(17–20): 2096–2098.
35. Seo, S.W., S.Y. Cho, and N.M. Jokerst, Integrated thin film InGaAsP laser and I X 4 polymer multimode interference splitter on silicon, *Optics Letters*, 2007, **32**(5): 548–550.
36. Balslev, S. et al., Micro-fabricated single mode polymer dye laser, *Optics Express*, 2006, **14**(6): 2170–2177.
37. Seo, J. and L.P. Lee, Disposable integrated microfluidics with self-aligned planar microlenses, *Sensors and Actuators B—Chemical*, 2004, **99**(2–3): 615–622.
38. Llobera, A. et al., Multiple internal reflection poly(dimethylsiloxane) systems for optical sensing, *Lab on a Chip*, 2007, **7**(11): 1560–1566.
39. Chediak, J.A. et al., Heterogeneous integration of CdS filters with GaN LEDs for fluorescence detection microsystems, *Sensors and Actuators A—Physical*, 2004, **111**(1): 1–7.
40. Llobera, A. et al., Monolithic PDMS passband filters for fluorescence detection, *Lab on a Chip*, 2010, **10**(15): 1987–1992.
41. Ligler, F.S., Perspective on optical biosensors and integrated sensor systems, *Analytical Chemistry*, 2009, **81**(2): 519–526.

42. Zhang, C.S., D. Xing, and Y.Y. Li, Micropumps, microvalves, and micromixers within PCR microfluidic chips: Advances and trends, *Biotechnology Advances*, 2007, **25**(5): 483–514.

43. Oh, K.W. and C.H. Ahn, A review of microvalves, *Journal of Micromechanics and Microengineering*, 2006, **16**(5): R13–R39.

44. Beebe, D.J., G.A. Mensing, and G.M. Walker, Physics and applications of microfluidics in biology, *Annual Review of Biomedical Engineering*, 2002, **4**: 261–286.

45. Iverson, B.D. and S.V. Garimella, Recent advances in microscale pumping technologies: A review and evaluation, *Microfluidics and Nanofluidics*, 2008, **5**(2): 145–174.

46. Wang, X.Y. et al., Electroosmotic pumps and their applications in microfluidic systems, *Microfluidics and Nanofluidics*, 2009, **6**(2): 145–162.

47. Mansur, E.A. et al., A state-of-the-art review of mixing in microfluidic mixers, *Chinese Journal of Chemical Engineering*, 2008, **16**(4): 503–516.

48. Chang, C.C. and R.J. Yang, Electrokinetic mixing in microfluidic systems, *Microfluidics and Nanofluidics*, 2007, **3**(5): 501–525.

49. Wang, Y.C. and J.Y. Han, Pre-binding dynamic range and sensitivity enhancement for immuno-sensors using nanofluidic preconcentrator, *Lab on a Chip*, 2008, **8**(3): 392–394.

50. Yu, H. et al., A simple, disposable microfluidic device for rapid protein concentration and purification via direct-printing, *Lab on a Chip*, 2008, **8**(9): 1496–1501.

51. Weigl, B.H. and P. Yager, Microfluidics—Microfluidic diffusion-based separation and detection, *Science*, 1999, **283**(5400): 346–347.

52. Gossett, D.R. et al., Label-free cell separation and sorting in microfluidic systems, *Analytical and Bioanalytical Chemistry*, 2010, **397**(8): 3249–3267.

53. Di Carlo, D. et al., Continuous inertial focusing, ordering, and separation of particles in microchannels, *Proceedings of the National Academy of Sciences of the United States of America*, 2007, **104**(48): 18892–18897.

54. Reed, G.T. and A.P. Knights, *Silicon Photonics—An Introduction*, 2004, John Wiley & Sons Ltd, England, U.K.

55. Intel Corporation, Silicon Photonics Research [cited 2011 4/8/2011]; available from: http://techresearch.intel.com/ResearchAreaDetails.aspx?Id=26

56. University of Washington, OpSIS (Optoelectronic Systems Integration in Silicon), 2011 [cited 2011 4/8/2011]; available from: http://depts.washington.edu/uwopsis/

57. Hewitt, P.D. and G.T. Reed, Improving the response of optical phase modulators in SOI by computer simulation, *Journal of Lightwave Technology*, 2000, **18**(3): 443–450.

58. Luxtera, 2011 [cited 2011 4/8/2011]; available from: http://www.luxtera.com/

59. Soref, R.A., J. Schmidtchen, and K. Petermann, Large single-mode rib wave-guides in Ge Si–Si and Si-on-Sio$_2$, *IEEE Journal of Quantum Electronics*, 1991, **27**(8): 1971–1974.

60. Luchansky, M.S. et al., Characterization of the evanescent field profile and bound mass sensitivity of a label-free silicon photonic microring resonator biosensing platform, *Biosensors & Bioelectronics*, 2010, **26**(4): 1283–1291.

61. Erickson, D. et al., Nanobiosensors: Optofluidic, electrical and mechanical approaches to biomolecular detection at the nanoscale, *Microfluidics and Nanofluidics*, 2008, **4**(1–2): 33–52.

62. Newman, J.D. and A.P.F. Turner, Home blood glucose biosensors: A commercial perspective, *Biosensors & Bioelectronics*, 2005, **20**(12): 2435–2453.

63. Pejcic, B. and R. De Marco, Impedance spectroscopy: Over 35 years of electrochemical sensor optimization, *Electrochimica Acta*, 2006, **51**(28): 6217–6229.

64. Daniels, J.S. and N. Pourmand, Label-free impedance biosensors: Opportunities and challenges, *Electroanalysis*, 2007, **19**(12): 1239–1257.

65. Patolsky, F., G.F. Zheng, and C.M. Lieber, Nanowire-based biosensors, *Analytical Chemistry*, 2006, **78**(13): 4260–4269.

66. Sadik, O.A., A.O. Aluoch, and A.L. Zhou, Status of biomolecular recognition using electrochemical techniques, *Biosensors & Bioelectronics*, 2009, **24**(9): 2749–2765.
67. Bellan, L.M., D. Wu, and R.S. Langer, Current trends in nanobiosensor technology, *Wiley Interdisciplinary Reviews: Nanomedicine and Nanobiotechnology*, 2011, **3**: 229–246.
68. Yu, X.B. et al., An impedance array biosensor for detection of multiple antibody–antigen interactions, *Analyst*, 2006, **131**(6): 745–750.
69. Lodes, M.J. et al., Use of semiconductor-based oligonucleotide microarrays for influenza A virus subtype identification and sequencing, *Journal of Clinical Microbiology*, 2006, **44**(4): 1209–1218.
70. Maurer, K. et al., The removal of the t-BOC group by electrochemically generated acid and use of an addressable electrode array for peptide synthesis, *Journal of Combinatorial Chemistry*, 2005, **7**(5): 637–640.
71. Maurer, K. et al., Use of a multiplexed CMOS microarray to optimize and compare oligonucleotide binding to DNA probes synthesized or immobilized on individual electrodes, *Sensors*, 2010, **10**(8): 7371–7385.
72. Cooper, J. et al., Targeted deposition of antibodies on a multiplex CMOS microarray and optimization of a sensitive immunoassay using electrochemical detection, *Plos One*, 2010, **5**(3): e9781.
73. Patolsky, F. et al., Electrical detection of single viruses, *Proceedings of the National Academy of Sciences of the United States of America*, 2004, **101**(39): 14017–14022.
74. Li, C. et al., Complementary detection of prostate-specific antigen using In(2)O(3) nanowires and carbon nanotubes, *Journal of the American Chemical Society*, 2005, **127**(36): 12484–12485.
75. Cui, Y. et al., Nanowire nanosensors for highly sensitive and selective detection of biological and chemical species, *Science*, 2001, **293**(5533): 1289–1292.
76. Hahm, J. and C.M. Lieber, Direct ultrasensitive electrical detection of DNA and DNA sequence variations using nanowire nanosensors, *Nano Letters*, 2004, **4**(1): 51–54.
77. Zhang, G.J. et al., Highly sensitive measurements of PNA–DNA hybridization using oxide-etched silicon nanowire biosensors, *Biosensors & Bioelectronics*, 2008, **23**(11): 1701–1707.
78. Malhotra, B.D., A. Chaubey, and S.P. Singh, Prospects of conducting polymers in biosensors, *Analytica Chimica Acta*, 2006, **578**(1): 59–74.
79. Hu, P. et al., Self-assembled nanotube field-effect transistors for label-free protein biosensors, *Journal of Applied Physics*, 2008, **104**(7): 074310–074315.
80. Star, A. et al., Label-free detection of DNA hybridization using carbon nanotube network field-effect transistors, *Proceedings of the National Academy of Sciences of the United States of America*, 2006, **103**(4): 921–926.
81. Zheng, G.F. et al., Multiplexed electrical detection of cancer markers with nanowire sensor arrays, *Nature Biotechnology*, 2005, **23**(10): 1294–1301.
82. Stern, E. et al., Label-free immunodetection with CMOS-compatible semiconducting nanowires, *Nature*, 2007, **445**(7127): 519–522.
83. Yang, Y.T. et al., Zeptogram-scale nanomechanical mass sensing, *Nano Letters*, 2006, **6**(4): 583–586.
84. Marx, K.A., Quartz crystal microbalance: A useful tool for studying thin polymer films and complex biomolecular systems at the solution–surface interface, *Biomacromolecules*, 2003, **4**(5): 1099–1120.
85. Cooper, M.A. and V.T. Singleton, A survey of the 2001 to 2005 quartz crystal microbalance biosensor literature: Applications of acoustic physics to the analysis of biomolecular interactions, *Journal of Molecular Recognition*, 2007, **20**(3): 154–184.
86. Seker, S., Y.E. Arslan, and Y.M. Elcin, Electrospun nanofibrous PLGA/fullerene-C60 coated quartz crystal microbalance for real-time gluconic acid monitoring, *IEEE Sensors Journal*, 2010, **10**(8): 1342–1348.

87. Hianik, T. et al., Detection of aptamer–protein interactions using QCM and electrochemical indicator methods, *Bioorganic & Medicinal Chemistry Letters*, 2005, **15**(2): 291–295.
88. Hook, F. et al., Characterization of PNA and DNA immobilization and subsequent hybridization with DNA using acoustic-shear-wave attenuation measurements, *Langmuir*, 2001, **17**(26): 8305–8312.
89. Su, X.D. et al., Detection of point mutation and insertion mutations in DNA using a quartz crystal microbalance and MutS, a mismatch binding protein, *Analytical Chemistry*, 2004, **76**(2): 489–494.
90. Liebau, M., A. Hildebrand, and R.H.H. Neubert, Bioadhesion of supramolecular structures at supported planar bilayers as studied by the quartz crystal microbalance, *European Biophysics Journal with Biophysics Letters*, 2001, **30**(1): 42–52.
91. Shen, Z.H. et al., Nonlabeled quartz crystal microbalance biosensor for bacterial detection using carbohydrate and lectin recognitions, *Analytical Chemistry*, 2007, **79**(6): 2312–2319.
92. Mahon, E., T. Aastrup, and M. Barboiu, Dynamic glycovesicle systems for amplified QCM detection of carbohydrate-lectin multivalent biorecognition, *Chemical Communications*, 2010, **46**(14): 2441–2443.
93. Briand, E. et al., Combined QCM-D and EIS study of supported lipid bilayer formation and interaction with pore-forming peptides, *Analyst*, 2010, **135**(2): 343–350.
94. Linden, M.V. et al., Characterization of phosphatidylcholine/polyethylene glycol-lipid aggregates and their use as coatings and carriers in capillary electrophoresis, *Electrophoresis*, 2008, **29**(4): 852–862.
95. Cooper, M.A. et al., Direct and sensitive detection of a human virus by rupture event scanning, *Nature Biotechnology*, 2001, **19**(9): 833–837.
96. Dickert, F.L. et al., Bioimprinted QCM sensors for virus detection—Screening of plant sap, *Analytical and Bioanalytical Chemistry*, 2004, **378**(8): 1929–1934.
97. Su, X.L. and Y.B. Li, A self-assembled monolayer-based piezoelectric immunosensor for rapid detection of *Escherichia coli* O157: H7, *Biosensors & Bioelectronics*, 2004, **19**(6): 563–574.
98. Su, X.L. and Y.B. Li, A QCM immunosensor for *Salmonella* detection with simultaneous measurements of resonant frequency and motional resistance, *Biosensors & Bioelectronics*, 2005, **21**(6): 840–848.
99. Ma, Z.W., Z.W. Mao, and C.Y. Gao, Surface modification and property analysis of biomedical polymers used for tissue engineering, *Colloids and Surfaces B—Biointerfaces*, 2007, **60**(2): 137–157.
100. Fawcett, N.C. et al., QCM response to solvated, tethered macromolecules, *Analytical Chemistry*, 1998, **70**(14): 2876–2880.
101. Rabe, J. et al., Monolithic miniaturized quartz microbalance array and its application to chemical sensor systems for liquids, *IEEE Sensors Journal*, 2003, **3**(4): 361–368.
102. Fritz, J. et al., Translating biomolecular recognition into nanomechanics, *Science*, 2000, **288**(5464): 316–318.
103. Backmann, N. et al., A label-free immunosensor array using single-chain antibody fragments, *Proceedings of the National Academy of Sciences of the United States of America*, 2005, **102**(41): 14587–14592.
104. Wu, G.H. et al., Bioassay of prostate-specific antigen (PSA) using microcantilevers, *Nature Biotechnology*, 2001, **19**(9): 856–860.
105. Gupta, A.K. et al., Anomalous resonance in a nanomechanical biosensor, *Proceedings of the National Academy of Sciences of the United States of America*, 2006, **103**(36): 13362–13367.
106. Ilic, B., Y. Yang, and H.G. Craighead, Virus detection using nanoelectromechanical devices, *Applied Physics Letters*, 2004, **85**(13): 2604–2606.

107. Gupta, A., D. Akin, and R. Bashir, Single virus particle mass detection using microresonators with nanoscale thickness, *Applied Physics Letters*, 2004, **84**(11): 1976–1978.

108. Johnson, L. et al., Characterization of vaccinia virus particles using microscale silicon cantilever resonators and atomic force microscopy, *Sensors and Actuators B—Chemical*, 2006, **115**(1): 189–197.

109. Ilic, B. et al., Single cell detection with micromechanical oscillators, *Journal of Vacuum Science & Technology B*, 2001, **19**(6): 2825–2828.

110. Park, K. et al., 'Living cantilever arrays' for characterization of mass of single live cells in fluids, *Lab on a Chip*, 2008, **8**(7): 1034–1041.

111. Ilic, B. et al., Enumeration of DNA molecules bound to a nanomechanical oscillator, *Nano Letters*, 2005, **5**(5): 925–929.

112. Lee, J.H. et al., Immunoassay of prostate-specific antigen (PSA) using resonant frequency shift of piezoelectric nanomechanical microcantilever, *Biosensors & Bioelectronics*, 2005, **20**(10): 2157–2162.

113. Burg, T.P. et al., Weighing of biomolecules, single cells and single nanoparticles in fluid, *Nature*, 2007, **446**(7139): 1066–1069.

114. Barton, R.A. et al., Fabrication of a nanomechanical mass sensor containing a nanofluidic channel, *Nano Letters*, 2010, **10**(6): 2058–2063.

115. Godin, M. et al., Using buoyant mass to measure the growth of single cells, *Nature Methods*, 2010, **7**(5): 387–390.

116. Bryan, A.K. et al., Measurement of mass, density, and volume during the cell cycle of yeast, *Proceedings of the National Academy of Sciences of the United States of America*, 2010, **107**(3): 999–1004.

117. von Muhlen, M.G. et al., Label-free biomarker sensing in undiluted serum with suspended microchannel resonators, *Analytical Chemistry*, 2010, **82**(5): 1905–1910.

118. Rich, R.L. and D.G. Myszka, Grading the commercial optical biosensor literature—Class of 2008: 'The Mighty Binders', *Journal of Molecular Recognition*, 2010, **23**(1): 1–64.

119. Fan, X.D. et al., Sensitive optical biosensors for unlabeled targets: A review, *Analytica Chimica Acta*, 2008, **620**(1–2): 8–26.

120. Liedberg, B., C. Nylander, and I. Lundstrom, Surface-plasmon resonance for gas-detection and biosensing, *Sensors and Actuators*, 1983, **4**(2): 299–304.

121. Homola, J., Surface plasmon resonance sensors for detection of chemical and biological species, *Chemical Reviews*, 2008, **108**(2): 462–493.

122. Boozer, C. et al., Looking towards label-free biomolecular interaction analysis in a high-throughput format: A review of new surface plasmon resonance technologies, *Current Opinion in Biotechnology*, 2006, **17**(4): 400–405.

123. Qavi, A.J. et al., Label-free technologies for quantitative multiparameter biological analysis, *Analytical and Bioanalytical Chemistry*, 2009, **394**(1): 121–135.

124. Wassaf, D. et al., High-throughput affinity ranking of antibodies using surface plasmon resonance microarrays, *Analytical Biochemistry*, 2006, **351**(2): 241–253.

125. Usui-Aoki, K. et al., A novel approach to protein expression profiling using antibody microarrays combined with surface plasmon resonance technology, *Proteomics*, 2005, **5**(9): 2396–2401.

126. Dhayal, M. and D.A. Ratner, XPS and SPR analysis of glycoarray surface density, *Langmuir*, 2009, **25**(4): 2181–2187.

127. Corn, R.M. et al., Fabrication of DNA microarrays with poly(L-glutamic acid) monolayers on gold substrates for SPR imaging measurements, *Langmuir*, 2009, **25**(9): 5054–5060.

128. Spangler, B.D. et al., Comparison of the Spreeta (R) surface plasmon resonance sensor and a quartz crystal microbalance for detection of *Escherichia coli* heat-labile enterotoxin, *Analytica Chimica Acta*, 2001, **444**(1): 149–161.

129. Chinowsky, T.M. et al., Performance of the Spreeta 2000 integrated surface plasmon resonance affinity sensor, *Sensors and Actuators B—Chemical*, 2003, **91**(1–3): 266–274.
130. Codner, E.P. and Corn, R.M., Portable Surface Plasmon Resonance Imaging Instrument, 2006, Wisconsin Alumni Research Foundation, Madison, WI.
131. Fu, E. et al., SPR imaging-based salivary diagnostics system for the detection of small molecule analytes, *Oral-Based Diagnostics*, 2007, **1098**: 335–344.
132. Chinowsky, T.M. et al., Portable 24-analyte surface plasmon resonance instruments for rapid, versatile biodetection, *Biosensors & Bioelectronics*, 2007, **22**(9–10): 2268–2275.
133. Lee, C.Y. et al., Surface coverage and structure of mixed DNA/alkylthiol monolayers on gold: Characterization by XPS, NEXAFS, and fluorescence intensity measurements, *Analytical Chemistry*, 2006, **78**(10): 3316–3325.
134. Laibinis, P.E. et al., Comparison of the structures and wetting properties of self-assembled monolayers of normal-alkanethiols on the coinage metal-surfaces, Cu, Ag, Au, *Journal of the American Chemical Society*, 1991, **113**(19): 7152–7167.
135. Ulman, A., Formation and structure of self-assembled monolayers, *Chemical Reviews*, 1996, **96**(4): 1533–1554.
136. Rich, R.L. and D.G. Myszka, Survey of the year 2006 commercial optical biosensor literature, *Journal of Molecular Recognition*, 2007, **20**(5): 300–366.
137. Rich, R.L. and D.G. Myszka, Survey of the year 2007 commercial optical biosensor literature, *Journal of Molecular Recognition*, 2008, **21**(6): 355–400.
138. Kersey, A.D. et al., Fiber grating sensors, *Journal of Lightwave Technology*, 1997, **15**(8): 1442–1463.
139. Chryssis, A.N. et al., Detecting hybridization of DNA by highly sensitive evanescent field etched core fiber Bragg grating sensors, *IEEE Journal of Selected Topics in Quantum Electronics*, 2005, **11**(4): 864–872.
140. DeLisa, M.P. et al., Evanescent wave long period fiber Bragg grating as an immobilized antibody biosensor, *Analytical Chemistry*, 2000, **72**(13): 2895–2900.
141. Chen, X. et al., Dual-peak long-period fiber gratings with enhanced refractive index sensitivity by finely tailored mode dispersion that uses the light cladding etching technique, *Applied Optics*, 2007, **46**(4): 451–455.
142. Washburn, A.L. and R.C. Bailey, Photonics-on-a-chip: Recent advances in integrated waveguides as enabling detection elements for real-world, lab-on-a-chip biosensing applications, *Analyst*, 2011, **136**(2): 227–236.
143. Szekacs, A. et al., Development of a non-labeled immunosensor for the herbicide trifluralin via optical waveguide lightmode spectroscopic detection, *Analytica Chimica Acta*, 2003, **487**(1): 31–42.
144. Adanyi, N. et al., Development of immunosensor based on OWLS technique for determining Aflatoxin B1 and Ochratoxin A, *Biosensors & Bioelectronics*, 2007, **22**(6): 797–802.
145. Wittmer, C.R. and P.R. Van Tassel, Probing adsorbed fibronectin layer structure by kinetic analysis of monoclonal antibody binding, *Colloids and Surfaces B—Biointerfaces*, 2005, **41**(2–3): 103–109.
146. Blattler, T.M. et al., High salt stability and protein resistance of poly(L-lysine)-g-poly(ethylene glycol) copolymers covalently immobilized via aldehyde plasma polymer interlayers on inorganic and polymeric substrates, *Langmuir*, 2006, **22**(13): 5760–5769.
147. Horvath, R. et al., Structural hysteresis and hierarchy in adsorbed glycoproteins, *Journal of Chemical Physics*, 2008, **129**(7): 071102.
148. Adrian, J. et al., Wavelength-interrogated optical biosensor for multi-analyte screening of sulfonamide, fluoroquinolone, beta-lactam and tetracycline antibiotics in milk, *Trac-Trends in Analytical Chemistry*, 2009, **28**(6): 769–777.

149. Heideman, R.G., R.P.H. Kooyman, and J. Greve, Performance of a highly sensitive optical wave-guide Mach–Zehnder interferometer immunosensor, *Sensors and Actuators B—Chemical*, 1993, **10**(3): 209–217.

150. Heideman, R.G. and P.V. Lambeck, Remote opto-chemical sensing with extreme sensitivity: Design, fabrication and performance of a pigtailed integrated optical phase-modulated Mach–Zehnder interferometer system, *Sensors and Actuators B—Chemical*, 1999, **61**(1–3): 100–127.

151. Shew, B.Y., Y.C. Cheng, and Y.H. Tsai, Monolithic SU-8 micro-interferometer for biochemical detections, *Sensors and Actuators A-Physical*, 2008, **141**(2): 299–306.

152. Sánchez del Río, J., L.G. Carrascosa, F.J. Blanco, M. Moreno, J. Berganzo, A. Calle, C. Domínguez, and L.M. Lechuga, Lab-on-a-chip platforms based on highly sensitive nanophotonic Si biosensors for single nucleotide DNA testing, in *Proceedings of the SPIE: Silicon Photonics II*, Kubby, J.A. and Reed, G.T., ed., 2007, SPIE, San Jose, CA.

153. Densmore, A. et al., Silicon photonic wire biosensor array for multiplexed real-time and label-free molecular detection, *Optics Letters*, 2009, **34**(23): 3598–3600.

154. Brandenburg, A. and R. Henninger, Integrated optical Young interferometer, *Applied Optics*, 1994, **33**(25): 5941–5947.

155. Brandenburg, A., Differential refractometry by an integrated-optical Young interferometer, *Sensors and Actuators B—Chemical*, 1997, **39**(1–3): 266–271.

156. Ymeti, A. et al., Realization of a multichannel integrated Young interferometer chemical sensor, *Applied Optics*, 2003, **42**(28): 5649–5660.

157. Ymeti, A. et al., An ultrasensitive Young interferometer handheld sensor for rapid virus detection, *Expert Review of Medical Devices*, 2007, **4**(4): 447–454.

158. Hoffmann, C. et al., Interferometric biosensor based on planar optical waveguide sensor chips for label-free detection of surface bound bioreactions, *Biosensors & Bioelectronics*, 2007, **22**(11): 2591–2597.

159. Griffel, G. et al., Morphology-dependent resonances of a microsphere-optical fiber system, *Optics Letters*, 1996, **21**(10): 695–697.

160. Vollmer, F., S. Arnold, and D. Keng, Single virus detection from the reactive shift of a whispering-gallery mode, *Proceedings of the National Academy of Sciences of the United States of America*, 2008, **105**(52): 20701–20704.

161. Vollmer, F. et al., Protein detection by optical shift of a resonant microcavity, *Applied Physics Letters*, 2002, **80**(21): 4057–4059.

162. Hanumegowda, N.M. et al., Refractometric sensors based on microsphere resonators, *Applied Physics Letters*, 2005, **87**(20).

163. Hanumegowda, N.M. et al., Label-free protease sensors based on optical microsphere resonators, *Sensor Letters*, 2005, **3**(4): 315–319.

164. Vollmer, F. et al., Multiplexed DNA quantification by spectroscopic shift of two microsphere cavities, *Biophysical Journal*, 2003, **85**(3): 1974–1979.

165. Armani, A.M. et al., Label-free, single-molecule detection with optical microcavities, *Science*, 2007, **317**(5839): 783–787.

166. Arnold, S., S.I. Shopova, and S. Holler, Whispering gallery mode bio-sensor for label-free detection of single molecules: Thermo-optic vs. reactive mechanism, *Optics Express*, 2010, **18**(1): 281–287.

167. Squires, T.M., R.J. Messinger, and S.R. Manalis, Making it stick: Convection, reaction and diffusion in surface-based biosensors, *Nature Biotechnology*, 2008, **26**(4): 417–426.

168. Carlborg, C.F. et al., A packaged optical slot-waveguide ring resonator sensor array for multiplex label-free assays in labs-on-chips, *Lab on a Chip*, 2010, **10**(3): 281–290.

169. Chao, C.Y., W. Fung, and L.J. Guo, Polymer microring resonators for biochemical sensing applications, *IEEE Journal of Selected Topics in Quantum Electronics*, 2006, **12**(1): 134–142.

170. Huang, Y.Y. et al., Fabrication and replication of polymer integrated optical devices using electron-beam lithography and soft lithography, *Journal of Physical Chemistry B*, 2004, **108**(25): 8606–8613.
171. Yalcin, A. et al., Optical sensing of biomolecules using microring resonators, *IEEE Journal of Selected Topics in Quantum Electronics*, 2006, **12**(1): 148–155.
172. Ramachandran, A. et al., A universal biosensing platform based on optical micro-ring resonators, *Biosensors & Bioelectronics*, 2008, **23**(7): 939–944.
173. Barrios, C.A. et al., Label-free optical biosensing with slot-waveguides, *Optics Letters*, 2008, **33**(7): 708–710.
174. Hosseini, E.S. et al., Systematic design and fabrication of high-Q single-mode pulley-coupled planar silicon nitride microdisk resonators at visible wavelengths, *Optics Express*, 2010, **18**(3): 2127–2136.
175. Xu, D.X. et al., Label-free biosensor array based on silicon-on-insulator ring resonators addressed using a WDM approach, *Optics Letters*, 2010, **35**(16): 2771–2773.
176. De Vos, K. et al., SOI optical microring resonator with poly(ethylene glycol) polymer brush for label-free biosensor applications, *Biosensors & Bioelectronics*, 2009, **24**(8): 2528–2533.
177. Washburn, A.L., L.C. Gunn, and R.C. Bailey, Label-free quantitation of a cancer biomarker in complex media using silicon photonic microring resonators, *Analytical Chemistry*, 2009, **81**(22): 9499–9506.
178. Gylfason, K.B. et al., On-chip temperature compensation in an integrated slot-waveguide ring resonator refractive index sensor array, *Optics Express*, 2010, **18**(4): 3226–3237.
179. Bailey, R.C., A robust silicon photonic platform for multiparameter biological analysis, in *Proceedings of the SPIE: Silicon Photonics IV*, Kubby, J.A. and Reed, G.T. eds., The International Society for Optical Engineering, San Jose, CA, Vol. 7220, 2009.
180. Kirk, J.T. et al., Multiplexed inkjet functionalization of silicon photonic biosensors, *Lab Chip*, 2011, **11**(7): 1372–1377.
181. Luchansky, M.S. and R.C. Bailey, Silicon photonic microring resonators for quantitative cytokine detection and T-cell secretion analysis, *Analytical Chemistry*, 2010, **82**(5): 1975–1981.
182. Qavi, A.J. and R.C. Bailey, Multiplexed detection and label-free quantitation of microRNAs using arrays of silicon photonic microring resonators, *Angewandte Chemie-International Edition*, 2010, **49**(27): 4608–4611.
183. Washburn, A.L. et al., Quantitative, label-free detection of five protein biomarkers using multiplexed arrays of silicon photonic microring resonators, *Analytical Chemistry*, 2010, **82**(1): 69–72.

4 Quartz Crystal Microbalance–Based Biosensors

Şükran Şeker and Y. Murat Elçin

CONTENTS

ABBREVIATIONS

AW	Acoustic wave
BAW	Bulk acoustic wave
F	Frequency
FPW	Flexure plate wave
GOx	Glucose oxidase
LW	Love wave
PLGA	Poly(lactic-co-glycolic acid)

PS   Poly(styrene)
PZ   Piezoelectric
QCM   Quartz crystal microbalance
QCR   Quartz crystal resonator
R   Resistance
SAM   Self-assembled monolayer
SAW   Surface acoustic wave
SEM   Scanning electron microscopy
SH-APW   Shear horizontal acoustic plate wave
TSM   Thickness shear mode

4.1 INTRODUCTION

A biosensor is an analytical device that determines the specific interactions between biological molecules. Biosensors are commonly used in various fields such as food industry, environmental field, medicine, and drug discovery regarding its advantages such as label-free detection of molecules, fast response time, minimum sample pretreatment, and high sample throughput. Biosensors which convert a biological response into an electrical signal are integrated by a physicochemical transducer in various forms such as optical, electrochemical, thermometric, magnetic, or piezoelectric (PZ) [1]. Among the current biosensor systems, PZ biosensors play an important role in biochemical sensing. The quartz crystal microbalance (QCM) as a PZ sensor is fundamentally a mass sensing device with the ability to measure very small amounts of mass changes on a quartz crystal resonator (QCR) in real time. It operates on the principle that the resonance frequency of the quartz crystal changes with the amount of mass deposited. One of the major advantages of this technique is that it allows for a label-free detection of molecules.

The QCM, often called the QCR, is one of the most common acoustic-wave (AW) sensors operating with mechanical AWs as the transduction mechanism [2]. It works with a thickness shear mode (TSM) and has become a mature, commercially available, robust, and affordable technology [3]. The sensor's transducer element is based upon a solid plate of PZ material, usually a quartz crystal, which can generate AWs in the substrate in order to detect small mass uptakes in the nanogram level. Typically in QCM applications, a thin piece of quartz is compressed between two metal plates used as electrodes that have evaporated onto both sides.

It is apparent that the performance of a mechanically resonating sensor is largely influenced by the properties of the sensory coating or a film interacting with target molecules. Thus, surface coating and modification method is one of the most important factors for the improvement of the sensitivity, selectivity, and time response of a sensor system.

QCM techniques have been widely applied to investigate the interaction of biomolecules; thus, their types are categorized according to biorecognition component. QCM immunosensors have been used to detect antigen–antibody binding on the quartz crystal [4]. Enzyme-based QCM measures the product of the substrate conversion by means of the enzyme functioning in the enzymatic reaction [5]. Nucleic acid–based QCM biosensors are used to measure hybridization reaction between a probe oligonucleotide on the quartz with complementary single-stranded nucleotide in sample solution [6]. The QCM cell biosensors are used to detect the growth rate of cells on the crystal surface [7]. Moreover, cell–drug interaction in the drug discovery field [8] and the cell–materials interaction in biomaterials studies can be measured by means of the QCM [9].

4.2 PIEZOELECTRIC RESONATORS

4.2.1 FUNDAMENTALS OF PIEZOELECTRICITY AND QUARTZ CRYSTALS

PZ sensing is fundamentally based on PZ effect as described by Jacques and Pierre Curie in 1880. They found that when subjected to mechanical stress, a voltage occurred on the surface of certain types of crystals including quartz, tourmaline, and Rochelle salt. A year after their discovery, the opposite effect, the so-called the inverse PZ effect, was predicted from fundamental thermodynamic principles by Lippmann [10]. According to these phenomena, application of an electrical field on a PZ material causes mechanical deformation.

Figure 4.1 shows a schematic illustration of PZ effect. In a PZ material, the positive and negative charges are randomly distributed in each part of the surface. When some pressure on the material is applied, positive and negative charges in the molecules are separated from each other. This polarization generates an electrical field and can be used to transform the mechanical energy used in the material deformation into electrical energy [11].

A large number of crystals exhibit PZ effect, but the electrical, thermal, mechanical, and chemical properties of the quartz make it the most common crystal type used in analytical applications [12]. The quartz crystal is an SiO_2 monocrystal with a zinc-blend structure type, present in nature with various forms, α-quartz and

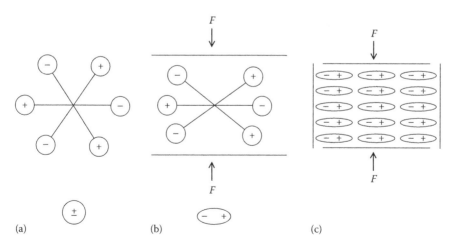

FIGURE 4.1 Simple molecular model for explaining PZ effect: (a) unperturbed molecule; (b) molecule subjected to an external force; (c) and polarizing effect on the material surfaces. (Adapted and redrawn from Springer Science+Business Media: *Piezoelectric Transducers and Applications*, 1st edn., 2004, Arnau, A.)

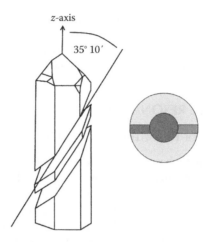

FIGURE 4.2 AT-cut quartz crystal. A quartz plate is cut with an angle of 35° 10′ with respect to the optical z-axis. Deviating just 5′ from the proper cut angle leads to a temperature coefficient that is different from zero (between 0°C and 50°C). (Adapted and redrawn from Janshoff, A. et al., *Angew. Chem. Int. Ed. Engl.*, 39, 4004, 2000.)

β-quartz. Alpha-quartz form is used for PZ applications since this form is insoluble in water and resistant to high temperatures.

The resonant frequency of quartz crystal vibration depends on the physical properties of the crystal (e.g., size, cut, density, and shear modulus). The quartz crystal is a precisely cut slab from a natural or synthetic crystal. Different types of quartz cuts have different properties and vibration modes. AT- and BT-cut crystals have commonly been used as PZ sensors. AT-cut quartz crystals are used as QCMs due to near zero frequency changes at the room temperature range.

Two crystal orientations oscillate exclusively in the TSM resonators, AT- and BT-cut crystals [13]; AT-cut crystals are in use as PZ sensors for bioanalytical applications. The AT-cut is made with an +35°10′ angle from the z-axis [14] (Figure 4.2). The advantage of the AT-cut quartz crystal is that it has a temperature coefficient of nearly zero at around room temperature.

In QCM sensor applications, the quartz crystal consists of a thin piece of AT-cut quartz crystal compressed between two metal plates used as electrodes on both sides as can be seen in Figure 4.3. When the electrodes are exposed to an AC current, the quartz crystal starts to oscillate at its resonance frequency due to the PZ effect.

4.2.2 THICKNESS SHEAR MODE RESONATORS (QUARTZ CRYSTAL MICROBALANCES)

There are several types of AW devices (Figure 4.4) depending on their wave propagation modes [14]. For example, TSM or bulk acoustic wave (BAW), flexure plate wave (FPW), surface acoustic wave (SAW), Love wave (LW), and shear horizontal acoustic wave mode resonators are the most commonly used resonators in sensor systems. BAW using QCMs travel through the interior of the substrate. By contrast, SAWs propagate on the surface of the crystal.

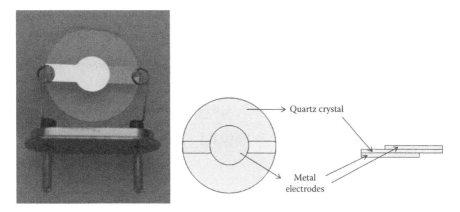

FIGURE 4.3 AT-cut quartz crystal used in QCM applications. The QCM consists of a thin piece of quartz crystal compressed between two metal plates coated on both sides, used as electrodes.

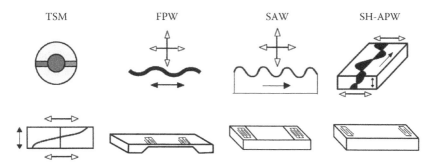

FIGURE 4.4 Schematic illustration of the four common kinds of AW sensors. TSM, thickness shear mode; FPW, flexure plate wave; SAW, surface acoustic wave; SH-APW, shear horizontal acoustic plate wave. (Adapted and redrawn from Janshoff, A. et al., *Angew. Chem. Int. Ed. Engl.*, 39, 4004, 2000.)

The QCM is a TSM-type resonator consisting of a thin disc of a AT-cut quartz with metal electrodes on both sides. The quartz crystal has a number of different resonator types depending on the cut angle determining the mode of mechanical vibration. AT-cut crystal operates in the TSM mode and is used for QCM systems. The oscillation in this mode creates a displacement parallel to the surface of the quartz wafer. A voltage applied between the electrodes causes a shear deformation of the quartz crystal. The TSM oscillation responds very sensitively to any mass changes on the crystal surfaces [15].

4.3 MASS LOADING EQUATIONS

Deposition of a mass on a thin quartz crystal surface induces a decrease in the resonant frequency for a rigid substance. The relationship between the change of the resonant frequency of a quartz crystal and adsorption of mass on the surface of a PZ resonator was formulated by Sauerbrey in 1959 [16]. Application of an alternating voltage potential across the quartz crystal surfaces causes the crystal to oscillate at a characteristic resonant frequency. According to this equation, the mass of a thin layer accumulated on the surface of a crystal can be calculated through measuring the changes in the resonant frequency:

$$\Delta F = \frac{-2F_o^2 \Delta M}{\left[A\left(\mu_q \rho_q\right)^{1/2} \right]} \tag{4.1}$$

where
   ΔF is the change in frequency in Hz
   F_o is the initial resonant frequency of the quartz crystal
   ΔM is the mass change in g cm^{-2}
   A is the area of the crystal in cm^2
   ρ_q is the density of quartz (2.648 g cm^{-3})
   μ_q is the shear modulus of quartz (2.947 × 10^{11} dyn cm^{-2})

The Sauerbrey equation is valid for oscillation in air and only applies to rigid masses attached to the crystal surface. Kanazawa and Gordon showed that when a crystal is dipped into a solution, the frequency shift also depends on the density and the viscosity of the liquid in contact with the QCM at one surface [17]. In this case, the viscosity and density of liquid affect the propagation of the shear wave that radiates from the resonator into the liquid media. Kanazawa equation, as shown in the following, was developed for the QCM measurements in the liquid phase:

$$\Delta F = -F_o^{3/2} \left[\frac{\left(\rho_L \eta_L\right)}{\left(\pi \rho_q \mu_q\right)} \right]^{1/2} \tag{4.2}$$

where
   ΔF is the measured frequency shift in this non-gravimetric regime
   ρ_L is the density of the liquid in contact with the crystal
   η_L is the viscosity of the liquid in contact with the crystal

4.4 METHODS OF QUARTZ CRYSTAL SURFACE MODIFICATIONS

The quality and sensitivity of the resonance frequency of crystal affect surface characteristics, such as the coating material, roughness, and hydrophobicity. Therefore, modification of the sensing surface is an important process in QCM biosensor studies. A variety of methods exist for surface coating or modification. The thin film deposition methods such as self-assembled monolayers (SAMs), electrochemical deposition, spin coating, and electrospinning have been commonly used to improve the functionalization of quartz crystal surfaces.

4.4.1 SELF-ASSEMBLED MONOLAYERS

Functionalized alkanethiolate SAMs are important in preparing biosensor surfaces. Modifying a quartz crystal surface with SAMs generates a suitable recognition layer with a specific property or function, for the use of a single sensor sensitive to a single compound. There are generally two kinds of methods depending on self-assembly for use in biosensor studies. The most commonly used are the gold-alkylthiolate monolayers and alkylsilane monolayers. Gold-alkylthiolate monolayer was first produced by Nuzzo and Allara in 1983 [18], demonstrating that alkanethiolates could be ordered on gold by adsorption of di-n-alkyl disulfides from dilute solutions.

In order to obtain well-ordered, defect-free SAMs, the quartz crystals are immersed into low concentrations of thiol solution (typically 1–2 mM) in ethanol overnight at room temperature, forming a monolayer bearing many active tails (–COOH, –NH$_2$, –OH, etc.), with improved analyte attachment features (Figure 4.5). The most commonly used solvent in preparation of SAMs is ethanol. One of the advantages of the gold-alkylthiolate monolayer is that it is stable when exposed to air and aqueous or ethanolic solutions for several months. The reason of this could be ascribed to the fact that SAMs on gold surfaces adsorb very strongly due to the formation of a covalent bond between the gold and the sulfur atoms [19,20].

The SAMs of functionalized alkanethiols on gold surfaces are used in various applications such as molecular recognition, selective immobilization of enzymes to surfaces, corrosion protection, patterned surfaces in micrometer-scale. The basic principle is that the thiol molecules adsorb readily from solution onto the gold surface, forming an ordered monolayer with the tail group. Different thiol molecules or mixing of thiol molecules can create SAM surfaces of desired chemical surface functionality and sizes. By performing some activation reactions, the tail groups can be chemically functionalized following the assembly of the SAM. And also, a thiol-derivatized probe can be covalently linked to a gold electrode of a QCM by thiol groups [21].

4.4.2 ELECTROCHEMICAL DEPOSITION

Electropolymerization technique has been used to form thin polymeric films on electrode surfaces. This electrochemical method has many advantages over the other coating methods for creating biosensors [22]. One major advantage is that the reproducible and precise formation of a polymeric coating over surfaces of desired

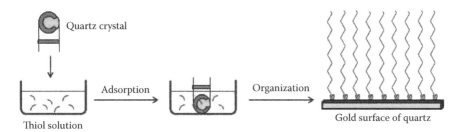

FIGURE 4.5   Schematic illustration of the preparation of SAMs.

size and geometry provides polymeric coating with an electrochemically controlled thickness.

The electrochemical quartz crystal microbalance (EQCM) system is a very convenient and powerful tool for creating electropolymerized thin films on the electrode surface of the quartz crystal (the working electrode) using cyclic voltammetry (CV); thus, the investigation of mass and viscoelastic properties is possible [23].

The electropolymerization of a wide range of different monomer types has been performed, such as pyrrole and substituted pyrrole, a large 32-unit ferrocenyle-dendrimer, reversible fullerene C_{60} derivatives, phenols, and biomimetic tyrosine derivatives [24].

4.4.3 ELECTROSPINNING AND SPIN COATING

Electrospinning [25] technique has become a popular method with the potential to produce nanofibrous nonwoven surfaces for a variety of applications, such as the modification of sensor surfaces [26–28]. Nanofibers fabricated via electrospinning have a surface area of approximately 10–100 times larger than that of continuous films, further increasing the adsorption rate and also the sensor sensitivity. The most important advantage of using nanofibers for coating is that it increases surface area by the highly porous topography for adhesion. In principle, by the help of the syringe pump, a polymer solution alone or with its content is transferred along the glass pipe to the needle where positive charge is applied. The charged polymer solution moves toward the grounded target, the solvent evaporates, attaining the three-dimensional nanofiber deposits [27] (Figure 4.6). Electrospun fibers with controllable membrane

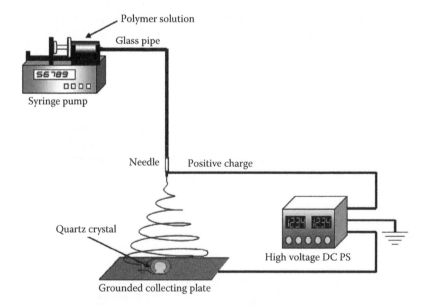

FIGURE 4.6 Schematic illustration of the electrospinning process used to coat quartz crystals with polymer solutions. (From Şeker, Ş. et al., *IEEE Sens. J.*, 10, 1342, 2010. With permission.)

FIGURE 4.7 Representative SEM images of (a) newly prepared and (b) enzyme (glucose oxidase) immobilized electrospun polymer coatings on coverslips used for the enzymatic oxidation of β-D-glucose. Images demonstrate the nonwoven mesh with randomly oriented nanofibers. Scale bars: (A) 10 μm and (B) 5 μm. (From Şeker, Ş. et al., *IEEE Sens. J.*, 10, 1342, 2010. With permission.)

thickness, fine structures, diversity of materials, and large specific surface can be prepared by this technique.

Recently, the electrospinning method has been used to produce polymeric nanofibers for sensing applications [5,29] (Figure 4.7). The fibrous membranes obtained by electrospinning technique have a strong potential application for sensor systems since the fibrous membranes have a larger surface area than that of continuous films.

Spin coating is an efficient, relatively simple and low-cost way to produce thin, uniform polymer coating a planar substrate [30]. In the spin coating process, the solution is first deposited onto a substrate which is then accelerated rapidly to the desired spin speed (Figure 4.8). The centrifugal force provides the spreading of

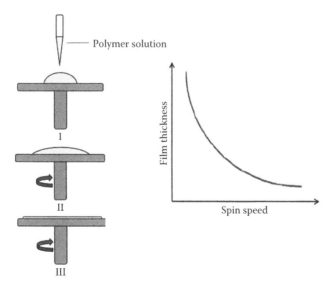

FIGURE 4.8 Schematic illustration of the spin coating process.

solution, resulting in the formation of homogeneous film on the substrate surface. The coating thickness depends on polymer concentration, solvent, and spin coating speed. Using spin coating technique, it is possible to reduce the surface roughness of the commercially produced quartz, in order to protect the electrode from oxidation [31].

4.5 BIOLOGICAL APPLICATIONS OF THE QCM

4.5.1 ENZYME BIOSENSORS

Enzyme biosensors based on highly specific enzymatic reactions have been widely exploited (in clinical and food analysis). In general, the QCM enzyme biosensors are used for the measurement of mass deposition of the product molecule from the enzymatic reaction.

Several QCM enzyme biosensors using immobilized urease [32] and glucose oxidase [33] have been studied. Wei and Shih [32] have developed a fullerene-cryptand-coated PZ urea sensor by measuring the ammonium ion, a product of the catalytic hydrolysis of urea by urease. They have found that the fullerene C_{60}-cryptand22 PZ crystal detection system exhibited good sensitivity and selectivity for urea with respect to other biological species in aqueous solutions. In the last case, a PZ sensor coated with a nanofibrous layer of PLGA containing saturated fullerene C_{60} was developed to detect gluconic acid, the oxidation product of β-D-glucose by glucose oxidase [5] (Figure 4.9). Fullerene-C_{60} containing nanofibrous poly(DL-lactide-co-glycolide) coatings with a thickness of ~625 nm were prepared by electrospinning on PZ quartz crystals. The sensor was able to monitor D-gluconic acid in real time, in the glucose concentration range between 1.4 and 14.0 mM quite linearly.

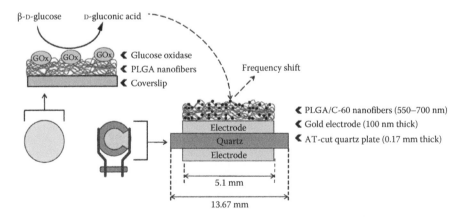

FIGURE 4.9 Schematic illustration of the real-time monitoring of gluconic acid by QCM. (From Şeker, Ş. et al., *IEEE Sens. J.*, 10, 1342, 2010. With permission.)

4.5.2 NUCLEIC ACID–BASED BIOSENSORS

QCM DNA biosensors (genosensors) are analytical devices consisting of an immobilized sequence-specific probe to detect the complimentary sequence by hybridization reaction. Analysis of specific DNA sequences in clinical, food, and environmental samples facilitates detection and identification of infectious diseases from biological species or living systems such as viruses and bacteria in real time, without the use of any labels such as radioisotopes, enzymes, and fluorophores. This has been presented as an alternative to traditional methods of detecting specific DNA sequences, where labeled probes are required [6]. The basis of operation for a PZ DNA biosensor is the monitoring of the decrease in oscillation frequency due to hybridization reaction on the quartz surface. The hybridization reaction is detected following the frequency change, resulting from the interaction between the single-stranded oligonucleotides (probe) immobilized on the quartz crystal and the target complementary strand in the solution. Many types of nucleic acid–based biosensors have already found use in several analytical fields, for instance, gene mutation [34], detection of genetically modified organisms [21], bacteria [35], or viruses [36], and in toxicology studies [37].

The probe is usually a short synthetic oligonucleotide that is immobilized onto the quartz crystal. In general, the probe sequence is 18–25 nucleotide in size; longer capture oligonucleotides often exhibit particularly unfavorable hybridization specificity, yielding to intramolecular hydrogen bonding and consequent formation of nonreactive hairpin structures [38].

Immobilization of probe DNA on the quartz crystal surface is a fundamental step in DNA biosensor development since the affinity and specificity of the biosensor can be greatly improved by choosing the proper immobilization procedure. The probe DNA should be attached to the crystal surface without losing its native conformation and activity. A number of immobilization methods have been employed to suitably bind the probe strand to quartz surfaces. The 5′-phosphate residues of the probe strand can be easily modified with thiol [39] or biotin [40]. Immobilization of oligonucleotide probes to quartz crystal surface is achieved most commonly by using the biotin–avidin interaction. In this method, the probe DNA biotinylated at the 5′ end is immobilized onto streptavidin-coated gold electrodes. The thiolated probe can be directly immobilized onto the gold surface of quartz crystal by SAM formation. The immobilization of thiol-modified probes is easily performed in one step.

4.5.3 QCM IMMUNOSENSORS

Nowadays, QCM immunosensors are the most widely used analysis tool for detecting antibody–antigen reactions. The method is based on the detection of the frequency shift, resulting from the highly specific interaction of the antibody immobilized on the quartz crystal surface and the antigen inside the solution. The QCM immunosensor has many potential applications, ranging from clinical diagnosis and food control to environmental analysis for the detection of bacteria or organic compounds by using an antigen–antibody reaction (Table 4.1). The advantage of the QCM immunosensor

TABLE 4.1

Immunosensor Applications of Piezoelectric Quartz Crystal Microbalance

Target	Immobilization Method	Detection Limit
Microorganism detection		
Escherichia coli [41]	SAM, Protein A	1.0×10^3 CFU mL^{-1}
Pseudomonas aeruginosa [42]	SAM via sulfo-LC-SPDP	$1.3 \times 10^7 - 1.3 \times 10^8$ cells mL^{-1}
Listeria monocytogenes [4]	SAM	1.0×10^7 cells mL^{-1}
Staphylococcal enterotoxins [43]	PEI, SAM, Protein A	$2.7 - 12.1$ μg mL^{-1}
Salmonella enteritidis [44]	SAM via MPA	1.0×10^5 cells mL^{-1}
Herpes viruses [45]	Protein A	5.0×10^4 cells mL^{-1}
Protein detection		
Ferritin [46]	SAM	2.4 nmol L^{-1}
α-Fetoprotein [47]	SAM	1.5 nmol L^{-1}
Albumin [48]	SAM/Protein A	$1-5$ μg mL^{-1}
Human chorionic gonadotropin [49]	SAM via sulfo-LC-SPD	$2.5 - 500$ mIU mL^{-1}

SAM, self-assembled monolayer; MPA, 3-mercaptopropionic acid; Sulfo-LC-SPDP, sulfosuccin-imidyl 6-[3-(2-pyridyldithio)propionamido] hexanoate.

is that it allows for the direct measurement of immunointeraction without using any labels and additional chemicals. By means of this interaction, the QCM immunosensors can qualitatively or quantitatively detect antigen or antibody on quartz crystal which is caused by a change in the resonant frequency.

Protein A is a widely used component for immobilizing an antibody, such as for immunosensor development. Gao et al. [50] developed a *Staphylococcal enterotoxin* C$_2$ (SEC$_2$) immunosensor which was coated with different immobilization methods of SEC$_2$ antibody on a gold electrode of the PZ crystal. In this study, the electrode coated with protein A demonstrated the best result for SEC$_2$ detection. Covalent immobilization to organic polymer layers is also used for immunosensor development. Polyethylenemine (PEI) crosslinked with glutaraldehyde is the most commonly used polymer in immunosensor development studies. Tsai and Lin [51] showed that the amount and the reaction activity of bound antibody on PEI film were better than those on SAM. The interaction between thiols and gold surface of PZ crystals has been used for the immobilization of antibodies. A monolayer of the thiol onto the gold surface of crystal is facilitated by covalent binding due to high affinity between the gold and sulfur atoms. Vaughan et al. [4] have developed an immunosensor to detect *Listeria monocytogenes* against antibody immobilized on an SAM of thiosalicylic acid. They have shown that the sensor could detect *L. monocytogenes* cells in real time in solution of 1.0×10^7 cells mL^{-1}.

4.5.4 QCM Mammalian Cell Biosensors

4.5.4.1 Detection of Cell–Surface Interactions

The detection of cell attachment onto various substrates is very important in determining mammalian cell behavior toward biomaterials and substances that are of biological importance. Determining materials that favor cell adhesion (a desirable condition for many medical implants) and developing materials that prevent cell adhesion are crucial [52]. One of the major shortcomings in biomaterials research and development studies is the lack of suitable tools to detect the cell–materials interactions under in vitro culture conditions in real time [53]. In order to detect cell attachment onto biomaterials, a variety of techniques have been used [54], that is, counting of labeled cells directly, measurement of the cellular zones and the cell density using optical techniques, and fluorescence density measurements to analyze cell adhesion strength. Following cell adhesion, the removal of cells from the substrate surface and the staining or fixation stages are regarded as disadvantages of such techniques [52]. The QCM system, however, is a useful technique to detect cell adhesion without removing cells from the surface and without using any labels [55].

PZ biosensors incorporating living cells as the active sensing element provide useful information regarding the properties of cultured cells, such as attachment, proliferation, cell–substrate and cell–drug interactions under different conditions. The behavior of cells toward materials is important in determining the compatibility of biomaterial surfaces (Figure 4.10).

QCM mammalian cell biosensor is a less intensely studied area than other QCM biosensor applications due to added difficulties [56]. Sterile culture conditions are required in these experiments. The cells must be attached to the quartz surface under cell culture conditions (e.g., 5% CO_2–95% air and 37°C) to create QCM cell biosensors. The quartz crystal surface should attain a certain level of hydrophilicity for cell attachment. Therefore, some modifications or coating techniques have been applied to improve the hydrophilic character of the hydrophobic gold surface of quartz crystal. In many studies, the gold surface of quartz crystal is treated chemically to render its hydrophilicity [57].

Many studies have focused on the process of cellular attachment as this behavior is important for biomaterial development. In QCM cell biosensors, endothelial cells (ECs) [58], osteoblasts [59], MDCK I and II cells [60], 3T3 cells [60], CHO cells [52], Neuro-2A cells [61], McCoy human fibroblasts cells [9], MC3T3-E1 cells [55], MCF-7 cells [62], and VERO cells [63] have been used as a cell source. These studies have indicated that mammalian cell attachment and spreading on quartz surface can be monitored by the QCM sensor. Overall, the results have shown that adherent cells on the surfaces cause a decrease in frequency. However, most studies have suggested

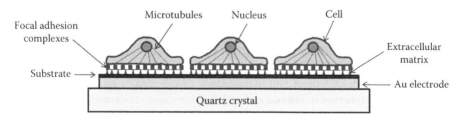

FIGURE 4.10 Schematic representation of the QCM cell biosensor. (Adapted and redrawn from Marx, K.A. et al., *Anal. Biochem.*, 361, 77, 2007.)

that the cell layer formed on the crystal surface does not act as a rigid mass, and in this case, the Sauerbrey equation should not be valid.

In the past decade, QCM-based mammalian cell biosensors have been successfully developed to monitor living cell proliferation and cell attachment to quartz electrode surface. For instance, Khraiche et al. [61] demonstrated the use of an acoustic sensor to measure adhesion of neuro-blastoma cells (Neuro-2A) on uncoated gold electrodes and poly-L-lysine (PLL) coatings. They have shown that the acoustic sensor has sufficient sensitivity to monitor neuronal adhesion on the PLL coating in real time [61]. Fohlerová et al. [64] investigated the attachment of rat epithelial cells (WB F344) and lung melanoma cells (B16F10) to QCM electrode coated with extracellular matrix proteins, vitronectin, and laminin. Their results have demonstrated that the QCM cell sensor is suitable for the evaluation of different cell adhesion processes. Redepenning et al. [59] investigated the attachment and spreading of osteoblasts to quartz crystal surface, by recording the shift of the resonance frequency responding proportionally to the surface cell coverage. Fredriksson et al. [52] have shown that cell adhesion on polystyrene (PS) surfaces was dependent on wettability. Wegener et al. [65] revealed that different mammalian cell types generated individual responses to QCM when contacting with the quartz crystal surface. Marx et al. [66] studied the changes in cellular viscoelastic properties of the cells based on frequency and resistance shift measured by the QCM cell biosensor with living ECs or human breast cancer cells (MCF-7) on the gold electrode surface. Guillou-Buffello et al. [9] disclosed that PMMA-based bioactive polymers exhibiting either carboxylate and/or sulfonate functional groups inhibited adhesion of the McCoy fibroblastic cells when compared to nonfunctionalized PMMA and PMMA-based copolymers, which comprised only one of these functional groups.

4.5.4.2 Detection of Cell–Drug Interactions

A new area of drug discovery research is the evaluation of pharmaceutical compounds and drugs using the QCM mammalian cell biosensors. Cell-based label-free technologies play a fundamental role in preclinical drug development processes. The QCM technology approach offers several advantages over the methods currently in use. The QCM is a technique that does not require cell removal from the surfaces or a labeling step for monitoring cell response. By measuring the resonance frequency shifts of quartz crystals, important biological information such as cell adhesion, cell proliferation, and cytotoxicity can be obtained.

Microtubules are long tubular polymers composed of dimeric subunits consisting of α- and β-tubulin protein which form the tubulin heterodimer. In microtubule formation (Figure 4.11), α- and β-tubulin heterodimers polymerize to form a short microtubule nucleus. The second step is the elongation of the microtubule at both ends [(+) or (−) end] to form a cylinder [67].

Microtubules are involved in the process of cell division and mitosis, in cell signaling, in the development and maintenance of cell shape, and in intercellular transport. Their function in mitosis and cell division makes microtubules important agents in cancer chemotherapy [68].

There are a number of chemically diverse compounds binding to tubulin in microtubules (Figure 4.12). These compounds with diverse structures disrupt the

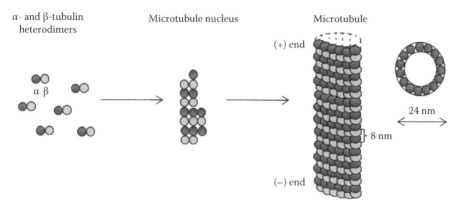

FIGURE 4.11 Polymerization of microtubules. (Adapted and redrawn from Jordan, M.A. and Wilson, L., *Nat. Rev. Cancer,* 4, 253, 2004.)

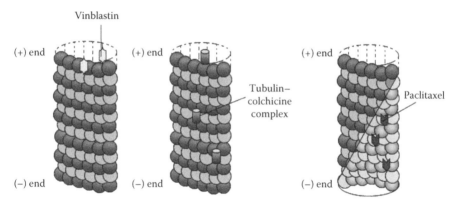

FIGURE 4.12 Antimicrotubule agents binding to microtubules at diverse sites. (Adapted and redrawn from Jordan, M.A. and Wilson, L., *Nat. Rev. Cancer,* 4, 253, 2004.)

polymerization dynamics of microtubules and cause inhibition of cell proliferation. These events make it possible to discover and develop important anticancer agents for the treatment of cancer in clinical and preclinical stages. Microtubule-binding antimitotic drugs are specified in two major groups [67]. First group causes destabilization of the microtubule, inhibiting microtubule polymerization at high drug concentrations, namely, the *Vinca* alkaloids (vinblastine, vincristine, vinorelbine, vindesine, and vinflunine), cryptophycins, halichondrins, estramustine, colchicine, and combretastatins. Second group of these compounds is called the microtubule-stabilizing agents which induce microtubule polymerization and stabilize microtubules at high concentrations. These drugs include paclitaxel, docetaxel, epothilones, discodermolide, eleutherobins, sarcodictyins, laulimalide, and rhazinalam. The antimitotic drugs are able to bind to diverse sites on tubulin and at different positions in the microtubule and show different effects on microtubule dynamics through different chemical mechanisms. For example, vinblastine binds to the microtubule

(+) end, suppressing microtubule dynamics. Colchicine binds tubulin dimers and copolymerizes into the microtubule lattice, suppressing microtubule dynamics. Paclitaxel binds the interior region of the microtubule, thus suppressing microtubule dynamics [67].

The cancer cells can develop resistance to microtubule binding compounds following long-term treatment. Acquired resistance to microtubule binding drugs loses the function on cancer cells. Thus, there is a need to develop more antimitotic drugs that can be used for the treatment of cancer [68,69]; the discovery of novel microtubule binding drugs is considered invaluable for effective cancer treatment.

In the pharmaceutical industry, drug discovery studies are searching for novel methods to determine pharmaceutical compounds and drugs, using PZ crystal coated with a cell layer. There are a few number of studies based on QCM drug biosensor for detecting the effect of drugs on cells. Detailed evaluation of cell–drug interactions was performed by Marx et al. [70] using bovine aortic ECs. They investigated the behavior of ECs on QCM biosensor by adding different nocodazole concentrations (in the range of $0.11–15\,\mu M$) to cell growth media and determined the frequency and resistance shift effects–based nocodazole doses. Following this study, effects of taxol and nocodazole on two different cell types, namely, ECs and metastatic human mammary cancer cell line (MDA-MB-231), were investigated by PZ whole-cell biosensors [71]. In this particular study, the behavior of cells to varying concentrations of microtubule binding drugs—taxol and nocodazole—was detected by measuring changes in frequency and resistance of quartz crystal [71]. Braunhut et al. [8] investigated the response of two human breast cancer cell lines (MCF-7 and MDA-MB-231) to docetaxel and paclitaxel drugs. The results accurately predicted that docetaxel was more effective than paclitaxel, and MCF-7 cells were more resistant to taxanes compared to MDA-MB-231 cells.

These results indicate that QCM mammalian cell biosensor can be used for the detection of cytoskeletal alterations in microtubules as well as the changes in cell shape, cell attachment, and viscoelastic properties of cells. To sum up, the QCM cell biosensor may be useful in the discovery of new antimitotic drugs that affect cellular attachment in real time.

ACKNOWLEDGMENT

YME acknowledges the support of the Turkish Academy of Sciences, TÜBA (Ankara, Turkey).

REFERENCES

1. J. M. Cooper and A. E. G. Cass, *Biosensors: A Practical Approach*, Oxford University Press, Oxford, U.K., 2004.
2. R. Lucklum and P. Hauptmann, The quartz crystal microbalance: Mass sensitivity, viscoelasticity and acoustic amplification, *J. Sens. Actuators B*, 70: 30–36, 2000.
3. M.-I. Rocha-Gaso, C. March-Iborra, A. Montoya-Baides, and A. Arnau-Vives, Surface generated acoustic wave biosensors for the detection of pathogens: A review, *Sensors*, 9: 5740–5769, 2009.

4. R. D. Vaughan, C. K. O'Sullivan, and G. G. Guilbault, Development of a quartz crystal microbalance (QCM) immunosensor for the detection of *Listeria monocytogenes*, *Enzyme Microb. Technol.*, 29: 635–638, 2001.

5. Ş. Şeker, Y. E. Arslan, and Y. M. Elçin, Electrospun nanofibrous PLGA/Fullerene-C_{60} coated quartz crystal microbalance for real-time gluconic acid monitoring, *IEEE Sens. J.*, 10: 1342–1348, 2010.

6. M. Minunni, S. Tombelli, R. Scielzi, I. Mannelli, M. Macsini, and C. Gaudiano, Detection of β-Thalassemia by a DNA piezoelectric biosensor coupled with polymerase chain reaction, *Anal. Chim. Acta*, 481: 55–64, 2003.

7. X. Jia, L. Tan, Q. Xie, Y. Zhang, and S. Yao, Quartz crystal microbalance and electrochemical cytosensing on a chitosan/multiwalled carbon nanotubes/Au electrode, *Sens. Actuators B*, 134: 273–280, 2008.

8. S. J. Braunhut, D. McIntosh, E. Vorotnikova, T. Zhou, and K. A. Marx, Detection of apoptosis and drug resistance of human breast cancer cells to taxane treatments using quartz crystal microbalance biosensor technology, *ASSAY Drug Dev. Technol.*, 3: 77–88, 2005.

9. D. L. Guillou-Buffello, G. Helary, M. Gindre, G. Pavon-Djavid, P. Laugier, and V. Migonney, Monitoring cell adhesion processes on bioactive polymers with the quartz crystal resonator technique, *Biomaterials*, 26: 4197–4205, 2005.

10. G. Lippmann, Principe de conservation de l'électricité, *Annales de Physique et de Chimie*, 5a Serie 24: 145–178, 1881.

11. A. Arnau, *Piezoelectric Transducers and Applications*, 1st edn., Springer Verlag, Berlin, Germany, 2004.

12. M. R. Deakin and D. A. Buttry, Electrochemical applications of the quartz crystal microbalance, *Anal. Chem.*, 61: 1147A–1154A, 1989.

13. C. Lu and A. W. Czanderna, *Methods and Phenomena 7: Application of Piezoelectric Quartz Crystal Microbalance*, Elsevier, New York, 393 pp., 1984.

14. A. Janshoff, H.-J. Galla, and C. Steinem, Piezoelectric mass-sensing devices as biosensors—An alternative to optical biosensors?, *Angew. Chem. Int. Ed. Engl.*, 39: 4004–4032, 2000.

15. G. G. Guilbault and J. M. Jordan, Analytical uses of piezoelectric crystals: A review, *CRC Crit. Rev. Anal. Chem.*, 19: 1–28, 1988.

16. G. Sauerbrey, Verwendung von Schwingquarzen zur Wägung dünner Schichten und zur Mikrowägung, *Zeitschrift Physik*, 155: 206–212, 1959.

17. K. K. Kanazawa and J. G. Gordon, Frequency of a quartz microbalance in contact with liquid, *Anal. Chem.*, 57: 1770–1771, 1985.

18. R. G. Nuzzo and D. L. Allara, Adsorption of bifunctional organic disulfides on gold surfaces, *J. Am. Chem. Soc.*, 105: 4481–4483, 1983.

19. R. G. Nuzzo, B. R. Zegarski, and L. H. Dubois, Fundamental studies of the chemisorption of organosulfur compounds on Au(111). Implications for molecular self-assembly on gold surfaces, *J. Am. Chem. Soc.*, 109: 733–740, 1987.

20. H. O. Finklea, S. Avery, M. Lynch, and T. Furtsch, Blocking oriented monolayers of alkyl mercaptans on gold electrodes, *Langmuir*, 3: 409–413, 1987.

21. I. Mannelli, M. Minunni, S. Tombelli, and M. Mascini, Quartz crystal microbalance (QCM) affinity biosensor for genetically modified organisms (GMOs) detection, *Biosens. Bioelectron.*, 18: 129–140, 2003.

22. S. Cosnier, Biosensors based on electropolymerized films: New trends, *Anal. Bioanal. Chem.*, 377: 507–520, 2003.

23. K. A. Marx, T. Zhou, D. McIntosh, and S. J. Braunhut, Electropolymerized tyrosine-based thin films: Selective cell binding via peptide recognition to novel electropolymerized biomimetic tyrosine RGDY films, *Anal. Biochem.*, 384: 86–95, 2009.

24. K. A. Marx, The quartz crystal microbalance and the electrochemical QCM: Applications to studies of thin polymer films, electron transfer systems, biological macromolecules, biosensors, and cells, *Springer Ser. Chem. Sens. Biosens.*, 5: 371–424, 2007.
25. G. I. Taylor, Disintegration of water drops in an electric field, *P. R. Soc. Lond. Ser. A-Math. Phys. Sci.*, 280(1382): 383–397, 1964.
26. B. Ding, J. Kima, Y. Miyazaki, and S. Shiratori, Electrospun nanofibrous membranes coated quartz crystal microbalance as gas sensor for NH_3 detection, *J. Sens. Actuators B*, 101: 373–380, 2004.
27. R. Ramaseshan, S. Sundarrajan, and R. Jose, Nanostructured ceramics by electrospinning, *J. Appl. Phys.*, 102(11): 111101–111118, 2007.
28. M. W. Frey, A. J. Baeumner, D. Li, and P. Kakad, Electrospun nanofiber-based biosensor assemblies, U.S. Patent 7485591, Mar. 2, 2009.
29. B. Ding, M. Yamazaki, and S. Shiratori, Electrospun fibrous polyacrylic acid membrane-based gas sensors, *Sens. Actuators B*, 106: 477–483, 2005.
30. D. B. Hall, P. Underhill, and J. M. Torkelso, Spin coating of thin and ultrathin polymer films, *Polym. Eng. Sci.*, 38: 2039–2045, 1998.
31. S. P. Sakti, S. Rösler, R. Lucklum, P. Hauptmann, F. Bühling, and S. Ansorge, Thick polystyrene-coated quartz crystal microbalance as a basis of a cost effective immunosensor, *Sens. Actuators A: Phys.*, 76: 98–102, 1999.
32. L.-F. Wei and J.-S. Shih, Fullerene-cryptand coated piezoelectric crystal urea sensor based on urease, *Anal. Chim. Acta*, 437: 77–85, 2001.
33. S. M. Reddy, J. P. Jones, T. J. Lewis, and P. M. Vadgama, Development of an oxidase-based glucose sensor using thickness-shear-mode quartz crystals, *Anal. Chim. Acta*, 363: 203–213, 1998.
34. D. Dell'Atti, S. Tombelli, M. Minunni, and M. Mascini, Detection of clinically relevant point mutations by a novel piezoelectric biosensor, *Biosens. Bioelectron.*, 21: 1876–1879, 2006.
35. X. Mao, L. Yang, X.-L. Su, and Y. Li, A nanoparticle amplification based quartz crystal microbalance DNA sensor for detection of *Escherichia* coli O157:H7, *Biosens. Bioelectron.*, 21: 1178–1185, 2006.
36. P. Skládal, C. dos Santos Riccardi, H. Yamanaka, and P. I. da Costa, Piezoelectric biosensors for real-time monitoring of hybridization and detection of hepatitis C virus, *J. Virol. Methods*, 117: 145–151, 2004.
37. L. H. Pope, S. Allen, M. C. Davies, C. J. Roberts, S. J. B. Tendler, and P. M. Williams, Probing DNA duplex formation and DNA–drug interactions by the quartz crystal microbalance technique, *Langmuir*, 17: 8300–8304, 2001.
38. F. Lucarelli, S. Tombelli, M. Minunni, G. Marrazza, and M. Mascini, Electrochemical and piezoelectric DNA biosensors for hybridisation detection, *Anal. Chim. Acta*, 609: 139–159, 2008.
39. S. Tombelli, M. Minunni, and M. Mascini, Piezoelectric biosensors: Strategies for coupling nucleic acids to piezoelectric devices, *Methods*, 37: 48–56, 2005.
40. S. Tombelli, M. Mascini, and A. P. F. Turner, Improved procedures for immobilisation of oligonucleotides on goldcoated piezoelectric quartz crystals, *Biosens. Bioelectron.*, 17: 929–936, 2002.
41. X.-L. Su and Y. Li, Self-assembled monolayer-based piezoelectric immunosensor for rapid detection of *Escherichia coli* O157:H7, *Biosens. Bioelectron.*, 19: 563–574, 2004.
42. N. Kim, I.-S. Park, and D.-K. Kim, Characteristics of a label-free piezoelectric immunosensor detecting *Pseudomonas aeruginosa*, *Sens. Actuators B*, 100: 432–438, 2004.
43. H.-C. Lin and W.-C. Tsai, Piezoelectric crystal immunosensor for the detection of staphylococcal enterotoxin B, *Biosens. Bioelectron.*, 18: 1479–1483, 2003.
44. S. H. Si, X. Li, Y. S. Fung, and D. R. Zhu, Rapid detection of *Salmonella enteritidis* by piezoelectric immunosensor, *Microchem. J.*, 68: 21–27, 2001.

45. B. Konig and M. Gratzel, A novel immunosensor for herpes viruses, *Anal. Chem.*, 66: 341–344, 1994.

46. S. F. Chou, W. L. Hsu, J. M. Hwang, and C. Y. Chen, Development of an immunosensor for human ferritin, a nonspecific tumor marker, based on a quartz crystal microbalance, *Anal. Chim. Acta*, 453: 181–189, 2002.

47. S. F. Chou, W. L. Hsu, J. M. Hwang, and C. Y. Chen, Determination of α-fetoprotein in human serum by a quartz crystal microbalance based immunosensor, *Clin. Chem.*, 48: 913–918, 2002.

48. I. Navrátilová, P. Skládal, and V. Viklickỳ, Development of piezoelectric immunosensors for measurement of albuminuria, *Talanta*, 55: 831–839, 2001.

49. B. Zhang, Q. Mao, X. Zhang, T. Jiang, M. Chen, F. Yu, and W. Fu, Novel piezoelectric quartz micro-array immunosensor based on self-assembled monolayer for determination of human chorionic gonadotropin, *Biosens. Bioelectron.*, 19: 711–720, 2004.

50. Z. Gao, F. Chao, Z. Chao, and G. Li, Detection of staphylococcal enterotoxin C_2 employing a piezoelectric crystal immunosensor, *Sens. Actuators B*, 66: 193–196, 2000.

51. W.-C. Tsai and I.-C. Lin, Development of a piezoelectric immunosensor for the detection of alpha-fetoprotein, *Sens. Actuators B*, 106: 455–460, 2005.

52. C. Fredriksson, S. Khilman, B. Kasemo, and D. M. Steel, In vitro real-time characterization of cell attachment and spreading, *J. Mater. Sci. Mater. Med.*, 9: 785–788, 1998.

53. M. S. Lord, C. Modin, M. Foss, M. Duch, A. Simmons, F. S. Pedersen, B. K. Milthorpe, and F. Besenbacher, Monitoring cell adhesion on tantalum and oxidised polystyrene using a quartz crystal microbalance with dissipation, *Biomaterials*, 27: 4529–4537, 2006.

54. K. Anselme, Osteoblast adhesion on biomaterials, *Biomaterials*, 21: 667–681, 2000.

55. C. Modin, A.-L. Stranne, M. Foss, M. Duch, J. Justesen, J. Chevallier, L. K. Andersen, A. G. Hemmersam, F. S. Pedersen, and F. Besenbacher, QCM-D studies of attachment and differential spreading of pre-osteoblastic cells on Ta and Cr surfaces, *Biomaterials*, 27: 1346–1354, 2006.

56. K. A. Marx, Quartz crystal microbalance: A useful tool for studying thin polymer films and complex biomolecular systems at the solution–surface interface, *Biomacromolecules*, 4(5): 1099–1120, 2003.

57. K. A. Marx, T. Zhou, M. Warren, and S. J. Braunhut, Quartz crystal microbalance study of endothelial cell number dependent differences in initial adhesion and steady-state behavior: Evidence for cell–cell cooperativity in initial adhesion and spreading, *Biotechnol. Prog.*, 19: 987–999, 2003.

58. T. Zhou, K. A. Marx, M. Warren, H. Schulze, and S. J. Braunhut, The quartz crystal microbalance as a continuous monitoring tool for the study of endothelial cell surface attachment and growth, *Biotechnol. Prog.*, 16: 268–277, 2000.

59. J. Redepenning, T. K. Schlesinger, E. J. Mechalke, D. A. Puleo, and R. Bizios, Osteoblast attachment monitored with a quartz crystal microbalance, *Anal. Chem.*, 65: 3378–3381, 1993.

60. J. Wegener, A. Janshoff, and H.-J. Galla, Cell adhesion monitoring using a quartz crystal microbalance: Comparative analysis of different mammalian cell lines, *Eur. Biophys. J.*, 28: 26–37, 1998.

61. M. L. Khraiche, A. Zhou, and J. Muthuswamy, Acoustic sensor for monitoring adhesion of Neuro-2A cells in real-time, *J. Neurosci. Methods*, 144: 1–10, 2005.

62. M. Guo, J. Chen, Y. Zhang, K. Chen, C. Pan, and S. Yao, Enhanced adhesion/spreading and proliferation of mammalian cells on electropolymerized porphyrin film for biosensing applications, *Biosens. Bioelectron.*, 23: 865–871, 2008.

63. D. M. Gryte, M. D. Ward, and W.-S. Hu, Real-time measurement of anchorage-dependent cell adhesion using a quartz crystal microbalance, *Biotechnol. Prog.*, 9: 105–108, 1993.

64. Z. Fohlerová, P. Skládal, and J. Turánek, Adhesion of eukaryotic cell lines on the gold surface modified with extracellular matrix proteins monitored by the piezoelectric sensor, *Biosens. Bioelectron.*, 22: 1896–1901, 2007.

65. J. Wegener, J. Seebach, A. Janshoff, and H. J. Galla, Analysis of the composite response of shear wave resonators to the attachment of mammalian cells, *Biophys. J.*, 78: 2821–2833, 2000.

66. K. A. Marx, T. Zhou, A. Montrone, D. McIntosh, and S. J. Braunhut, Quartz crystal microbalance biosensor study of endothelial cells and their extracellular matrix following cell removal: Evidence for transient cellular stress and viscoelastic changes during detachment and the elastic behavior of the pure matrix, *Anal. Biochem.*, 343: 23–34, 2005.

67. M. A. Jordan and L. Wilson, Microtubules as a target for anticancer drugs, *Nat. Rev. Cancer*, 4: 253–265, 2004.

68. M. Kavallaris, Microtubules and resistance to tubulin-binding agents, *Nat. Rev. Cancer*, 10: 194–204, 2010.

69. S. Drukman and M. Kavallaris, Microtubule alterations and resistance to tubulin-binding agents, *Int. J. Oncol.*, 21: 621–628, 2002.

70. K. A. Marx, T. Zhou, A. Montrone, H. Schulze, and S. J. Braunhut, Quartz crystal microbalance cell biosensor: Detection of microtubule alterations in living cells at nM nocodazole concentrations, *Biosens. Bioelectron.*, 16: 773–782, 2001.

71. K. A. Marx, T. Zhou, A. Montrone, D. McIntosh, and S. J. Braunhut, Comparative study of the cytoskeleton binding drugs nocodazole and taxol with a mammalian cell quartz crystal microbalance biosensor: Different dynamic responses and energy dissipation effects, *Anal. Biochem.*, 361: 77–92, 2007.

5 Toward Printable Lab-on-a-Chip Technologies for Cell Analytics

Martin Brischwein, Giuseppe Scarpa, Helmut Grothe, Bernhard Wolf, and Stefan Thalhammer

CONTENTS

5.1 BACKGROUND AND INTRODUCTION

Approximately two decades ago, a handful of researchers began discussing an intriguing idea: Could the equipment needed for everyday chemistry and biology procedures possibly be shrunk to fit on a chip the size of a fingernail? Miniature devices for, say, analyzing DNA and proteins should be faster and cheaper than conventional versions. Nowadays, the "lab on a chip" represents a quite mature and advanced technology that integrates a microfluidic system on a microscale device. Such a "laboratory" is created by means of channels, mixers, reservoirs, diffusion chambers, integrated electrodes, pumps, valves, and so on. During the past years, microfluidics, micrometer-scale total analysis systems (µTASs), or the so-called lab-on-a-chip devices have revised interest in the scaling laws and countless groups for downscaling purposes [1,2]. The term "microfluidic" refers to the ability to manipulate fluids in one or more channels with dimensions of 5–500 µm [3,4].

The development of microfluidics already started 20 years ago in ink-jet printer manufacturing. The ink-jet mechanism involves very small tubes carrying the ink for printing. The same etching techniques known from semiconductor technology are employed to create channels, tubes, and chambers in silicon or glass substrates, which can be layered on top of one another to result in more complex 2-D and 3-D structures. Chip designs can also be stamped, molded, or cut into plastic sheets. While the earliest reported microscale devices consisted of such channels etched in solid substrates such as silicon [5,6], glass [7], and plastic [8], microelectromechanical systems (MEMSs), fabrication technologies have been increasingly applied to fabricate highly sophisticated devices from a variety of materials, including soft elastomers such as polydimethylsiloxane (PDMS) [9] with hundreds of microchannels and integrated sensors to measure physiological parameters. Cell analysis and clinical diagnosis are now becoming the largest fields of application for microchip-based analytical systems [10]. The advantages of miniaturization include small sample and reagent volumes, smaller space requirements and size, easy operation and fast analysis time, as well as easy adaptation to custom needs. A variety of applications for use in environmental, food, and cell analysis (e.g., food quality evaluation, drugs, pharmacokinetics studies, and clinical diagnosis) have been envisaged. Furthermore, point-of-care (POC) testing is increasingly required but is difficult and impractical using conventional methods. Microchip-based systems have the potential to overcome most of the problems there, such as long assay time, as well as expensive handling procedures. Another important advantage of the microchip-based cell analysis system is the ability to integrate successive steps in an analytical process (e.g., cell culture, sample preparation, chemical reaction, separation and detection) on a single microchip. Using microchip-based analysis systems, the required number of cells required for sensitive testing can be reduced consistently [11]. This is useful, for instance, in tumor therapy when drug sensitivity testing using a patient's own cancer cells is required. Drug sensitivity testing using standard laboratory test involves invasive procedures to obtain the relatively large number of cells required. An example in this direction is presented later.

The aim of this contribution is to give a general overview on the state-of-the-art in cellular assays on chips. The emphasis will be on technologies and applications related to the authors' research objectives and developments. This includes the current microsensor strategies for cellular assays but also approaches for manipulation of cells on chips. Challenges, perspectives, and future developments based on recent achievements are also included.

5.2   CELL CULTURING AND MONITORING IN MICRODEVICES

5.2.1   CELLULAR SYSTEMS

Regardless of whether cells are analyzed as biological signal transducers and amplifiers in a biosensor or as the actual object of interest in a cellular assay, a consideration of the marked sensitivity of cells toward almost any perturbation of its environment is advisable. Living cells, prokaryotes or eukaryotes, are organic microsystems with functional compartments at the nano-size level interconnected by complex signal chains. Individual subunits are coupled to each other by biochemical and physical signaling pathways constituting a complex signaling network. The ground reason for development, tissue repair, and immunity comes from the ability of the cells to perceive and respond correctly to their environment. Cellular signal processing is non-linear and it works in parallel pathways. Cells are continuously integrating different sources of chemical and physical signals from the external environment, most often through a class of proteins called receptors. Molecules that activate receptors can be classified as hormones, growth factors, etc., but under a common name of "receptor ligands." While many receptors are cell surface proteins, some are also found inside cells. Cellular life is characterized by dynamic molecular processes whose highly interconnected regulation is essential for the whole organism. The "output" of this cellular signaling network may then be manifested as a decision about growth and mitosis, as a reorganization of the cell cytoskeleton and cell morphology, as the activation of distinct metabolic pathways, the production and release of proteins, the initiation of programmed cell death, or the transmission of another chemical or electrical signal to neighboring cells.

Keeping these basic facts in mind, a very careful ex vivo treatment of fragile cells in microsystems—particularly of primary cells prepared from living organs—appears appropriate to avoid artifacts and misinterpretation of results.

5.2.2 DETECTING CELLULAR OUTPUT SIGNALS

The signal-processing capabilities of living cells are frequently assessed by various functional cell assays. Most often, fluorescence techniques are used to dissect the molecular basis of cell functions, either with fluorescent protein analogs or with synthetic dyes. But no truly appropriate fluorescence techniques have been developed so far for the analysis of cell morphology, metabolic rates, and electric activity. In contrast to microelectric or microelectronic structures, optic setups with light sources, light guides, and detectors are less easily integrated into small, portable lab-on-a-chip systems. Moreover, labeling cells with dyes is costly and it involves an additional step of cell manipulation with unclear physiological side effects.

"Cell-on-a-chip" systems have already been reviewed [12–16]. Changes of cell morphology in adherently growing monolayers are monitored with electric impedance sensors [17,18] or surface acoustic wave (SAW) transducers [19,20]. Rates of extracellular acidification are recorded with light-activated potentiometric sensors and with pH-ISFETs [21–24]. Electrophysiologic activity of cells, that is, rapid changes in the membrane potential of electrically active cells or currents of ion channels are detected with microelectrode arrays or with patch-clamp chips [25,26]. In most systems, only one parameter is recorded on a given chip. However, there are efforts to develop multiparametric chips combining several complementary sensors for a comprehensive understanding of dynamic cellular behavior [16]. The principle of a cell monitoring microdevice is shown in Figure 5.1.

The coupling between the living microstructure of cells and technical microstructures of transducers is usually achieved by growing cells directly on the surface of inert and nontoxic materials, such as silicon- or glass-based sensor chips, which are accepted by many cells as substrate for stable adhesion and growth. The

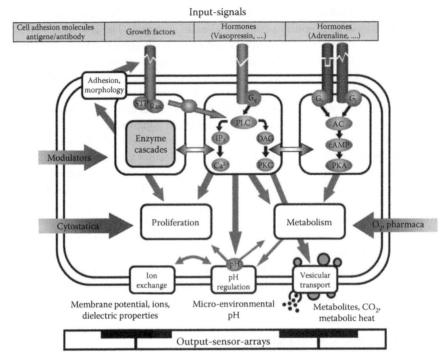

FIGURE 5.1 Schematic course of the mitogen signal transfer in a cell and multiparametric bioelectronic chip as an interface for signal reading.

cell-on-a-chip setup may be regarded as a cell–transducer hybrid. Many of the parameters detected (cell morphology, cell metabolism, and electric activity) are closely linked to the intracellular signal network and thereby can be used as sensitive indicators of any perturbation of cell physiology.

5.2.3 CHIP FABRICATION AND PACKAGING

In the following, a brief description of the basics of sensor chip design and processing is provided. It is confined to the technology used in the authors' clean room facilities and treats the construction of planar transducer structures.

Thin-film technology is used as a process to produce platinum electrode structures on substrates such as glass or ceramics. A liftoff photoresist process defines a resist mask for sputtering Ti/Pt (20/200 nm) on the wafer surface. The process steps are similar for glass and ceramic substrate. Platinum electrode structures are used for resistive temperature, amperometric dissolved oxygen, and electrochemical impedance sensors. The latter ones may be galvanically plated with gold to decrease the electrode impedance. Platinum is also used for electrically contacting the pH-sensitive ruthenium oxide sensors. After removal of the resist mask, ruthenium oxide is deposited in a reactive sputter process using a shadow mask. In a second photo process, SU-8 resist is structured for insulating all parts of the wafer which shall not be in direct contact with liquid or do not serve as contact pads. Finally, the wafers are diced to 7.55×7.55 mm chips. The diced chips are glued and bonded on a 24×24 mm^2 printed circuit board. An encapsulation forming a cell culture well for connection is fixed on the board by molding a two-component adhesive. Finally, the encapsulated chips are shrink wrapped into bags and plasma sterilized (Figure 5.2).

Glass chips with electrode structures are used as the bottom of 24-well cell test plates (Figure 5.3). Here, there is no dicing of single chips to facilitate the adhesive bonding to the polymeric plate corpus and contacting by pushpins. Besides, electrochemical sensors for dissolved oxygen and pH are replaced by optochemical sensors spotted on the glass surface. This solution was preferred in order to avoid a large number of electric contacts and the need for numerous pH reference electrodes. Besides, the optochemical sensors are read out by a process microscope that is already integrated in the system for cell imaging.

Another type of glass chip is fabricated as part of a 24-well "neuro" test plate. In general, multi-electrode arrays consist of a conductive layer deposited on a glass substrate and isolated on top by a dielectric layer. Only a subarea of the electrode ($\approx 10\,\mu$m in dimension) is de-isolated in order to record the capacitively coupled cell

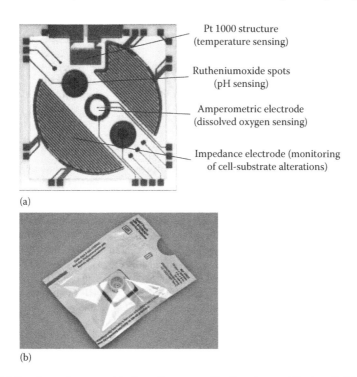

(a)

(b)

FIGURE 5.2 Ceramic sensor chip for cell assay applications in a mobile, handheld platform which is currently developed in cooperation with cellasys GmbH, Germany. (a) Layout with amperometric, potentiometric, and impedimetric transducer elements as it is fabricated in cooperation with Heraeus Sensor Technology GmbH, Germany; (b) chip in package, sterile and ready to use for cell culturing or connecting to a test platform.

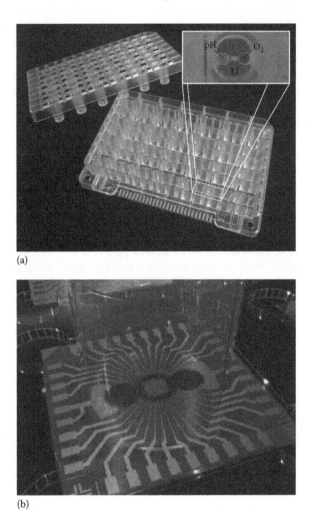

FIGURE 5.3 Glass chip–based functional test units. (a) 24-well cell test plate with bonded glass chip bottom. Plate corpus and cover lid are made from polycarbonate. The outer dimensions conform to standards for microtest plates. Each cell culture well has 7.7 mm in diameter. The insert shows a top view on a well with sensors for pH, dissolved O_2, and electric impedance (EI). This test plate has been constructed and fabricated together with partners from industry (Quarder Systemtechnik GmbH, Germany, and HP Medizintechnik GmbH, Germany). (b) Glass chip for a multiwell "neuro" test plate. Here, the single chips are diced and bonded separately to the cell culture containers.

electric activity. Each chip has an array of such electrodes to allow a cell network analysis. Currently, laser structuring of metalized glass and polymer sheet substrates is tested as a fabrication method alternative to conventional thin-film technology. Twenty-four chips bonded to culture vessels again are arranged to form the disposable part of the "neuro" test plate, which also provides electric signal filtering, amplification, and processing [27].

Common to all the described chip- and sensor-processing technologies are the comparatively high fabrication costs, even in industrial fabrication lines. Screen printing may be an alternative, but the achievable structural resolution is suboptimal for some of the required sensor functions [28]. Moreover, the functional properties of the metal-based active electrode surfaces (typically platinum and gold) may be improved. There are novel materials tailored for specific sensing purposes that could clearly enhance the performance of the sensors. For example, carbon nanomaterials are known to have superior electrocatalytic properties for many redox reactions [29,30] and yield low-impedance surfaces for electric stimulation and signal recording [31]. The challenge will be to deposit those materials as an active, durable sensor surface (e.g., by a printing process), and to combine these modified surfaces with low-cost substrates integrating the necessary conducting pathways.

5.2.4 SYSTEM ENGINEERING

A sensor chip is hardly useful as a stand-alone microdevice. Rather it is at the core of a system integrating sensorics, fluidics, cell culture maintenance, and signal processing. Where appropriate, this may result in a relatively bulky desktop or laboratory system, albeit with a smart and disposable central piece.

For many screening applications, the monitoring sites have to be arranged in high-density arrays to create a basis for sound data statistics. There are now increasingly successful attempts toward single-cell analysis in microdevices [14,32], allowing a statistical treatment of data derived from single cells. On the other hand, a disintegration of living tissue to yield single-cell preparations always involves a severe trauma to the cells with unpredictable effects on cell signaling. To design a personalized predictive test on specimens derived from human solid cancer, for example, an arraying of culture and monitoring sites packaged in the format of a microtest plate appears to be more appropriate.

A complete device for live cell monitoring combining parallel sensor-based monitoring, liquid handling, and imaging capabilities is shown in Figure 5.4. It operates with both types of test plates presented in Figure 5.3.

The demand for miniaturization may originate from several reasons. These are (1) scaling considerations from fluidics and the specific requirements of metabolic assays, (2) the saving of costly drug compounds, and (3) the saving of primary tissue which may be available only in very small amounts, for example, when cancer biopsy specimens are to be tested. Thus, the challenge is to establish a robust microscaled cell culturing along with the required sensing capabilities.

In cell-based biosensors, cellular recognition and signaling capabilities can also be exploited to detect a physiological response to possibly hazardous substances. In this case, cells are used for providing the appropriate receptor proteins and for the generation and amplification of an output signal which is detectable by a physical transducer. In fact, one of the most important advantages of using living cells as biological signal discriminators and amplifiers is their ability to regenerate the signal receptors and amplifying signaling cascades permanently in a native state. Such sensors that could be specific for toxins might be used, for example, in deterring biological warfare weapons or they could serve as a biomonitor of wastewater effluents [33–35].

FIGURE 5.4 Photograph showing the assay platform with the test plate in front position (A), the pipetting robot head (B) which is the interface between medium/drug solution containers (C) and the microfluidics of the test plate; (D) integrated process microscope.

For the design and engineering of almost any kind of microdevice, the aid of computer modeling is now state-of-the-art. Here, we use finite element modeling to simulate the combined effects of cell metabolic activity, diffusion, buffer capacity, material effects, and microfluidics to end up at optimized microculturing and monitoring devices.

5.2.5   CASE STUDY: ANALYSIS OF HUMAN TUMOR EXPLANTS

In oncological practice, there is an urgent need for personalized POC methods predicting the outcome of chemotherapeutic treatment regimens [36,37]. The approach that has been pursued almost since the beginning of chemotherapy is to establish assays for the ex vivo determination of cellular chemosensitivity. Unlike molecular markers, they have the intrinsic advantage of integrating the extremely complex cellular signaling network summing up into a detectable response parameter. More recently, methods relying on a sensor chip–based cell analytics have been described [38–40]. However, most of the reports are on measurements on monolayer cultures of tumor cell lines and not on primary cultures.

The rationale of the approach outlined here is based on previous work with human tumor cells and tissues [41–43]. In contrast to many other assays, sensor-based

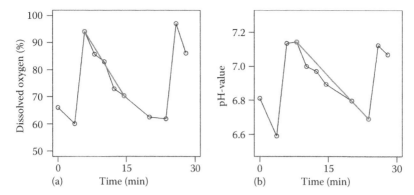

FIGURE 5.5   Sensor raw data plot comprising a single culture well with tissue slice during a single fluidic time interval. (a) Dissolved oxygen vs. time. (b) pH vs. time. The light path to read out the optochemical sensors is guided through the integrated process microscope. The excited area of the sensor spot is less than 1% of the total sensor area, leaving chance for a considerable further miniaturization. At $t \approx 5\,h$, the pipetting robot effects an exchange of medium in the microvolume, pH, and dissolved oxygen resume values of fresh media. In the following rest interval, the slopes of pH and dissolved oxygen provide estimates for the required metabolic rates. The changes of pH and dissolved oxygen are to be kept sufficiently small to avoid large deviations from normal values.

monitoring technology offers a fast, label-free cell preparation and the advantage of continuously following up individual specimens before, during, and after an ex vivo treatment.

It is realized with the platform shown in Figure 5.4. Within each well of the test plate, a microvolume of $\approx 23\,\mu L$ is created to achieve a high biomass: medium volume ratio. The "biomass" is represented by $\approx 200\,\mu m$ vibratome slices of the cancerous test specimens. The tissue preparation is guided by the maxim that traumatic, manipulative steps should be minimized.

Metabolic activity, that is, oxygen consumption and acid release, is monitored during the rest intervals of the fluidic system, typically lasting for 20 min (Figure 5.5).

The primary result of raw data evaluation is a pair of rate values assigned to each measurement cycle. Consequently, a pair of rate curves for each of the 24 microcultures characterizes the entire measurement typically extending over 288 cycles for a 96 h experiment. Higher-level data evaluation, that is, normalization of the plots, introduction of weight factors according to signal noise levels, or statistics on the kinetic sensor graphs for testing of hypotheses—to come up finally with a cutoff criterion for "sensitivity"—is not substantially included in the current stadium of this explorative study.

5.3   CELL MANIPULATION USING LAB-ON-A-CHIP APPROACHES

In the past 10 years, there has been an increased interest in analysis of even more complex biological systems such as living cells with the use of microfabricated structures. The focus is on noncontact cell-sorting approaches in the field of lab-on-a-chip

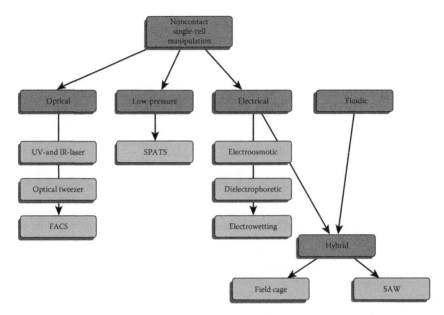

FIGURE 5.6 Overview of microdevice approaches for noncontact single-cell manipulation.

systems, which include single-cell electroosmotic and dielectrophoretic cell sorting and pressure-based and optical sorting. Recent advances in the field of genomics and proteomics have catalyzed a strong interest in the understanding of molecular processes at the cellular level like gene expression and cell biomechanics. Microscale devices are being applied to both manipulation and the interrogation of single cells in nanoliter volumes. At the single cell level, several noncontact manipulation approaches are briefly summarized (see Figure 5.6).

The ability to manipulate single cells and cell components has been accomplished over the past few decades using a variety of techniques, including mechanical approaches, for example, micropipettes and robotic micromanipulators [44], optical tweezers [45,46], and microelectrodes [47,48]. Adapting these general classes of techniques to nanofluidic systems, using microchannels or microfabricated structures incorporating nanoliters of fluid, offers an unprecedented level of control over positioning, handling, and patterning of single cells.

Several groups have recently reported on the development of nanofluidic systems to mechanically manipulate and isolate single cells or small groups of cells in microscale tubing and culture systems. The Quake Group used multilayer soft lithography, a technology to create stacked 2-D microscale channel networks from elastomers to fabricate integrated PDMS-based devices for programmable cell–based assays [49]. They applied the microdevice for the isolation of single *Escherichia coli* bacteria in subnanoliter chambers and assayed them for cytochrome c peroxidase activity. Khademhosseini et al. reported on the use of polyethylene glycol (PEG)-based microwells within microchannels to dock small groups of cell in predefined locations. The cells remained viable in the array format and were stained for cell surface receptors by sequential flow of antibodies and secondary fluorescent probes [50].

Trapping of cells using biomolecules in nanofluidic systems has been demonstrated using antibodies and proteins with high affinity to the target cell (for review, see Refs. [51,52]). Chang et al. used square silicon micropillars in a channel coated with the target protein, an E-selectin-IgC chimera, to mimic the rolling and tethering behavior of leukocyte recruitment to blood vessel walls [53]. Using electric fields to both induce flow and separate molecules is widely adapted to microscale devices to separate nucleic acids and proteins [54,55]. For cell-capture dielectrophoresis has been adapted to microscale devices, in which a nonuniform alternating current is applied to separate cells on the basis of their polarizability [56].

In an effort to miniaturize the cell-sorting process, microscale devices have been fabricated that use different strategies, including electrokinetic, pressure, and optical deflection-based methods [57]. While the first prototypes developed in the late 1990s were capable of tens of cells per second, state-of-the-art microscale fluorescent-activated cell sorter (FACS) is quite comparable to conventional sorters [58]. Wang et al. implemented a fluorescence-activated microfluidic cell sorter and evaluated its performance on live, stably transfected HeLa cells expressing a fused histone-green fluorescent protein [59]. Fu et al. developed a simple PDMS T-microchannel sorter device consisting of a single input and two electrode-connected collection outputs [57]. Sorting of cells was accomplished by switching the applied voltage potential between the two outputs. Flow switching in pressure-based cell sorting to the collection and waste outputs has been achieved by integrated elastomeric valves in multilayer PDMS-based devices and hydrodynamic switches [58]. Ozkan et al. used vertical cavity surface-emitting laser arrays to trap and manipulate individual cells in PDMS microchannels [60]. While field-trapping strengths are characteristically weak for larger objects like mammalian cells, the use of photonic pressure to repel cells has been shown to be an effective tool for cell-sorting microfabricated devices. Besides, multicellular sorting single-cell analysis plays an important role in molecular diagnosis.

Here, it is not the intention to provide an entire comprehensive list of microdevices to monitor single-cell physiology. A list of the major fields of noncontact research is summarized in Figure 5.6, and the present status and novel achievements in the monitoring cell physiology field are reviewed (for history, theory, and technology, see Ref. [61]; for analytical standard operations and applications, see Ref. [62]; and for latest trends, see Ref. [63]).

5.3.1 SMALLEST DROPLET ACTUATION

Microscale devices offer the possibility of solving system integration issues for cell biology, while minimizing the necessity for external control hardware. Many applications, such as single-gene library screening, are currently carried out as a series of multiple, labor-intensive steps required in the array process from, for example, DNA amplification, reporter molecule labeling, and hybridization. While the industrial approach to complexity has been to develop elaborate mechanical high-throughput workstations, this technology comes at a price, requiring considerable expense, space, and labor in the form of operator training and maintenance. For small laboratories or research institutions, this technology is basically out of reach. Devices consisting of addressable microscale fluidic networks can dramatically simplify the screening process, providing a compartmentalized platform for nanoliter aliquots. But making miniature labs is not just a question of scaling down conventional equipment. Nanoliter volumes of liquid behave in different ways than their macroscopic counterparts. For example, we are used to seeing liquids mix by turbulent flow, as illustrated by the way cream swirls into coffee. But such turbulence does not occur in closed channel or tubing systems just a few micrometers wide. They are governed by the rules of laminar flow or low Reynold's numbers [64]. Here, we present a completely different approach to liquid handling on small scales. Instead of closed tubing, a chemically modified surface providing virtual fluid confinements was developed [65]. The shape of a droplet on a surface is given by the properties of the substrate and is controlled by the surface free energy and not by channel walls. Depending on the chemical modification, droplets with high contact angle will form (hydrophobic) or the liquid will preferentially wet the surface (hydrophilic). This is one of the basic ideas behind the chip system described in this section: Small amounts of liquids do not need to be confined in tubes and trenches. They form their own test tubes, held together by surface tension effects. The droplets, fluidic tracks, and reaction sites are then defined by a monolayer chemical modification of the chip surface. The technology to create such fluidic tracks very much resembles the one used to define conducting paths on an electronic semiconductor device. A true lab on a chip, however, requires more than just test tubes. More importantly, their cargo has to be moved around, mixed, stirred, or processed in general. In comparison to conventional closed microfluidic systems with external pumping, faced with the difficulty to further miniaturize, SAWs are employed to agitate and actuate these little virtual test tubes along predetermined trajectories [65]. The interaction between the SAW and the fluid on the chip leads to acoustic streaming within the fluid, which can be used to pump and mix within a closed volume but also to actuate the small droplets as a whole. Liquid volumes in the range from $1\,\mu L$ down to $100\,pL$ are precisely moved on these fluidic "tracks" without any tubing system. The agitating SAWs are generated by high-frequency electrical signals on microstructured interdigital transducers (IDTs) embedded into the lab on a chip [66].

Well-defined analyses, controlled in the submicroliter regime, can be quickly and gently conducted on such an acoustically driven lab on a chip. Apart from its nearly unlimited applicability for many different biological assays, its programmability and extremely low manufacturing costs are another definite advantage of our "freely programmable lab-on-a-chip (LOC)" [67] (Figure 5.7). In fact, they can be produced at such low cost that their use as disposables in many areas of diagnosis can be envisioned.

In most conventional microfluidic systems, liquids are confined and moved in tubes or capillaries. Usually, the application of such systems is restricted to continuous flow processes. Most of the work in macroscopic laboratories, however, is carried out as a batch process. To realize the concept of microfluidics as a closed system, some major problems need to be overcome. For example, the pressure required for actuating the liquid scales with the channel dimension [68]. Furthermore, the shrinking of liquid-handling systems to the micron- and submicron-size range entails moving into the area of small Reynolds numbers. The fluid dynamics in this regime are very different from the macroscale. In microdevices with channel diameter smaller than $1\,\mu m$, the

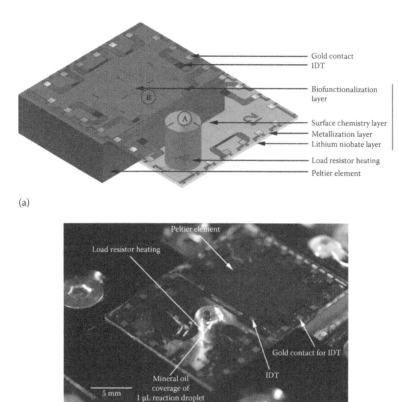

(a)

(b)

FIGURE 5.7 (a) Sketch of LOC functionality. The host substrate ($LiNbO_3$) is covered by a laterally patterned layer of Pt, Ni, and Au for transducers and sensor metallization. Subsequent silanization of the surface accounts for a hydrophilic/hydrophobic surface chemistry, facilitating a planar tracking system, which could be further functionalized by a question-specific detection array. Reaction center A is controlled by a load resistor heating, reaction point B by a peltier element. (b) Optical image of the chip installed in the heating unit; on reaction center A a 1 μL aqueous solution is covered by 5 μL mineral oil; scale bar 5 mm.

flow is usually governed by low Reynolds-number conditions, where turbulence and inertia effects are nonexistent. The Reynolds number (*Re*) is a dimensionless quantity, giving the ratio between viscosity- and inertia-related effects and is defined as

$$Re = \frac{\nu l \rho}{\eta}$$

where
   ν is the fluid velocity
   l is the characteristic channel dimension
   ρ is the fluid density
   η is the fluid viscosity

At high Reynolds numbers (i.e., $Re > 2000$), fluid flow is turbulent. At low Reynolds numbers (i.e., $Re < 2000$), fluid flow is laminar.

Mixing, for example, is typically only driven by diffusion under these conditions, and hence diffusion cannot be neglected in the design of these devices. Moreover, forces due to surface tension at liquid–air interfaces are dominant at this scale. Gravity- and inertia-related effects are usually negligible. As the magnitude of the physical effects in the microfluidic world is very much different from that at macroscopic scale, fluid-integrated microdevices must be designed from first principles rather than simply by miniaturizing macroscopic devices. Careful examination of a microscale fluid reveals that the effects of surface tension, for instance, by far exceed those of gravity. As a result of this, the power of pumps has to be typically increased while reducing the size, which complicates the integration into a complete system. Furthermore, when a biological solution is pumped through a narrow tube, the risk of reagent loss by adhesion to the wall is large due to the unfavorable surface to volume ratio. Other problems arise by small channels becoming easily clogged and by surface modification and functionalization being difficult to control. We use "virtual" beakers and channels to confine smallest possible amounts of liquids to the planar surface of piezoelectric substrates. SAWs are excited by interdigital transducers on the chip to act as electrically addressable and programmable nanopumps [69]. The reagents can be manipulated either as discrete droplets or by streaming patterns induced in somewhat larger volumes. The technology allows for both batch and continuous processes to be carried out at high speed. The most important feature, however, is the programmability of the chip, as different assay protocols can be realized with the same chip layout and hardware.

The driving force behind this interaction and the resulting acoustically driven flow is an effect called "acoustic streaming" [70]. It is a consequence of the pressure dependence of the mass density, leading to a nonvanishing time average of the acoustically induced pressure. Although acoustic streaming has been a well-known effect for a long time in macroscopic, classical systems, little attention has been paid to it, so far, in terms of miniaturization [71]. The acoustic wave entering the droplet is diffracted under a Rayleigh angle ΘR into the fluid (Figure 5.8a), where it generates a longitudinal pressure wave. For an infinite half-space, this diffraction angle is given by the ratio of the sound velocities in the substrate and in the fluid, respectively. Sound traveling through a liquid is viscously attenuated along its transmission through the medium. If the intensity is high, this attenuation creates an acoustic pressure gradient along the propagation of the wave. The gradient induces a force in the same direction, which in turn causes a flow in the fluid [72]. The upper boundary of a real, spatially confined fluid, defined by its surface, bends the streaming lines, resulting in a continuous flow within the droplet (Figure 5.8b). At larger SAW amplitudes, the internal streaming develops into a movement of the droplet as a whole into the desired direction on the chip. Velocities close to 1 m/s can be achieved in this way. Depending on the actual layout of the chip, biochemical droplets can be merged, mixed, and processed in almost any fashion. As the SAW nanopumps are electrically addressable, a complete sequence of different steps of biochemical reactions or biological assays can be computer controlled. Moreover, the simplicity of the fabrication process of the freely programmable

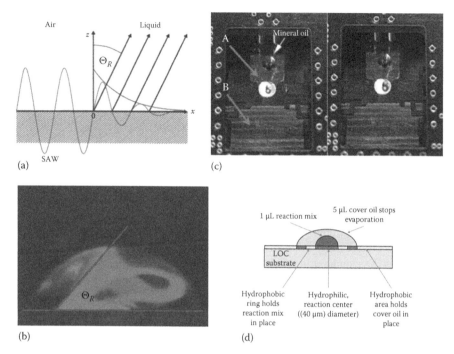

FIGURE 5.8 (a) A schematic illustration of the interaction between a surface acoustic wave (SAW) and a liquid at the surface of the LiNbO$_3$ substrate. The SAW is propagating from left to right, impinging the liquid at $x=0$. A longitudinal sound wave is radiated into the fluid under a refraction angle ΘR. (b) Acoustic streaming induced in a 50 nL droplet on the surface of a chip. The SAW propagating on the chip surface hits the droplet from left to right. Absorption of the wave creates a streaming pattern by coupling to a pressure wave in the liquid, which is excited under the Rayleigh angle ΘR. The streaming pattern is visualized by the dissolution of a fluorescent dye2. (c) The SAW-driven microfluidic lab on a chip. Four droplets (≈ 100 nL each) and a 5 μL droplet of mineral oil are moved in a "remotely controlled" manner and independently by the nanopumps. (d) Sketch of the planar hydrophilic/hydrophobic structured glass slide with 1 μL PCR reaction mixture covered with 5 μL mineral oil.

"bioprocessor" in combination with the possibility to operate and control the entire process via computer automation makes the developed system an ideal candidate for question- and patient-specific truly miniaturized laboratories on a chip (Figure 5.8c). The fluid actuation is controlled by a deliberately addressable manifold of IDTs, implemented in the metallization layer of the chip surface. The typically narrow-band IDTs are excited via a high-frequency source at the respective operating frequency. This will allow the complete automatic control of the reactants actuation in the subsequent steps. The analysis of genetic material is fundamental to various medical and scientific applications. These include diagnosis of genetic disorders from prenatal diagnosis, investigation of oncogenes, forensic analysis, disease diagnosis and screening, paternity testing, gene therapy formation, and detection of pathological specimen. But the primary reason for the analysis of genetic material, particularly in medicine, is the diagnosis of genetic disorders.

Reducing polymerase chain reaction (PCR) to the microliter level (Figure 5.8d) is not only of interest for portable detection technologies and high-throughput and parallel analytical systems. Also, as a consequent next step toward an individualized therapy or POC diagnosis, it is essential to dissipate the complexity in heterogeneous tissue sections [73,74].

5.4 CHALLENGES OF ORGANIC ELECTRONICS

The discovery of conducting polymers in the 1970s paved the way for a variety of new research fields. This achievement was recognized with the 2000 Nobel Prize in Chemistry indicating its importance to science and technology. One of the most significant fields emerging as a consequence of this discovery is now referred to as organic electronics. Using organic materials as an alternative for (or in combination with) conventional inorganic semiconductors gives rise to new electronic, optical, and mechanical properties for novel electronic devices. In recent years, research and development in the field of organic electronics have experienced great advances. The interest in novel electronic devices based on organic semiconducting materials has dramatically increased due to the great potential of this new technology for a wide range of engineering applications. Components based on organic semiconductors are diverse and include light-emitting diodes, photodetectors, thin-film transistors, sensors, and so on. OLED displays for different portable electronic devices represent the first commercial products arising from the field and more and more are destined to come. Figure 5.9 shows some examples of organic devices demonstrating at the same time their capability of being processed on flexible substrates.

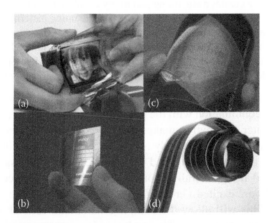

FIGURE 5.9 Examples of flexible products based on organic semiconductors. (a) Sony flexible OLED at SID 2007 (upper left, from http://www.sony.net). (b) Model of a polymer flexible RFID tag (lower left, Photo courtesy of PolyIC, Fuerth, Bavaria, Germany). (c) Organic flexible sheet-image scanner (right, from Prof. Takao Someya, Sheet-image scanner, http://www.ntech.t.u-tokyo.ac.jp/Archive/Archive_download/Archive_download_en.html#scanner). (d) Flexible solar cell from Konarka.

Certainly, one of the strongest advantages of organic technologies is their promise for ultralow costs. Being able to manufacture electronic components on flexible plastic substrates by simple coating and printing techniques is not only a fascinating vision, but also interesting from an economic point of view. Solution-processable polymers have the potential to fulfill these requirements in addition to providing the possibility of chemical tailoring, allowing flexibility in the design of organic semiconductors with desired properties. As mentioned at the beginning, small, portable, and inexpensive sensors, which can be used as disposables in life science applications, cover a wide range of sensing applications, that is, food and environmental monitoring or detection of biological hazardous material. Organic materials being suitable for large-area, low-cost, flexible, and maybe even disposable electronics can provide a promising answer in the aforementioned field of applications, and the field at the interface between organic electronics and biology represents a rare example of an almost perfect link between fundamental research and clearly defined applications dictated by the industry. Indeed, the term "organic bioelectronics," coined by Berggren and Richter-Dahlfors in their seminal review [75], has come to describe a new highly interdisciplinary research theme, which holds promise to open new possibilities in the realization of biomedical test systems in life sciences, help discover fundamental principles for the medicine of the future, increase the quality of our lives, and protect our environment. Sensors based on organic semiconductors will combine the addressed features. They are flexible and electrically conductive and can be produced in large areas with low cost [76]. In recent years, organic biosensors based on electrical (or electronic) transduction mechanisms came into the focus of several groups. A remarkable number of examples can already be found in the literature. Among these transistor-based devices (such as gas, vapor, and fluid sensors) [77–80], ion-sensitive field-effect transistors (ISFETs) [81–84], sensors used in liquids [85–88], and organic electrode materials used in amperometry/voltammetry [89–93] are in use.

A detailed description of the performance of the organic thin-film transistors (OTFTs) produced in our laboratories along with theoretical analysis has recently been reported [94]. Both spin-coating and spray-coating [95,96] deposition techniques were successfully applied.

Low-operating voltage OTFT devices have been fabricated, which can be used as sensors in electrolytes [97]. The devices are based on regioregular poly(3-hexylthiophene) (P3HT), being both a reasonably conductive and optically active polymer. For applications to monitor the cell physiology in life sciences, biocompatibility and biofunctionalization of conducting and semiconducting polymers is mandatory. Recently, we reported on the biocompatibility of solution-processable organic semiconducting polymers. Thiophene derivates, such as P3HT, represent a promising organic material. Normally, they are characterized by a good solubility, ease of processing, and good environmental and thermal stability. Therefore, P3HT biocompatibility and biofunctionalization have been investigated. P3HT solutions in toluene or chloroform were deposited and cured at temperatures higher than the boiling point of the used solvent, yielding nontoxic, solvent-free uniform layers. For biocompatibility and cell adhesion, and to evaluate the effect of surface treatments and protein-based coatings, cell-growth studies and adhesion experiments on

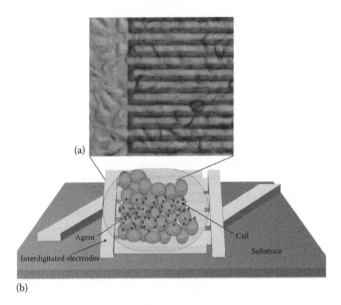

FIGURE 5.10 (a) Optical microscopic image of cellular growth on P3HT transistor after biofunctionalization of the polymer surface. Cells were grown for 24 h on sterilized P3HT prefunctionalized with collagen. (b) Schematic of one possible sensor application scenario, in which external agents modify the cell metabolism. (From Scarpa, G. et al., *IEEE Trans. Nanotechnol.*, 9, 527, 2010.)

the modified P3HT thin-film layers have been carried out with fibroblast cell lines. The spindle shape states the adherent growth of the cells on the underlying surface (see Figure 5.10) [98]. To overcome biocompatibility problems, protein-based coatings and oxygen-plasma treatments have been adopted to enable growth of adherent living cells on those modified surfaces.

Thereby, the surface characteristics have a significant impact on the adhesion, differentiation, and proliferation of surrounding cells. For biocompatibility studies, glass substrates were coated with thin P3HT layers, by spinning and casting the polymer under nitrogen atmosphere. The latter method was applied to obtain films in the range of 100 nm to 10 μm thickness. These hydrophobic coatings have not been changed further. The details on the biocompatibility and biofunctionalization studies of P3HT are published, along with all relevant experimental details [99]. With the demonstration of biocompatible semiconducting polymeric layers, a hurdle for the realizations of low-cost and mass-produced sensors in life science has been overcome. This opens new possibilities and applications of biological sensing with organic electronic devices.

5.5   FUTURE PERSPECTIVES

The key factors for commercial success of organic electronic components are efficiency, lifetime, and cost. Efficiency and lifetime are inferior when compared to traditional inorganic devices and are therefore at the same time the major drawbacks.

A lot of research is now done trying to overcome these problems, and for some applications state-of-the-art organic devices are already sufficient. However, in the field of biology and medicine, unprecedented applications can open up by using the possibilities offered by organic devices. On the basis of what has been discussed so far, multianalyte and disposable analytical formats can be realized. Multiplexed analysis can be obtained by developing "multispot" analytical systems: different bio-specific probes can be immobilized in different positions of a device and the signals from the different positions (corresponding to the binding of the specific analytes) will be separately detected. Therefore, different analytes in the same sample will be simultaneously detected and quantified. This aspect is very important in clinical diagnosis and POC monitoring and home diagnostics, because the devices could be used for the detection of "panels" of analytes related to a given pathology. Development of devices in disposable formats will also be considered, because disposable devices represent a valid alternative to overcome the problem of biosensors regeneration, avoid the risk of cross-contamination between different samples, and reduce the hazards related to the manipulation of potentially infected clinical samples. In particular, going for low cost and tests close to clinics, large volumes and low costs for single-use applications, and distributed diagnostics, a series of enabling technologies need to be developed. Device stability in an industrial environment and the know-how of industrial actors will be a key issue here. However, still a lot of research efforts are needed. Besides the biocompatibility, the liability (especially on flexible plastic substrates) and the stability in terms of cell culture and liquid media are also critical issues for the suitability of the individual materials and processes. Finally, the integration of all analytical process, from sample collection to analyte detection, on the same device, lab-on-a-chip system, can be sought. Sensors platform able to perform sample collection and analysis on the same device could be designed. The ability to precisely control parameters such as substrate, flow rate, buffer composition, and surface chemistry in these microscale devices would make them ideal for a broad spectrum of cell biology–based applications. It ranges from high-throughput screening of single cells and 3-D scaffolds for tissue engineering to complex biochemical reactions like PCR, POC diagnosis, and drug detection. Potential applications could be assays for cancer detection at an early stage, cardiovascular disease, biomarkers for Alzheimer's disease, virus and bacteria detection, and online cell analysis like cell activity.

REFERENCES

1. A. Poghossian, J.W. Schultze, and M.J. Schöning, Multi-parameter detection of (bio-) chemical and physical quantities using an identical transducer principle, *Sensors and Actuators B: Chemical*, 91, 2003, 83–91.
2. A. Manz, N. Graber, and H.M. Widmer, Miniaturized total chemical analysis systems: A novel concept for chemical sensing, *Sensors and Actuators B: Chemical*, 1, Jan. 1990, 244–248.
3. T. Squires, Microfluidics: Fluid physics at the nanoliter scale, *Reviews of Modern Physics*, 77, 2005, 977–1026.
4. D.B. Weibel and G.M. Whitesides, Applications of microfluidics in chemical biology, *Current Opinion in Chemical Biology*, 10, Dec. 2006, 584–591.

5. A. Manz, Y. Miyahara, J. Miura, Y. Watanabe, H. Miyagi, and K. Sato, Design of an open-tubular column liquid chromatograph using silicon chip technology, *Sensors and Actuators B: Chemical*, 1, 1990, 249–255.

6. P. Wilding, M. Shoffner, and L. Kricka, PCR in a silicon microstructure, *Clinical Chemistry*, 40, 1994, 1815–1818.

7. D.J. Harrison, K. Fluri, K. Seiler, Z. Fan, C.S. Effenhauser, and A. Manz, Micromachining a miniaturized capillary electrophoresis-based chemical analysis system on a chip, *Science*, 261, Aug. 1993, 895–897.

8. L. Martynova, L.E. Locascio, M. Gaitan, G.W. Kramer, R.G. Christensen, and W.A. MacCrehan, Fabrication of plastic microfluid channels by imprinting methods, *Analytical Chemistry*, 69, Dec. 1997, 4783–4789.

9. D.C. Duffy, J.C. McDonald, O.J.A. Schueller, and G.M. Whitesides, Rapid prototyping of microfluidic systems in poly(dimethylsiloxane), *Analytical Chemistry*, 70, Dec. 1998, 4974–4984.

10. K. Sato, K. Mawatari, and T. Kitamori, Microchip-based cell analysis and clinical diagnosis system, *Lab on a chip*, 8, Dec. 2008, 1992–1998.

11. Y. Tanaka, K. Sato, M. Yamato, T. Okano, and T. Kitamori, Drug response assay system in a microchip using human hepatoma cells, *Analytical Sciences*, 20, Mar. 2004, 411–413.

12. J. El-Ali, P.K. Sorger, and K.F. Jensen, Cells on chips, *Nature*, 442, Jul. 2006, 403–411.

13. G.B. Salieb-Beugelaar, G. Simone, A. Arora, A. Philippi, and A. Manz, Latest developments in microfluidic cell biology and analysis systems, *Analytical Chemistry*, 82, Jun. 2010, 4848–4864.

14. L. Gac and A.V.D. Berg, Single cells as experimentation units in lab-on-a-chip devices, *Trends in Biotechnology*, 28(2), 2009, 55–62.

15. Y. Tanaka, K. Sato, T. Shimizu, M. Yamato, T. Okano, and T. Kitamori, Biological cells on microchips: New technologies and applications, *Biosensors & Bioelectronics*, 23, Nov. 2007, 449–458.

16. B. Wolf, M. Brischwein, H. Grothe, C. Stepper, J. Ressler, and T. Weyh, Lab-on-a-chip systems for cellular assays, *BioMEMS—Series: Microsystems*, 16, 2006, 269–307.

17. C.W. Scott and M.F. Peters, Label-free whole-cell assays: Expanding the scope of GPCR screening, *Drug Discovery Today*, 15, Jun. 2010, 704–716.

18. R. McGuinness, Impedance-based cellular assay technologies: Recent advances, future promise, *Current Opinion in Pharmacology*, 7, Oct. 2007, 535–540.

19. D. Le Guillou-Buffello, G. Hélary, M. Gindre, G. Pavon-Djavid, P. Laugier, and V. Migonney, Monitoring cell adhesion processes on bioactive polymers with the quartz crystal resonator technique, *Biomaterials*, 26, Jul. 2005, 4197–4205.

20. G.N.M. Ferreira, A.-C. Da-Silva, and B. Tomé, Acoustic wave biosensors: Physical models and biological applications of quartz crystal microbalance, *Trends in Biotechnology*, 27, Dec. 2009, 689–697.

21. P. Wang, G. Xu, L. Qin, Y. Xu, Y. Li, and R. Li, Cell-based biosensors and its application in biomedicine, *Sensors and Actuators*, 108, 2005, 576–584.

22. M. George, Highly integrated surface potential sensors, *Sensors and Actuators B: Chemical*, 69, Oct. 2000, 266–275.

23. B. Stein, M. George, H.E. Gaub, J.C. Behrends, and W.J. Parak, Spatially resolved monitoring of cellular metabolic activity with a semiconductor-based biosensor, *Biosensors & Bioelectronics*, 18, Jan. 2003, 31–41.

24. S.E. Eklund, R.M. Snider, J. Wikswo, F. Baudenbacher, A. Prokop, and D.E. Cliffel, Multianalyte microphysiometry as a tool in metabolomics and systems biology, *Journal of Electroanalytical Chemistry*, 587, 2006, 333–339.

25. A.F.M. Johnstone, G.W. Gross, D.G. Weiss, O.H.-U. Schroeder, A. Gramowski, and T.J. Shafer, Microelectrode arrays: A physiologically based neurotoxicity testing platform for the 21st century, *Neurotoxicology*, 31, Aug. 2010, 331–350.

26. M. Kreir, S. Stoelzle, A. Haythornthwaite, C. Haarmann, C. Farre, A. Brueggemann, D.R. Guinot, M. George, and N. Fertig, Automated patch clamp recordings of action potentials from stem cell derived cardiomyocytes, *Journal of Pharmacological and Toxicological Methods*, 62(2), Sep. 2010, e35.

27. F. Ilchmann, J. Meyer, V. Lob, C. Zhang, and H. Grothe, Automated multiparametric 24 well neuro screening system, *Proceedings of the 6th International Meeting on Substrate-Integrated Microelectrodes*, Reutlingen, Germany, 2008, pp. 302–303.

28. M. Brischwein, S. Herrmann, W. Vonau, F. Berthold, H. Grothe, E.R. Motrescu, and B. Wolf, Electric cell-substrate impedance sensing with screen printed electrode structures, *Lab on a Chip*, 6, Jun. 2006, 819–822.

29. G.A. Rivas, M.D. Rubianes, M.C. Rodríguez, N.F. Ferreyra, G.L. Luque, M.L. Pedano, S.A. Miscoria, and C. Parrado, Carbon nanotubes for electrochemical biosensing, *Talanta*, 74, Dec. 2007, 291–307.

30. C. Hu and S. Hu, Carbon nanotube-based electrochemical sensors: Principles and applications in biomedical systems, *Journal of Sensors*, 9, 2009, 1–40.

31. E.W. Keefer, B.R. Botterman, M.I. Romero, A.F. Rossi, and G.W. Gross, Carbon nanotube coating improves neuronal recordings, *Nature Nanotechnology*, 3, Jul. 2008, 434–439.

32. S. Lindström and H. Andersson-Svahn, Overview of single-cell analyses: Microdevices and applications, *Lab on a Chip*, 10, Oct. 2010, 3363–3372.

33. T.M. Curtis, M.W. Widder, L.M. Brennan, S.J. Schwager, W.H. van der Schalie, J. Fey, and N. Salazar, A portable cell-based impedance sensor for toxicity testing of drinking water, *Lab on a Chip*, 9, Aug. 2009, 2176–2183.

34. J. Pancrazio, A portable microelectrode array recording system incorporating cultured neuronal networks for neurotoxin detection, *Biosensors & Bioelectronics*, 18, Oct. 2003, 1339–1347.

35. J. Wiest, T. Stadthagen, M. Schmidhuber, M. Brischwein, J. Ressler, U. Raeder, H. Grothe, A. Melzer, and B. Wolf, Intelligent mobile lab for metabolics in environmental monitoring, *Analytical Letters*, 39, Jul. 2006, 1759–1771.

36. U. Abel, Chemotherapy of advanced epithelial cancer—A critical review, *Biomedicine & Pharmacotherapy*, 46, 1992, 439–452.

37. D.J. Demetrick, Targeting cancer treatment: The challenge of anatomical pathology to the analytical chemist, *Analyst*, 128, Jul. 2003, 995–997.

38. D. Wlodkowic and J.M. Cooper, Tumors on chips: Oncology meets microfluidics, *Current Opinion in Chemical Biology*, 14, Sep. 2010, 556–567.

39. M. Wu, A. Neilson, A.L. Swift, R. Moran, J. Tamagnine, S. Armistead, K. Lemire, J. Orrell, J. Teich, D.A. Ferrick, D. Parslow, and S. Chomicz, Multiparameter metabolic analysis reveals a close link between attenuated mitochondrial bioenergetic function and enhanced glycolysis dependency in human tumor cells, *American Journal of Physiology: Cell Physiology*, 292, Jan. 2007, C125–C136.

40. P. Mestres and A. Morguet, The Bionas technology for anticancer drug screening, *Expert Opinion on Drug Discovery*, 4, Jul. 2009, 785–797.

41. B. Wolf, M. Kraus, M. Brischwein, R. Ehret, W. Baumann, and M. Lehmann, Biofunctional hybrid structures—Cell-silicon hybrids for applications in biomedicine and bioinformatics, *Bioelectrochemistry and Bioenergetics*, 46, 1998, 215–225.

42. T. Henning, M. Brischwein, W. Baumann, R. Ehret, I. Freund, R. Kammerer, M. Lehmann, A. Schwinde, and B. Wolf, Approach to a multiparametric sensorchip-based tumor chemosensitivity assay, *Anti-Cancer Drugs*, 12, Jan. 2001, 21–32.

43. M. Brischwein, E.R. Motrescu, E. Cabala, A.M. Otto, H. Grothe, and B. Wolf, Functional cellular assays with multiparametric silicon sensor chips, *Lab on a Chip*, 3, Nov. 2003, 234–240.

44. H. Matsuoka, T. Komazaki, Y. Mukai, M. Shibusawa, H. Akane, A. Chaki, N. Uetake, and M. Saito, High throughput easy microinjection with a single-cell manipulation supporting robot, *Journal of Biotechnology*, 116, Mar. 2005, 185–194.

45. A. Ashkin and J. Dziedzic, Optical trapping and manipulation of viruses and bacteria, *Science*, 235, Mar. 1987, 1517–1520.

46. T.N. Buican, M.J. Smyth, H.A. Crissman, G.C. Salzman, C.C. Stewart, and J.C. Martin, Automated single-cell manipulation and sorting by light trapping, *Applied Optics*, 26, Dec. 1987, 5311–5316.

47. T. Schnelle, R. Hagedorn, G. Fuhr, S. Fiedler, and T. Muller, Three-dimensional electric field traps for manipulation of cells—Calculation and experimental verification, *Biochimica et Biophysica*, 1157, 1993, 127–140.

48. C. Duschl, P. Geggier, M. Jäger, M. Stelzle, T. Müller, T. Schnelle, and G.R. Fuhr, Versatile chip-based tool for the controlled manipulation of microparticles in biology using high frequency electromagnetic fields, *Lab-on-Chips for Cellomics: Micro and Nanotechnologies for Life Science*, H. Andersson and A. Berg, eds., Dordrecht, the Netherlands: Kluwer, 2004, pp. 83–122.

49. T. Thorsen, S.J. Maerkl, and S.R. Quake, Microfluidic large-scale integration, *Science*, 298, Oct. 2002, 580–584.

50. A. Khademhosseini, J. Yeh, S. Jon, G. Eng, K.Y. Suh, J.A. Burdick, and R. Langer, Molded polyethylene glycol microstructures for capturing cells within microfluidic channels, *Lab on a Chip*, 4, Oct. 2004, 425–430.

51. H. Andersson, Microfluidic devices for cellomics: A review, *Sensors and Actuators B: Chemical*, 92, Jul. 2003, 315–325.

52. H. Andersson and A. van den Berg, Microtechnologies and nanotechnologies for single-cell analysis, *Current Opinion in Biotechnology*, 15, Feb. 2004, 44–49.

53. W.C. Chang, L.P. Lee, and D. Liepmann, Biomimetic technique for adhesion-based collection and separation of cells in a microfluidic channel, *Lab on a Chip*, 5, Jan. 2005, 64–73.

54. N. Xu, Y. Lin, S.A. Hofstadler, D. Matson, C.J. Call, and R.D. Smith, A microfabricated dialysis device for sample cleanup in electrospray ionization mass spectrometry, *Analytical Chemistry*, 70, Sep. 1998, 3553–3556.

55. J. Bergkvist, S. Ekström, L. Wallman, M. Löfgren, G. Marko-Varga, J. Nilsson, and T. Laurell, Improved chip design for integrated solid-phase microextraction in on-line proteomic sample preparation, *Proteomics*, 2, Apr. 2002, 422–429.

56. S. Fiedler, S.G. Shirley, T. Schnelle, and G. Fuhr, Dielectrophoretic sorting of particles and cells in a microsystem, *Analytical Chemistry*, 70, May. 1998, 1909–1915.

57. A.Y. Fu, H.-P. Chou, C. Spence, F.H. Arnold, and S.R. Quake, An integrated microfabricated cell sorter, *Analytical Chemistry*, 74, Jun. 2002, 2451–2457.

58. A. Wolff, I.R. Perch-Nielsen, U.D. Larsen, P. Friis, G. Goranovic, C.R. Poulsen, J.P. Kutter, and P. Telleman, Integrating advanced functionality in a microfabricated high-throughput fluorescent-activated cell sorter, *Lab on a Chip*, 3, Feb. 2003, 22–27.

59. M.M. Wang, E. Tu, D.E. Raymond, J.M. Yang, H. Zhang, N. Hagen, B. Dees, E.M. Mercer, A.H. Forster, I. Kariv, P.J. Marchand, and W.F. Butler, Microfluidic sorting of mammalian cells by optical force switching, *Nature Biotechnology*, 23, Jan. 2005, 83–87.

60. M. Ozkan, M. Wang, C. Ozkan, R. Flynn, and S. Esener, Optical manipulation of objects and biological cells in microfluidic devices, *Biomedical Microdevices*, 5, 2003, 61–67.

61. D.R. Reyes, D. Iossifidis, P.-A. Auroux, and A. Manz, Micro total analysis systems. 1. Introduction, theory, and technology, *Analytical Chemistry*, 74, Jun. 2002, 2623–2636.

62. P.-A. Auroux, D. Iossifidis, D.R. Reyes, and A. Manz, Micro total analysis systems. 2. Analytical standard operations and applications, *Analytical Chemistry*, 74, Jun. 2002, 2637–2652.

63. P.S. Dittrich, K. Tachikawa, and A. Manz, Micro total analysis systems. Latest advancements and trends, *Analytical Chemistry*, 78, Jun. 2006, 3887–3908.

64. E.M. Purcell, Life at low Reynolds number, *American Journal of Physics*, 45, 1977, 3–11.

65. A. Wixforth, Acoustically driven planar microfluidics, *Superlattices and Microstructures*, 33, May. 2003, 389–396.

66. A. Renaudin, P. Tabourier, V. Zhang, J. Camart, and C. Druon, SAW nanopump for handling droplets in view of biological applications, *Sensors and Actuators B: Chemical*, 113, Jan. 2006, 389–397.

67. S. Thalhammer, Programmable lab-on-a-chip system for single cell analysis, *Proceedings of SPIE*, 7364, May 2009, 73640B.

68. J.P. Brody, P. Yager, R.E. Goldstein, and R.H. Austin, Biotechnology at low Reynolds numbers, *Biophysical Journal*, 71, Dec. 1996, 3430–3441.

69. T. Franke, A.R. Abate, D.A. Weitz, and A. Wixforth, Surface acoustic wave (SAW) directed droplet flow in microfluidics for PDMS devices, *Lab on a Chip*, 9, Sep. 2009, 2625–2627.

70. W. Nyborg, Acoustic streaming, *Physical Acoustics*, Volume II B, W. Mason, ed., New York: Academic Press, 1965, pp. 265–331.

71. C. Eckart, Vortices and streams caused by sound waves, *Physical Review*, 73, Jan. 1948, 68–76.

72. L.Y. Yeo and J.R. Friend, Ultrafast microfluidics using surface acoustic waves, *Biomicrofluidics*, 3, Jan. 2009, 012002, 1–23.

73. V. Mayer, U. Schoen, E. Holinski-Feder, U. Koehler, and S. Thalhammer, Single cell analysis of mutations in the APC gene, *Fetal Diagnosis and Therapy*, 26, Jan. 2009, 148–156.

74. D. Woide, A. Zink, and S. Thalhammer, Technical Note: PCR analysis of minimum target amount of ancient DNA, *American Journal of Physical Anthropology*, 42, Jun. 2010, 321–327.

75. M. Berggren and A. Richter-Dahlfors, Organic bioelectronics, *Advanced Materials*, 19, Oct. 2007, 3201–3213.

76. J.T. Mabeck and G.G. Malliaras, Chemical and biological sensors based on organic thin-film transistors, *Analytical and Bioanalytical Chemistry*, 384, Jan. 2006, 343–353.

77. L. Torsi, G.M. Farinola, F. Marinelli, M.C. Tanese, O.H. Omar, L. Valli, F. Babudri, F. Palmisano, P.G. Zambonin, and F. Naso, A sensitivity-enhanced field-effect chiral sensor, *Nature Materials*, 7, May. 2008, 412–417.

78. M.C. Tanese, D. Fine, A. Dodabalapur, and L. Torsi, Interface and gate bias dependence responses of sensing organic thin-film transistors, *Biosensors and Bioelectronics*, 21, Nov. 2005, 782–788.

79. M. Kaempgen and S. Roth, Transparent and flexible carbon nanotube/polyaniline pH sensors, *Journal of Electroanalytical Chemistry*, 586, Jan. 2006, 72–76.

80. C. Bartic, A. Campitelli, and S. Borghs, Field-effect detection of chemical species with hybrid organic/inorganic transistors, *Applied Physics Letters*, 82, 2003, 475–477.

81. C. Bartic, B. Palan, A. Campitelli, and G. Borghs, Monitoring pH with organic-based field-effect transistors, *Sensors and Actuators B: Chemical*, 83, Mar. 2002, 115–122.

82. A. Loi, I. Manunza, and A. Bonfiglio, Flexible, organic, ion-sensitive field-effect transistor, *Applied Physics Letters*, 86, 2005, 103512.

83. C. Bartic and G. Borghs, Organic thin-film transistors as transducers for (bio) analytical applications, *Analytical and Bioanalytical Chemistry*, 384, Sep. 2006, 354–365.

84. A. Sargent, T. Loi, S. Gal, and O.A. Sadik, The electrochemistry of antibody-modified conducting polymer electrodes, *Journal of Electroanalytical Chemistry*, 470, Jul. 1999, 144–156.

85. A. Sargent and O.A. Sadik, Monitoring antibody–antigen reactions at conducting polymer-based immunosensors using impedance spectroscopy, *Electrochimica Acta*, 44, Sep. 1999, 4667–4675.

86. M.G.H. Meijerink, D.J. Strike, N.F. de Rooij, and M. Koudelka-Hep, Reproducible fabrication of an array of gas-sensitive chemo-resistors with commercially available polyaniline, *Sensors and Actuators B: Chemical*, 68, Aug. 2000, 331–334.

87. Y. Sakurai, H.-S. Jung, T. Shimanouchi, T. Inoguchi, S. Morita, R. Kuboi, and K. Natsukawa, Novel array-type gas sensors using conducting polymers, and their performance for gas identification, *Sensors and Actuators B: Chemical*, 83, Mar. 2002, 270–275.

88. M.E. Roberts, S.C.B. Mannsfeld, N. Queraltó, C. Reese, J. Locklin, W. Knoll, and Z. Bao, Water-stable organic transistors and their application in chemical and biological sensors, *Proceedings of the National Academy of Sciences of the United States of America*, 105, Aug. 2008, 12134–12139.

89. B. Piro, L.A. Dang, M.C. Pham, S. Fabiano, and C. Tran-Minh, A glucose biosensor based on modified-enzyme incorporated within electropolymerised poly(3,4-ethylenedioxythiophene) (PEDT) films, *Journal of Electroanalytical Chemistry*, 512, Oct. 2001, 101–109.

90. S.K. Sharma, R. Singhal, B.D. Malhotra, N. Sehgal, and A. Kumar, Lactose biosensor based on Langmuir–Blodgett films of poly(3-hexyl thiophene), *Biosensors & Bioelectronics*, 20, Oct. 2004, 651–657.

91. S.K. Sharma, R. Singhal, B.D. Malhotra, N. Sehgal, and A. Kumar, Langmuir–Blodgett film based biosensor for estimation of galactose in milk, *Electrochimica Acta*, 49, Jun. 2004, 2479–2485.

92. L. Setti, A. Fraleoni-Morgera, B. Ballarin, A. Filippini, D. Frascaro, and C. Piana, An amperometric glucose biosensor prototype fabricated by thermal inkjet printing, *Biosensors & Bioelectronics*, 20, Apr. 2005, 2019–2026.

93. L. Setti, A. Fraleoni-Morgera, I. Mencarelli, A. Filippini, B. Ballarin, and M. Di Biase, An HRP-based amperometric biosensor fabricated by thermal inkjet printing, *Sensors and Actuators B: Chemical*, 126, Sep. 2007, 252–257.

94. S.M. Goetz, C.M. Erlen, H. Grothe, B. Wolf, P. Lugli, and G. Scarpa, Organic field-effect transistors for biosensing applications, *Organic Electronics*, 10, Jul. 2009, 573–580.

95. A. Abdellah, D. Baierl, B. Fabel, P. Lugli, and G. Scarpa, Exploring spray technology for the fabrication of organic devices based on poly(3-hexylthiophene), *Proceedings of the 9th IEEE Conference on Nanotechnology, IEEE-NANO*, Genoa, Italy, 2009, pp. 831–934.

96. A. Abdellah, B. Fabel, P. Lugli, and G. Scarpa, Spray deposition of organic semiconducting thin-films: Towards the fabrication of arbitrary shaped organic electronic devices, *Organic Electronics*, 11, Jun. 2010, 1031–1038.

97. G. Scarpa, A.-L. Idzko, A. Yadav, and S. Thalhammer, Organic ISFET based on poly (3-hexylthiophene), *Sensors*, 10, Mar. 2010, 2262–2273.

98. G. Scarpa, A.-L. Idzko, A. Yadav, E. Martin, and S. Thalhammer, Toward cheap disposable sensing devices for biological assays, *IEEE Transactions on Nanotechnology*, 9, Sep. 2010, 527–532.

99. G. Scarpa, A.-L. Idzko, S. Götz, and S. Thalhammer, Biocompatibility studies of functionalized regioregular poly(3-hexylthiophene) layers for sensing applications, *Macromolecular Bioscience*, 10, Apr. 2010, 378–383.

6 Construction of Enzyme Biosensors Based on a Commercial Glucose Sensor Platform

Yue Cui

CONTENTS

6.1 INTRODUCTION TO GLUCOSE BIOSENSOR

Biosensors are analytical devices incorporating a biological material (e.g., tissues, microorganisms, organelles, cell receptors, enzymes, antibodies, nucleic acids, natural products, etc.), a biologically derived material (e.g., recombinant antibodies, engineered proteins, aptamers, etc.), or a biomimetic material (e.g., synthetic receptors, biomimetic catalysts, combinatorial ligands, imprinted polymers, etc.) intimately within a physicochemical transducer or transducing microsystem, which may be optical, electrochemical, thermometric, piezoelectric, magnetic, or micromechanical [1], as shown in Figure 6.1. The generated electrical signal is related to the concentration of analytes through biological reactions. Diabetes is a worldwide public health problem. It is one of the leading causes of death and disability in the world. The diagnosis and management of diabetes mellitus requires a tight monitoring of blood glucose levels, which is the major diagnostic criterion for diabetes and is a useful tool for patient monitoring [2]. To date, the most commercially successful biosensors are amperometric enzyme glucose biosensors for monitoring diabetes, which account for approximately 85% of the current world market.

Glucose oxidase (GOD) [3] is the standard enzyme for glucose biosensors, and it has a relatively higher selectivity for glucose. GOD is a stable enzyme, which can

149

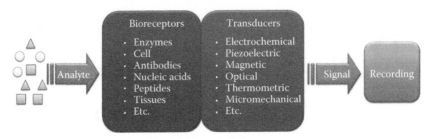

FIGURE 6.1 Schematic illustration for biosensors.

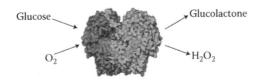

FIGURE 6.2 Enzymatic reactions with GOD for the detection of glucose. (From http://en.wikipedia.org/wiki/Glucose_oxidase)

withstand extreme pH, ionic strength, and temperature compared with many other enzymes, to allow less stringent conditions during the manufacturing process and relatively relaxed storage norms for use, and it is also cost-effective and commercially available [4]. Thus, GOD is widely used for glucose biosensor construction. As shown in Figure 6.2, glucose is catalyzed by GOD to consume oxygen and produce glucolactone and hydrogen peroxide.

Due to the advantages of being simple, small, cost-effective, and easy to handle, amperometric biosensors have been extensively employed. Commercial glucose biosensors are mainly constructed based on Clark-type electrode and screen-printed electrodes with the immobilization of GOD. As shown in Figure 6.3A, for a typical Clark-type glucose biosensor, the enzyme matrix of the biosensor is prepared by mixing the enzyme solution and curing agent. The enzyme mixture is spread over a Teflon membrane and dried at 4°C overnight. Then it is covered with a dialysis membrane to sandwich the enzyme layer between the two membranes and fixed with a membrane holder and an O-ring. The Teflon side facing the electrode received 15 μL of saturated KCl gel as electrolyte before the enzyme membrane being placed on top of the Clark-type electrode. The sensor is then screwed into the measuring cell, and rehydrated in buffer at room temperature before usage the Teflon side. In the presence of glucose, GOD catalyzes the specific oxidation of glucose. Dissolved oxygen acting as an essential material for the enzymatic activity of GOD is consumed proportionally to the concentration of glucose during the measurements. A detectable signal, caused by the consumption of dissolved oxygen by GOD, is monitored by the Clark electrode. Electroactive interferences are eliminated by the Teflon membrane in front of the Clark electrode. As shown in Figure 6.3B, the working electrode of the screen-printed electrode is covered by a mixture containing GOD and curing agent, followed by drying and storing at

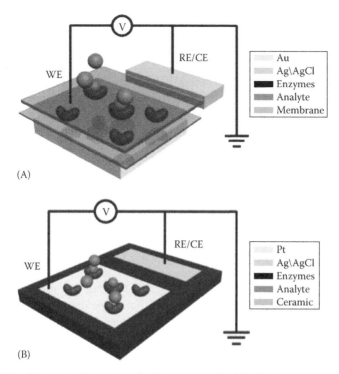

FIGURE 6.3 Schematic illustration for biosensors with (A) Clark-type oxygen electrode and (B) screen-printed electrode.

4°C overnight. The enzyme electrode is then screwed into the measuring cell and rehydrated at room temperature to allow the enzyme matrix to swell before use. Experiments were performed at room temperature by applying a specific potential for the screen-printed electrode for the detection of hydrogen peroxide. The current difference (nA) between the stationary currents is recorded, which is proportional to the concentration of glucose.

6.2 BIOSENSOR FOR METABOLIC COMPOUND: ATP

Metabolic compounds are important analytes for medical diagnosis and environmental monitoring, and metabolic biosensors are widely developed based on sequential and competitive enzymatic reactions on Clark-type electrode and screen-printed electrodes. Here, I describe the biosensor construction for a typical metabolic compound, ATP [5], which is based on the sequential enzymatic reactions and the detection of the substrate consumption and enzymatic product.

ATP, as a mediator of energy exchanges for all living cells, plays an important role in various vital biological processes [6,7]. Besides, it is also widely used as an index for biomass determinations in clinical microbiology, food quality control, and environmental analyses owing to the ubiquitous presence in living matter [8,9].

Due to the importance of ATP, various methods have been developed for its determination, including luminescent, colorimetric, spectrofluorimetric, chromatographic, patch-sniffing, and potentiometric biosensing methods [10–17]. However, these methods result in a long measuring time, a large amount of enzyme consumption, sophisticated procedures which need to be performed by skilled personnel, or a complex signal processing that is difficult to adapt for the handheld devices.

The importance of amperometric enzyme–based biosensors has increased considerably in recent years thanks to the advantages of being highly sensitive, rapid, accurate, economical, and easy to handle for specific measurement of target analyte in complex matrices such as blood, food products, and environmental samples. Various amperometric ATP biosensors have been developed by using different combinations of enzymes or enzymes with mediators [18–23]. Recently, the coimmobilizations of a NAD(P)$^+$-dependent dehydrogenase with p-hydroxybenzoate hydroxylase (HBH) in front of a Clark-type oxygen electrode and on a screen-printed electrode have been investigated for developing a general type of dehydrogenase-based biosensors [24–26], which show high-performance characteristics.

Here, the developments of two types of ATP biosensors are described which are based on new combinations of enzymes and electrodes by using the coimmobilizations of HBH, G6PDH, and HEX on a Clark-type oxygen electrode and on a screen-printed electrode. The schematic illustrations for the biosensor setup and the determination principles are shown in Figure 6.4. HEX transfers the phosphate group from ATP to glucose to form glucose-6-phosphate. G6PDH catalyzes the specific dehydrogenation of glucose-6-phosphate by consuming NADP$^+$. The product "NADPH" initiates the irreversible hydroxylation of p-hydroxybenzoate by HBH to consume dissolved oxygen and generate 3,4-dihydroxybenzoate. During the measurement of ATP, a detectable signal caused by the consumption of oxygen by HBH can be monitored at −0.6 V versus Ag/AgCl by the Clark-type electrode, and another detectable signal caused by the generation of 3,4-dihydroxybenzoate by HBH can be monitored at 0.42 V versus Ag/AgCl by the screen-printed electrode. The electronic

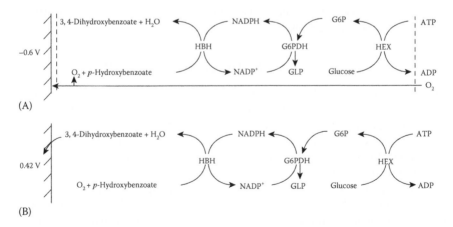

FIGURE 6.4 Schematic illustration for ATP biosensors with (A) the Clark-type electrode and (B) the screen-printed electrode. (From Cui, Y., *IEEE Sens. J.*, 10, 979, 2010.)

signals are monitored and processed with a potentiostat, and the data acquisitions are performed with a computer.

The Clark-type sensor performance for the determination of ATP was characterized. Figure 6.5A shows the current–time curve and the recovery study of the sensor obtained by the additions of various concentrations of ATP. The cathodic current decreased after the addition of ATP due to the consumption of dissolved oxygen through enzymatic reactions by HBH, G6PDH, and HEX, and the reduction of cathodic current was proportional to the concentration of ATP. The response of this biosensor was rapid (2 s) with high reproducibility and short recovery time (1 min). The steady-state background current decreased after the addition of ATP and reached a new stationary state in 20 s, and with a washing step to remove ATP in the buffer

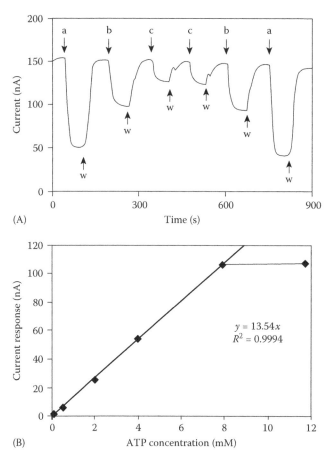

(A)

(B)

FIGURE 6.5 Characterization of the Clark-type sensor performance. (A) Current–time curve of the sensor to (a) 8 mM, (b) 4 mM, and (c) 2 mM ATP with (w) washing step. (B) Calibration curve for ATP by the sensor ($n=3$) (Sensor: 2 U HBH, 1.2 G6PDH, and 0.8 U HEX on a Clark-type oxygen electrode. Buffer: 100 mM Tris-SO$_4$ buffer at pH 8.0, 1 mM p-hydroxybenzoate, 0.2 mM NADP$^+$, 2 mM glucose, and 10 mM MgSO$_4$). (From Cui, Y., *IEEE Sens. J.*, 10, 979, 2010.)

solution for recovery, the current increased to the initial background current. The total measurement using the sensor took less than 3 min. Figure 6.5B shows the calibration curve for ATP with the sensor. A linear relationship was obtained between the current response and the concentration of ATP from 0.1 to 8 mM with a detection limit of 0.05 mM (slope: 13.5 nA mM^{-1}, $R^2 = 0.9994$, $n = 3$). Also, the Teflon membrane–covered Clark-type electrode was protected from contacting the electroactive interferences to produce unreliable signals. The sensor showed no signal response to the electroactive substances, such as ascorbic acid, uric acid, and 20 L-amino acids (L-alanine, L-valine, L-leucine, L-isoleucine, L-phenylalanine, L-tryptophan, L-methionine, L-glycine, L-glutamate, L-serine, L-threonine, L-cysteine, L-tyrosine, L-asparagine, L-glutamine, L-aspartic acid, L-lysine, L-arginine, L-histidine, L-proline). Therefore, the Clark-type sensor has a high specificity for the determination of ATP.

The screen-printed sensor performance for the determination of ATP was characterized. Figure 6.6A shows the current–time curve and the recovery study of the sensor obtained by the additions of various concentrations of ATP. The anodic current increased after the addition of ATP due to the oxidation of catechol generated by HBH, G6PDH, and HEX, and the increase of anodic current was proportional to the concentration of ATP. The response of this biosensor was rapid (2 s) with high reproducibility and short recovery time (1 min). The steady-state background current increased after the addition of ATP and reached a new stationary state in around 1 min, and with a washing step to remove ATP in the buffer solution for recovery, the current decreased to the initial background current. The total measurement using the sensor took less than 3 min. Figure 6.6B shows the calibration curve for ATP with the sensor. A linear relationship was obtained between the current response and the concentration of ATP from 5 μM to 4 mM with a detection limit of 4 μM (slope: 178.9 nA mM^{-1}, $R^2 = 0.9992$, $n = 3$). Also, some typical electroactive substances were investigated for their interference effects on the sensor response. The sensor showed almost no signal response ($<\pm0.5$ nA mM^{-1}) to most L-amino acids (L-alanine, L-valine, L-leucine, L-isoleucine, L-phenylalanine, L-tryptophan, L-glycine, L-glutamate, L-serine, L-threonine, L-cysteine, L-asparagine, L-glutamine, L-aspartic acid, L-lysine, L-arginine, L-histidine, L-proline), but a signal response of 13 nA mM^{-1} to L-methionine, 100 nA mM^{-1} to L-tyrosine, 780 nA mM^{-1} to ascorbic acid, and 400 nA mM^{-1} to uric acid due to the direct exposure of the electrode surface to buffer solution. While as in most real samples the concentrations of these substances are in the micromolar ranges and the concentrations of ATP are in the millimolar ranges [27,28], the screen-printed sensor could also determine the concentrations of ATP relatively accurately.

Compared with the screen-printed sensor, the Clark-type sensor showed a higher specificity of being free from electroactive interferences, while it also showed a lower sensitivity and a higher detection limit, which were probably mainly due to the smaller electrode area for signal transduction and the diffusion barriers from the dialysis membrane and the Teflon membrane. The determination performances of both types of sensors could be further enhanced by using other kinds of electrodes (e.g., another electrode with a larger diameter) for sensor constructions.

In summary, the developments of two types of amperometric trienzyme ATP biosensors are described based on new combinations of enzymes and electrodes by

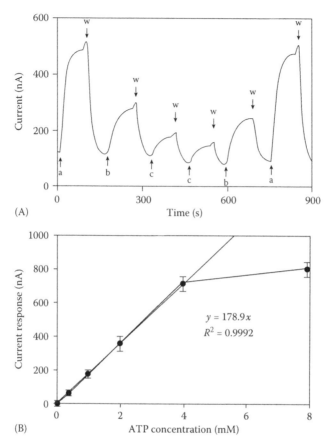

FIGURE 6.6 Characterization of the screen-printed sensor performance. (A) Current–time curve of the sensor to (a) 2 mM, (b) 1 mM, and (c) 0.5 mM ATP with (w) washing step. (B) Calibration curve for ATP by the sensor ($n=3$) (Sensor: 2 U HBH, 1.2 G6PDH, and 0.8 U HEX on a Clark-type oxygen electrode. Buffer: 100 mM Tris-SO$_4$ buffer at pH 8.0, 1 mM p-hydroxybenzoate, 0.2 mM NADP+, 2 mM glucose, and 10 mM MgSO$_4$). (From Cui, Y., *IEEE Sens. J.*, 10, 979, 2010.)

using the coimmobilizations of HBH, G6PDH, and HEX on a Clark-type oxygen electrode and on a screen-printed electrode. The sensors show high-performance characteristics with broad detection ranges, short measuring times, and good specificities. Thus, the methods provide new analytical approaches for the determinations of ATP which are rapid, sensitive, accurate, and easy to handle.

6.3 BIOSENSOR FOR ENZYME ACTIVITY: PHOSPHOGLUCOMUTASE

The determination of enzyme activity is of great importance for various biological and medical applications. The biosensors for enzyme activities are generally constructed based on the detection of the decrease of the substrates or the increase of products from the enzymatic reaction. Here, phosphoglucomutase (PGM) biosensor will be described as an example for the illustration the sensor construction for enzyme activities [29].

PGM is a ubiquitous enzyme that is expressed in all organisms from bacteria to plants to animals and controls a key branch point for carbohydrate metabolism. PGM catalyzes the interconversion of glucose-1-phosphate and glucose-6-phosphate. In this process, the enzyme links various catabolic pathways to yield energy ATP or reducing power NAD(P)H, and several anabolic pathways, such as to lead to the synthesis of polysaccharides [30–33].

Due to its importance, measurements for the enzymatic activity of PGM have been widely performed, by the use of optical methods with coupled enzyme system [32–40], or the use of a combination of ion/molecule reactions and FT-ICR mass spectrometry [41]. However, these measurements either result in a long detection time, a large amount of enzyme consumption, or complex procedures which need to be performed by skilled personnel. The importance of amperometric enzyme–based biosensors has increased considerably in recent years thanks to the advantages of being highly sensitive, rapid, accurate, economical, and easy to handle for specific measurement of target analyte in complex matrices such as blood, food product, and environmental sample. To the best of our knowledge, there is no report for the determination of PGM activity using an amperometric biosensor.

Here, an amperometric biosensor for PGM activity with a bienzyme screen-printed biosensor is described. As shown in Figure 6.7, the principle is as follows: PGM (EC 5.4.2.2, from rabbit muscle, Sigma) converts glucose-1-phosphate to glucose-6-phosphate. Glucose-6-phosphate dehydrogenase (G6PDH, EC 1.1.1.49, from Leuconostoc mesenteroides, Sigma) catalyzes the specific dehydrogenation of glucose-6-phosphate by consuming NAD^+. The product "NADH" initiates the irreversible decarboxylation and the hydroxylation of salicylate by salicylate hydroxylase (SHL, EC 1.14.13.1, from Pseudomonas sp., GDS Technology Inc., Elkhart, IN) in the presence of oxygen to produce catechol, which results in a detectable signal due to its oxidation at the working electrode.

The bienzyme electrode for the measurement of PGM activity is based on the detection of glucose-6-phosphate, the product of the PGM-catalyzed reaction. From literature, glucose-6-phosphate biosensors have been developed by several methods,

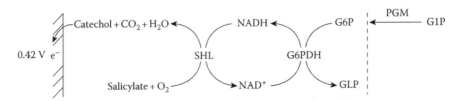

FIGURE 6.7 Schematic illustration of the amperometric screen-printed biosensor for the determination of PGM activity (G1P: glucose-1-phosphate; G6P: glucose-6-phosphate; GLP: D-glucono-1,5-lactone 6-phosphate; PGM: phosphoglucomutase; G6PDH: glucose-6-phosphate dehydrogenase; SHL: salicylate hydroxylase). (From Cui, Y. et al., *Anal. Biochem.*, 354, 162, 2006.)

including the combinations of G6PDH and various mediators, or the combination of phosphatase and GOD [42–44]. Here, the successful coupling of SHL and G6PDH on a screen-printed electrode can also serve as a glucose-6-phosphate biosensor.

The screen-printed electrode was covered with a mixture of SHL, G6PDH, and glutaraldehyde, followed by drying and storing at 4°C overnight. The biosensor was then screwed into the measuring cell, which was filled with buffer solution, and rehydrated for around 1 h at room temperature to allow the enzyme matrix swelling before usage. Experiments were performed by applying the specific potential for this type of screen-printed electrode at 0.42 V and magnetically stirring the solution at 300 rpm to obtain a uniform distribution of PGM. The measurement was started by adding PGM into the buffer solution contained in the measuring cell, and the current velocity (nA min^{-1}) was recorded for plotting calibration curve. A syringe was used between each measurement for sucking the buffer solution out of the measuring cell to remove PGM in the solution, and thus to remove catechol on the electrode surface.

A two-step optimization was needed before the sensor calibration in order to improve the sensor performance for the determination of PGM activity. The first step is the optimization of the enzyme matrix, including the enzyme loadings and the immobilization agent concentrations. Various loadings of SHL, G6PDH, and glutaraldehyde on screen-printed electrodes were investigated to obtain the maximum current velocity. Based on the optimization, a mixture containing 0.33 U SHL and 1.88 U G6PDH with 1% glutaraldehyde in 0.5 μL of enzyme matrix was used for further experiments. The second step was the optimization of the working condition, including pH value, substrate, and cofactor concentrations in the buffer solution. The effects of glucose-1,6-diphosphate and MgCl$_2$ were also studied due to being the activating cofactor and metal cofactor for the PGM-catalyzed reaction. To improve the sensor performance, various loadings of glucose-1-phosphate, salicylate, NAD$^+$, glucose-1,6-diphosphate, MgCl$_2$, and buffer pH were investigated. In order to obtain the maximum current velocity for the measurement of PGM activity, the substrate and cofactor concentrations in the buffer solution should be sufficient to avoid the signal saturation due to their inadequate loadings. The optimized working condition obtained was 100 mM Tris-HCl buffer solution containing 5 mM glucose-1-phosphate, 5 mM salicylate, 5 mM NAD$^+$, 50 μM glucose-1,6-diphosphate, and 5 mM MgCl$_2$ at pH 8.0.

Figure 6.8 shows the calibration curve for PGM activity using the bienzyme screen-printed sensor. The measurements of PGM activities were performed by the biosensor with the optimal enzyme matrix and working condition. As the PGM activity was proportional to the rate of production of glucose-6-phosphate, it was further proportional to the rate of production of catechol and the current velocity; thus, current velocity (nA min^{-1}) was used for the determination of PGM activity. As shown in the figure, a linear relationship was obtained between the current velocity and the PGM activity from 0.05 to 5 U mL^{-1} with a detection limit of 0.02 U mL^{-1} (slope: 76.36 (nA min^{-1})/(U mL^{-1}), $R^2 = 0.9988$, $n = 3$). The linear detection range and detection limit are decided by the recording method, and if the record of current velocity is changed from nA min^{-1} to nA (2 min)$^{-1}$ or nA (3 min)$^{-1}$, the linear detection range and detection limit will be changed and improved.

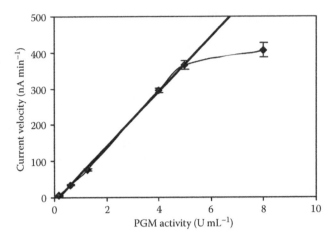

FIGURE 6.8 Calibration curve for PGM activity using the amperometric screen-printed biosensor (Sensor: 0.33 U SHL, 1.88 U G6PDH with 1% glutaraldehyde in 0.5 µL of enzyme matrix. Buffer: 100 mM Tris-HCl buffer containing 5 mM glucose-1-phosphate, 5 mM salicylate, 5 mM NAD$^+$, 50 µM glucose-1,6-diphosphate, and 5 mM MgCl$_2$ at pH 8.0). (From Cui, Y. et al., *Anal. Biochem.*, 354, 162, 2006.)

Also, the sensor has a fast measuring time (1 min) and a short recovery time (2 min) with high reproducibility. Hence, the total measurement of PGM activity takes less than 4 min with simple operations, which is more rapid and convenient than conventional methods.

In summary, a biosensor for PGM is described using an amperometric screen-printed biosensor based on the coimmobilization of SHL and G6PDH. The sensor shows high-performance characteristics with a broad detection range (0.05–5 U mL^{-1}) and a rapid measuring time (1 min). Thus, it provides a new analytical approach to the determination of PGM activity rapidly, sensitively, economically, and easy to handle.

6.4   BIOSENSOR FOR TOXIN: AZIDE

Toxins are harmful for various metabolic activities, and the determination of toxins are important for environmental and medical monitoring. The biosensor constructions are generally based on the inhibition of enzymatic activities on electrodes. Here, azide is used as an example for the construction of biosensors for toxins [45].

Sodium azide is a rapidly acting toxic chemical that exists as an odorless white solid. It prevents the cells from using oxygen, thus is very harmful to organs, especially the heart and the brain. In addition, sodium azide is used daily in the air bags of vehicles with large tonnage, in hospitals and laboratories as a chemical preservative, in agriculture for pest control, and in detonators and other explosives. Moreover, sodium azide is the starting material of heavy sodium azide, pure sodium metal, hydrazoic acid, and a variety of medicines [46–49].

Due to its importance, various methods have been developed for the determination of azide concentration, including ion chromatography [50,51], gas chromatography [52], high-performance liquid chromatography [53], and the combination of diffusion extraction and spectrophotometry [54]. However, these methods either require expensive equipment or need to be performed by skilled personnel. As the increase of the importance of amperometric biosensors due to the advantages of being highly sensitive, rapid, economical, and easy to handle for measurements, several types of biosensors have been constructed for the determination of azide based on the inhibition of enzymatic activity, including a laccase- or tyrosinase-immobilized mediated carbon electrode [55,56], and a catalase-immobilized Clark-type oxygen electrode [57].

Here, two methods to construct biosensors for the amperometric determination of azide are described here by using a disposable, screen-printed electrode immobilized with catalase or tyrosinase, as shown in Figure 6.9.

The first method for azide determination is based on the inhibition of the enzymatic consumption of H_2O_2 on the catalase-immobilized screen-printed electrode by azide, as shown in Figure 6.9A. H_2O_2 can result in a high current due to its oxidation at the working electrode; however, the immobilized catalase on electrode initiates the decomposition of H_2O_2 into H_2O and O_2 which consumes H_2O_2 at the electrode surface considerably, and therefore the current response to H_2O_2 is not apparent. The presence of azide inhibits the enzymatic activity of catalase, thus the consumption of H_2O_2 by catalase on the electrode decreases, followed by the significant increase of current signal, and the current difference after the injection of azide is proportional to the concentration of azide.

The second method for azide determination is based on the inhibition of the enzymatic consumption of catechol on the tyrosinase-immobilized screen-printed electrode by azide, as shown in Figure 6.9B. Catechol can result in a high current

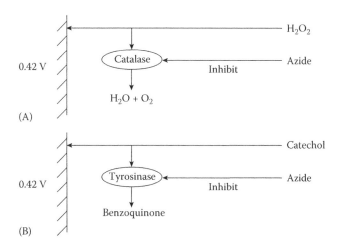

FIGURE 6.9 Schematic illustration for azide determination with (A) the catalase-immobilized screen-printed electrode and (B) the tyrosinase-immobilized screen-printed electrode. (From Cui, Y. et al., *Anal. Sci.*, 22, 1279, 2006.)

due to its oxidation at the working electrode; however, the immobilized tyrosinase on electrode converts it into benzoquinone in the presence of oxygen which consumes catechol at the electrode surface significantly, and therefore the current response to catechol using the tyrosinase-immobilized electrode is much smaller than that using a bare electrode. The presence of azide inhibits the enzymatic activity of tyrosinase; thus the consumption of catechol by tyrosinase on the electrode decreases, followed by the significant increase of current signal, and the current difference after the injection of azide is proportional to the concentration of azide.

A bare screen-printed electrode showed no current response to azide, but a high current response of 3480 nA to 1 mM H_2O_2. The catalase-immobilized screen-printed electrode exhibited no apparent current response to 1 mM H_2O_2 due to the enzymatic decomposition of H_2O_2 by catalase. Figure 6.10A shows a current–time curve with the catalase-immobilized electrode obtained by adding various amounts of azide. As shown in the figure, after the injection of azide, the anodic current increased due to the decrease of H_2O_2 consumption by catalase, and the increase of the anodic current was proportional to the concentration of azide. The response of the sensor was rapid (1 s) with a high reproducibility and a short recovery time (1 min). The steady background current increased after the addition of azide and reached a new steady state within 30 s. Therefore, the total measurement using the sensor took less than 3 min. Figure 6.10B shows the calibration curve for azide with the catalase-immobilized electrode. A linear relationship was obtained between the current response and the concentration of azide from 0.1 to 50 μM with a slope of 18.51 nA μM^{-1} and a correlation coefficient of 0.9923. This method also has a high reproducibility with a RSD of 6.6% for five different catalase-immobilized electrodes by testing the sensitivity. Compared to the biosensor methods reported previously [55–57], this method shows high-performance characteristics with a most sensitive detection range, a short measuring time, and an easy-to-handle operation.

A bare screen-printed electrode showed a high current response of 2350 nA to 1 mM catechol, and the tyrosinase-immobilized screen-printed electrode exhibited a current response of 350 nA to 1 mM catechol due to the incomplete consumption of catechol by tyrosinase. Figure 6.11A shows a current–time curve with the tyrosinase-immobilized electrode obtained by adding various amounts of azide. As shown in the figure, after the injection of azide, the anodic current increased due to the decrease of catechol consumption by tyrosinase, and the increase of the anodic current was proportional to the concentration of azide. The response of the sensor was rapid (1 s) with a high reproducibility and a short recovery time (1 min). The steady background current increased after the addition of azide and reached a new steady state within 30 s. Therefore, the total measurement using the sensor took less than 3 min. Figure 6.11B shows the calibration curve for azide with the tyrosinase-immobilized electrode. A linear relationship was obtained between the current response and the concentration of azide from 5 to 1000 μM with a slope of 0.45 nA μM^{-1} and a correlation coefficient of 0.9989. This method also has a high reproducibility with a RSD of 4.5% for five different tyrosinase-immobilized electrodes by testing the sensitivity. In this experimental

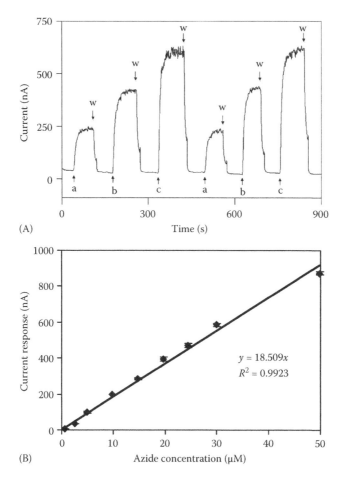

FIGURE 6.10 Determination performance with the catalase-immobilized electrode. (A) Current–time curve to (a) 10 μM, (b) 20 μM, and (c) 30 μM of azide with (w) washing step. (B) Calibration curve for azide ($n = 3$) (Sensor: 75 U catalase with 1% glutaraldehyde in 0.5 μL of enzyme matrix. Buffer: 50 mM K-PBS buffer containing 1 mM H_2O_2 at pH 7.0). (From Cui, Y. et al., *Anal. Sci.*, 22, 1279, 2006.)

condition, the determination of azide using the tyrosinase-immobilized electrode was not as sensitive as that using a catalase-immobilized electrode, which was probably due to the inhibition effect of azide on tyrosinase not as large as that on catalase.

In summary, the construction for amperometric azide biosensor is described based on the inhibition of enzymatic consumption of hydrogen peroxide or catechol with a disposable, screen-printed electrode immobilized with catalase or tyrosinase. Either of these methods provides a new analytical approach to the determination of azide that is rapid, sensitive, economical, and easy to handle. Besides, the two methods also provide new analytical methods for the determinations of some other toxic substances.

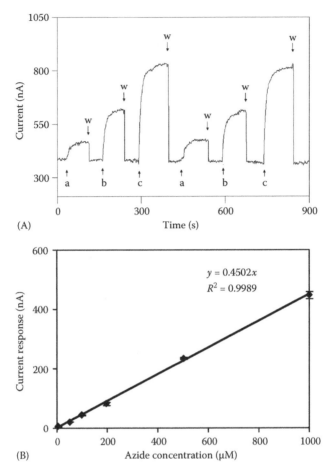

(A)

(B)

FIGURE 6.11 Determination performance with the tyrosinase-immobilized electrode. (A) Current–time curve to (a) 200 μM, (b) 500 μM, and (c) 1000 μM of azide with (w) washing step. (B) Calibration curve for azide ($n=3$) (Sensor: 75 U tyrosinase with 1% glutaraldehyde in 0.5 μL of enzyme matrix. Buffer: 50 mM K-PBS buffer containing 1 mM catechol at pH 7.0). (From Cui, Y. et al., *Anal. Sci.*, 22, 1279, 2006.)

6.5 BIOSENSOR FOR POLYMER: POLY(3-HYDROXYBUTYRATE)

Polymers play an important role in various metabolisms and applications in different fields. The biosensor constructions for polymers are generally based on the digestion of polymers into monomers and the detection of monomers with biosensors. Here, poly(3-hydroxybutyrate) (PHB) is used as an example for the polymer biosensor constructions [58].

PHB is a common intracellular biodegradable polymer involved in bacterial carbon and energy storage, and plays an important role in the course of metabolism [59,60]. As one of the most interesting biodegradable materials, it has promising applications in medicine, material science, agriculture, etc. [61–63] It has also been found in activated sludge samples from conventional wastewater treatment plants [64].

Due to its importance, various methods have been developed for measuring the PHB concentration, including the use of gravimetry [65], turbidimetry [66], and spectrophotometry [67], and the uses of chromatography [68,69], capillary isotachophoresis [70,71], and capillary zone electrophoresis [72] with pretreatments. However, these methods either result in an unreliable determination, a long measurement time (several hours), procedures with high temperature or pressure, or the need to be performed by skilled personnel.

The importance of amperometric enzyme–based biosensors has increased considerably in recent years thanks to the advantages of being highly sensitive, rapid, reliable, economical, and easy to handle for specific measurements of the target analyte in complex matrices, such as blood, food products, and environmental samples. In this work, we present a simple method for the determination of PHB using a combination of an amperometric enzyme–based biosensor and alkaline hydrolysis, which is the first report concerning the measurement of PHB with a biosensor technique.

As shown in Figure 6.12, PHB is first decomposed to produce its monomer 3-hydroxybutyrate (3-HB) through alkaline hydrolysis, followed by neutralization with acid. The product, 3-HB formed accordingly, is measured with an enzyme-based 3-HB biosensor. The 3-HB sensor [73] was constructed by immobilizing 3-hydroxybutyrate dehydrogenase (HBDH, EC 1.1.1.30) and SHL (EC 1.14.13.1) on a Clark-type oxygen electrode. The determination principle of the biosensor is as follows: HBDH catalyzes the specific dehydrogenation of 3-HB in the presence of NAD^+. The product, NADH, initiates the irreversible decarboxylation and hydroxylation of salicylate by SHL. Dissolved oxygen, acting as an essential material for the enzymatic activity of SHL, is consumed proportional to the concentration of 3-HB during the measurements. A detectable signal, caused by consumption of the dissolved oxygen by SHL, was monitored at $-0.6\,V$ versus Ag/AgCl by the Clark electrode. Both enzymes were entrapped by a poly(carbamoyl) sulfonate (PCS) hydrogel, which was sandwiched between a dialysis membrane and a Teflon membrane. Electroactive interferences were eliminated by a Teflon membrane.

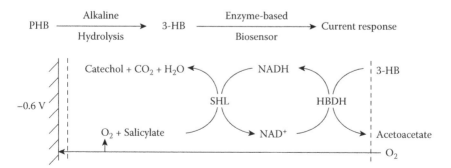

FIGURE 6.12 Schematic illustration for the determination of PHB, and enzyme-based 3-HB biosensor. (From Cui, Y. et al., *Anal. Sci.*, 22, 1323, 2006.)

PHB, with a concentration of 21.5 g L^{-1} (250 mM 3-HB Unit) in 6 M KOH, was decomposed into its monomer product, 3-HB, through alkaline hydrolysis at 50°C for 30 or 10 min, followed by neutralization with 6 M HCl to a pH of around 8 with a volume ratio of 1.1 for a 30 min hydrolysis or 1.07 for a 10 min hydrolysis. The volume ratios of KOH to HCl, being larger than 1.0, were due to the acid forms of the hydrolyzed products of PHB contained in KOH.

The production of 3-HB from PHB was measured with a 3-HB biosensor. One milliliter of K-PBS (100 mM, pH 7.5) buffer containing 0.5 mM sodium salicylate and 0.5 mM NAD$^+$ was added into the measuring chamber. After achieving a steady background current (<30 min), the measurement was started at room temperature (22°C) by adding a pretreated PHB solution, or its 10-times diluted solution (0.1–30 µL) into the measuring buffer. The concentration of PHB was determined by the decrease in the dissolved oxygen reduction current. Until a stationary current occurred, the current difference was recorded for plotting the calibration curve. The current response was proportional to the production of 3-HB from PHB, and was further proportional to the concentration of PHB.

Figure 6.13A shows the current response to various concentrations of the pretreated PHB using the enzyme-based 3-HB sensor with hydrolysis in 6 M KOH at 50°C for 30 min. The cathodic current decreased due to the consumptions of dissolved oxygen by the SHL and HBDH in the enzyme layer after the addition of the pretreated PHB solution. Also, the reduction of the cathodic current was proportional to the concentration of PHB. The response of the bienzyme 3-HB sensor was rapid (2 s) with a high reproducibility and short recovery time (2 min). The steady-state background current decreased after the addition of the pretreated PHB solution and reached a stationary state within 30 s. The total measuring time using the 3-HB sensor for the pretreated PHB was less than 4 min. Hence, the whole measurement for PHB took only several minutes, being less than 40 min with a saturated production of 3-HB using a 30 min hydrolysis. If the hydrolysis time is 10 min, the total time is less than 15 min with part production of 3-HB, which mostly depended on the hydrolysis time. They were much quicker than that using a conventional method.

Figure 6.13B shows the calibration curves for the pretreated PHB using the enzyme-based 3-HB sensor. A linear relationship is observed between the current response and the concentration of PHB from 0.5 to 110 mg L^{-1} (slope: 1.76 nA/ (mg L^{-1}), $R^2 = 0.9937$, $n = 3$) with hydrolysis in 6 M KOH at 50°C for 30 min. A sharp saturation at 120 mg L^{-1} was observed due to oxygen depletion by the enzymatic reaction, as indicated by a net current of zero at the Clark electrode. For calculating the detection limit, 3 M KCl (30 µL), a neutralized solution of 6 M KOH and 6 M HCl with a volume ratio of 1:1 was added as blank solution into the measuring cell in order to determine the blank signal. The detection limit of the system with a 30 min hydrolysis of PHB, calculated as the mean blank signal plus three times the standard derivation of the mean blank signal, was 0.3 mg L^{-1} of PHB. This method also has a high average reproducibility of 98.6%, which was obtained by repeated analysis of different concentrations of PHB ranging from 0 to 300 mg L^{-1} for three times.

The characteristics for the determination of PHB using the combination of enzyme-based biosensor and alkaline hydrolysis are summarized in Table 6.1. This

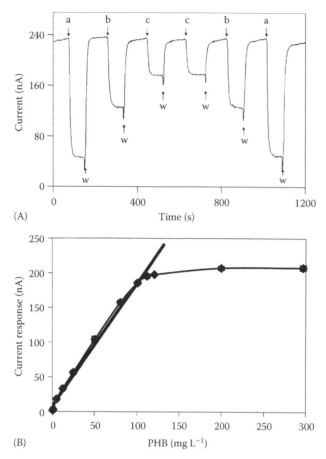

FIGURE 6.13 (A) Current response to (a) 100 mg L⁻¹, (b) 50 mg L⁻¹, and (c) 25 mg L⁻¹ of PHB, with (w) washing step. (B) Calibration curve for PHB using the combination of enzyme-based biosensor and alkaline hydrolysis in 6 M KOH at 50°C for 30 min. (From Cui, Y. et al., *Anal. Sci.*, 22, 1323, 2006.)

method shows a fast measurement, which takes several minutes due to the short hydrolysis time and the quick sensor measuring time. It also has a sensitive detection range and a low detection limit. Also, the determination of PHB using a 30 min hydrolysis has a smaller detection range, and takes a longer measurement time than those using a 10 min hydrolysis, while it has a higher sensitivity and a lower detection limit. This method also shows a high average reproducibility of larger than 97%. From a previous report [73], the enzyme-based sensor has good stability, which can retain above 50% of the initial response for 17 days using storage in a buffer solution at 4°C after a measurement. Due to the Teflon membrane, the determination for the pretreated PHB using the enzyme-based biosensor is also free from any electroactive interference. Since this method employs biosensor detection through specific enzymatic reactions and a relatively fast pretreatment, it is more rapid and reliable than conventional methods for the determination of PHB.

TABLE 6.1

Characteristics for the Determination of PHB Using the Combination of Enzyme-Based Biosensor and Alkaline Hydrolysis in 6 M KOH at 50°C for 30 or 10 min

Characteristics	With a 30 min Hydrolysis	With a 10 min Hydrolysis
Detection time	<40 min	<15 min
Linear range	0.5–110 mg L^{-1}	1.0–160 mg L^{-1}
Sensitivity	1.76 nA/(mg L^{-1})	1.20 nA/(mg L^{-1})
Detection limit	0.3 mg L^{-1}	0.5 mg L^{-1}
Reproducibility	98.6%	97.7%
Stability	17 days for the half-life of the 3-HB bienzyme sensor [15]	
Specificity	Specific, free from electroactive interferences [15]	

Source: From Cui, Y. et al., *Anal. Sci.*, 22, 1323, 2006.

In summary, the determination of PHB is described using a combination of enzyme-based biosensor and alkaline hydrolysis. It shows high-performance characteristics with a sensitive detection range, a short measurement time, and simple operation. Thus, the method provides a new analytical approach for the determination of PHB that is rapid, sensitive, and easy to handle.

6.6 CONCLUSIONS AND FUTURE PERSPECTIVES

The development of a general approach for the construction of enzyme biosensors could expand opportunities for sensors in both fundamental studies and a variety of device platforms. Glucose biosensors for the detection of diabetes have been widely commercialized. Here, the development of biosensors for several typical important analytes based on the commercial glucose biosensor platform are described, including metabolic compounds, toxins, polymers, and enzyme activities. These biosensors are constructed based on the immobilization of enzyme matrix with sequential enzymatic reactions or competitive enzymatic reactions on conventional Clark-type electrode or disposable screen-printed. The characterization shows that the sensors can detect the analytes sensitively, selectively, rapidly, and easy to handle. These sensors can be easily incorporated into the commercial sensor devices and applied to industry market. I anticipate that these methods could open exciting opportunities in the use of biosensors in fundamental bioanalytical research, as well as applications ranging from medical diagnosis to environmental monitoring.

REFERENCES

1. Turner, A.P.F., Karube, I., and Wilson, G.S. Biosensors: Fundamentals and Applications, Oxford University Press (1987).

2. Wang, J. Glucose biosensors: 40 years of advances and challenges. *Electroanalysis* **13**, 983–988 (2001).
3. Glucose Oxidase, http://en.wikipedia.org/wiki/Glucose-oxidase
4. Yoo, E. H. and Lee, S. Y. Glucose biosensors: An overview of use in clinical practice. *Sensors* **10**, 4558–4576 (2010).
5. Cui, Y. Amperometric ATP biosensors based on coimmobilizations of *p*-hydroxybenzoate hydroxylase, glucose-6-phosphate dehydrogenase, and hexokinase on Clark-type and screen-printed electrodes. *IEEE Sensors Journal* **10**, 979–983 (2010).
6. Higgins, C. F., Hiles, I. D., and Salmond, G. P. C. A family of related ATP-binding subunits coupled to many distinct biological processes in bacteria. *Nature* **323**, 448–450 (1986).
7. Stekhoven, F. S. Energy transfer factor A.D (ATP synthetase) as a complex Pi-ATP exchange enzyme and its stimulation by phospholipids. *Biochemical and Biophysical Research Communications* **47**, 7–14 (1972).
8. Jorgensen, P. E., Eriksen, T., and Jensen, B. K. Estimation of viable biomass in wastewater and activated sludge by determination of ATP, oxygen utilization rate and FDA hydrolysis. *Water Research* **26**, 1495–1501 (1992).
9. Pietrzak, E. M. and Denes, A. S. Comparison of luminol chemiluminescence with ATP bioluminescence for the estimation of total bacterial load in pure cultures. *Journal of Rapid Methods and Automation in Microbiology* **4**, 207–218 (1996).
10. Jose, D. A. et al. Colorimetric sensor for ATP in aqueous solution. *Organic Letters* **9**, 1979–1982 (2007).
11. Miao, Y., Liu, J., Hou, F., and Jiang, C. Determination of adenosine disodium triphosphate (ATP) using norfloxacin-Tb^{3+} as a fluorescence probe by spectrofluorimetry. *Journal of Luminescence* **116**, 67–72 (2006).
12. Ronner, P., Friel, E., Czerniawski, K., and Fränkle, S. Luminometric assays of ATP, phosphocreatine, and creatine for estimation of free ADP and free AMP. *Analytical Biochemistry* **275**, 208–216 (1999).
13. Brown, P. and Dale, N. Spike-independent release of ATP from Xenopus spinal neurons evoked by activation of glutamate receptors. *Journal of Physiology* **540**, 851–860 (2002).
14. Karatzaferi, C., De Haan, A., Offringa, C., and Sargeant, A. J. Improved high-performance liquid chromatographic assay for the determination of 'high-energy' phosphates in mammalian skeletal muscle: Application to a single-fibre study in man. *Journal of Chromatography B: Biomedical Sciences and Applications* **730**, 183–191 (1999).
15. Katsu, T. and Yamanaka, K. Potentiometric method for the determination of adenosine-5'-triphosphate. *Analytica Chimica Acta* **276**, 373–376 (1993).
16. Adachi, Y., Sugawara, M., Taniguchi, K., and Umezawa, Y. Na+/K+-ATPase-based bilayer lipid membrane sensor for adenosine 5'-triphosphate. *Analytica Chimica Acta* **281**, 577–584 (1993).
17. Gotoh, M., Tamiya, E., Karube, I., and Kagawa, Y. A microsensor for adenosine-5'-triphosphate pH-sensitive field effect transistors. *Analytica Chimica Acta* **187**, 287–291 (1986).
18. Compagnone, D. and Guilbault, G. G. Glucose oxidase/hexokinase electrode for the determination of ATP. *Analytica Chimica Acta* **340**, 109–113 (1997).
19. Cui, Y., Barford, J. P., and Renneberg, R. Amperometric trienzyme ATP biosensors based on the coimmobilization of salicylate hydroxylase, glucose-6-phosphate dehydrogenase, and hexokinase. *Sensors and Actuators, B: Chemical* **132**, 1–4 (2008).
20. Kueng, A., Kranz, C., and Mizaikoff, B. Amperometric ATP biosensor based on polymer entrapped enzymes. *Biosensors and Bioelectronics* **19**, 1301–1307 (2004).

21. Liu, S. and Sun, Y. Co-immobilization of glucose oxidase and hexokinase on silicate hybrid sol–gel membrane for glucose and ATP detections. *Biosensors and Bioelectronics* **22**, 905–911 (2007).

22. Llaudet, E., Hatz, S., Droniou, M., and Dale, N. Microelectrode biosensor for real-time measurement of ATP in biological tissue. *Analytical Chemistry* **77**, 3267–3273 (2005).

23. Yang, X., Johansson, G., Pfeiffer, D., and Scheller, F. Enzyme electrodes for ADP/ATP with enhanced sensitivity due to chemical amplification and intermediate accumulation. *Electroanalysis* **3**, 659–663 (1991).

24. Cui, Y., Barford, J. P., and Renneberg, R. Development of an L-glutamate biosensor using the coimmobilization of L-glutamate dehydrogenase and *p*-hydroxybenzoate hydroxylase on a Clark-type electrode. *Sensors and Actuators, B: Chemical* **127**, 358–361 (2007).

25. Cui, Y., Barford, J. P., and Renneberg, R. Development of a glucose-6-phosphate biosensor based on coimmobilized *p*-hydroxybenzoate hydroxylase and glucose-6-phosphate dehydrogenase. *Biosensors and Bioelectronics* **22**, 2754–2758 (2007).

26. Gajovic, N., Warsinke, A., Huang, T., Schulmeister, T., and Scheller, F. W. Characterization and mathematical modeling of a bienzyme electrode for L-malate with cofactor recycling. *Analytical Chemistry* **71**, 4657–4662 (1999).

27. Gribble, F. M. et al. A novel method for measurement of submembrane ATP concentration. *Journal of Biological Chemistry* **275**, 30046–30049 (2000).

28. Chenzhuo, L., Murube, J., Latorre, A., and Martin del Rio, R. The presence of high amounts of amino acid taurine in human tears. *Arch. Soc. Canar. Oftal.* **11**, 11–12 (2000).

29. Cui, Y., Barford, J. P., and Renneberg, R. Amperometric determination of phosphoglucomutase activity with a bienzyme screen-printed biosensor. *Analytical Biochemistry* **354**, 162–164 (2006).

30. Ray, W. J. and Peck, E. J. Phosphomutases. *The Enzymes* **6**, 407–477 (1972).

31. Lytovchenko, A., Sweetlove, L., Pauly, M., and Fernie, A. R. The influence of cytosolic phosphoglucomutase on photosynthetic carbohydrate metabolism. *Planta* **215**, 1013–1021 (2002).

32. Akutsu, J. I. et al. Characterization of a thermostable enzyme with phosphomannomutase/phosphoglucomutase activities from the hyperthermophilic archaeon *Pyrococcus horikoshii* OT3. *Journal of Biochemistry* **138**, 159–166 (2005).

33. Mesak, L. R. and Dahl, M. K. Purification and enzymatic characterization of PgcM: A β-phosphoglucomutase and glucose-1-phosphate phosphodismutase of *Bacillus subtilis*. *Archives of Microbiology* **174**, 256–264 (2000).

34. Frazier, D. M., Clemons, E. H., and Kirkman, H. N. Minimizing false positive diagnoses in newborn screening for galactosemia. *Biochemical Medicine and Metabolic Biology* **48**, 199–211 (1992).

35. Inoue, H., Kondo, S., Hinohara, Y., Juni, N., and Yamamoto, D. Enhanced phosphorylation and enzymatic activity of phosphoglucomutase by the Btk29A tyrosine kinase in *Drosophila*. *Archives of Biochemistry and Biophysics* **413**, 207–212 (2003).

36. Videira, P. A., Cortes, L. L., Fialho, A. M., and Sá-Correia, I. Identification of the pgmG gene, encoding a bifunctional protein with phosphoglucomutase and phosphomannomutase activities, in the gellan gum-producing strain *Sphingomonas paucimobilis* ATCC 31461. *Applied and Environmental Microbiology* **66**, 2252–2258 (2000).

37. Zhang, G. et al. Catalytic cycling in β-phosphoglucomutase: A kinetic and structural analysis. *Biochemistry* **44**, 9404–9416 (2005).

38. Howard, S. C., Deminoff, S. J., and Herman, P. K. Increased phosphoglucomutase activity suppresses the galactose growth defect associated with elevated levels of Ras signaling in *S. cerevisiae*. *Current Genetics* **49**, 1–6 (2006).

39. Masuda, C. A., Xavier, M. A., Mattos, K. A., Galina, A., and Montero-Lomelí, M. Phosphoglucomutase is an in vivo lithium target in yeast. *Journal of Biological Chemistry* **276**, 37794–37801 (2001).

40. Sergeeva, L. I. and Vreugdenhil, D. In situ staining of activities of enzymes involved in carbohydrate metabolism in plant tissues. *Journal of Experimental Botany* **53**, 361–370 (2002).

41. Gao, H. and Leary, J. A. Kinetic measurements of phosphoglucomutase by direct analysis of glucose-1-phosphate and glucose-6-phosphate using ion/molecule reactions and Fourier transform ion cyclotron resonance mass spectrometry. *Analytical Biochemistry* **329**, 269–275 (2004).

42. Bassi, A. S., Tang, D., and Bergougnou, M. A. Mediated, amperometric biosensor for glucose-6-phosphate monitoring based on entrapped glucose-6-phosphate dehydrogenase, Mg2+ ions, tetracyanoquinodimethane, and nicotinamide adenine dinucleotide phosphate in carbon paste. *Analytical Biochemistry* **268**, 223–228 (1999).

43. Hung Tzang, C., Yuan, R., and Yang, M. Voltammetric biosensors for the determination of formate and glucose-6-phosphate based on the measurement of dehydrogenase-generated NADH and NADPH. *Biosensors and Bioelectronics* **16**, 211–219 (2001).

44. Mazzei, F., Botrè, F., and Botrè, C. Acid phosphatase/glucose oxidase-based biosensors for the determination of pesticides. *Analytica Chimica Acta* **336**, 67–75 (1996).

45. Cui, Y., Barford, J. P., and Renneberg, R. A disposable, screen-printed electrode for the amperometric determination of azide based on the immobilization with catalase or tyrosinase. *Analytical Sciences* **22**, 1279–1281 (2006).

46. Betterton, E. A. Environmental fate of sodium azide derived from automobile airbags. *Critical Reviews in Environmental Science and Technology* **33**, 423–458 (2003).

47. Heeschen, W. H., Ubben, E. H., Gyodi, P., and Beer, P. Kieler Milchw. *ForschBer.* **45**, 109–136 (1993).

48. Chang, S. and Lamm, S. H. Human health effects of sodium azide exposure: A literature review and analysis. *International Journal of Toxicology* **22**, 175–186 (2003).

49. Hagenbuch, J. P. Opportunities and limits of the use of azides in industrial production. Implementation of safety measures. *Chimia* **57**, 773–776 (2003).

50. Annable, P. L. and Sly, L. A. Azide determination in protein samples by ion chromatography. *Journal of Chromatography* **546**, 325–334 (1991).

51. Kruszyna, R., Smith, R. P., and Kruszyna, H. Determining sodium azide concentration in blood by ion chromatography. *Journal of Forensic Sciences* **43**, 200–202 (1998).

52. Kage, S., Kudo, K., and Ikeda, N. Determination of azide in blood and urine by gas chromatography-mass spectrometry. *Journal of Analytical Toxicology* **24**, 429–432 (2000).

53. Vácha, J., Tkaczyková, M., and Rejholcová, M. Determination of sodium azide in the presence of proteins by high-performance liquid chromatography. *Journal of Chromatography B: Biomedical Sciences and Applications* **488**, 506–508 (1989).

54. Tsuge, K., Kataoka, M., and Seto, Y. Rapid determination of cyanide and azide in beverages by microdiffusion spectrophotometric method. *Journal of Analytical Toxicology* **25**, 228–236 (2001).

55. Daigle, F., Trudeau, F., Robinson, G., Smyth, M. R., and Leech, D. Mediated reagentless enzyme inhibition electrodes. *Biosensors and Bioelectronics* **13**, 417–425 (1998).

56. Leech, D. Optimisation of a reagentless laccase electrode for the detection of the inhibitor azide. *Analyst* **123**, 1971–1974 (1998).

57. Sezgintürk, M. K., Göktuğ, T., and Dinçkaya, E. A biosensor based on catalase for determination of highly toxic chemical azide in fruit juices. *Biosensors and Bioelectronics* **21**, 684–688 (2005).

58. Cui, Y., Barford, J. P., and Renneberg, R. Determination of poly(3-hydroxybutyrate) using a combination of enzyme-based biosensor and alkaline hydrolysis. *Analytical Sciences* **22**, 1323–1326 (2006).

59. Freier, T. et al. In vitro and in vivo degradation studies for development of a biodegradable patch based on poly(3-hydroxybutyrate). *Biomaterials* **23**, 2649–2657 (2002).

60. Mansfield, D. A., Anderson, A. J., and Naylor, L. A. Regulation of PHB metabolism in *Alcaligenes eutrophus. Canadian Journal of Microbiology* **41**, 44–49 (1995).

61. Koller, M. et al. Biotechnological production of poly(3-hydroxybutyrate) with Wautersia eutropha by application of green grass juice and silage juice as additional complex substrates. *Biocatalysis and Biotransformation* **23**, 329–337 (2005).

62. Pouton, C. W. and Akhtar, S. Biosynthetic polyhydroxyalkanoates and their potential in drug delivery. *Advanced Drug Delivery Reviews* **18**, 133–162 (1996).

63. Saad, B., Neuenschwander, P., Uhlschmid, G. K., and Suter, U. W. New versatile, elastomeric, degradable polymeric materials for medicine. *International Journal of Biological Macromolecules* **25**, 293–301 (1999).

64. Dircks, K., Henze, M., Van Loosdrecht, M. C. M., Mosbak, H., and Aspegren, H. Storage and degradation of poly-β-hydroxybutyrate in activated sludge under aerobic conditions. *Water Research* **35**, 2277–2285 (2001).

65. Tsuji, H. and Suzuyoshi, K. Environmental degradation of biodegradable polyesters. IV. The effects of pores and surface hydrophilicity on the biodegradation of poly(ε-caprolactone) and poly[(R)-3-hydroxybutyrate] films in controlled seawater. *Journal of Applied Polymer Science* **90**, 587–593 (2003).

66. Murase, T., Suzuki, Y., Doi, Y., and Iwata, T. Nonhydrolytic fragmentation of a poly[(R)-3-hydroxybutyrate] single crystal revealed by use of a mutant of polyhydroxybutyrate depolymerase. *Biomacromolecules* **3**, 312–317 (2002).

67. Yilmaz, M., Soran, H., and Beyatli, Y. Determination of poly-β-hydroxybutyrate (PHB) production by some *Bacillus* spp. *World Journal of Microbiology and Biotechnology* **21**, 565–566 (2005).

68. Jan, S. et al. Study of parameters affecting poly(3-hydroxybutyrate) quantification by gas chromatography. *Analytical Biochemistry* **225**, 258–263 (1995).

69. Saeki, T., Tsukegi, T., Tsuji, H., Daimon, H., and Fujie, K. Hydrolytic degradation of poly[(R)-3-hydroxybutyric acid] in the melt. *Polymer* **46**, 2157–2162 (2005).

70. Hudecova, D., Dudasova, S., and Sulo, P. Rapid and simple analysis of poly-beta-hydroxybutyrate content by capillary isotachophoresis. *Chemické Listy* **90**, 727–727 (1996).

71. Sulo, P., Hudecová, D., Propperová, A., and Bašnák, I. Rapid and simple analysis of poly-β-hydroxybutyrate content by capillary isotachophoresis. *Biotechnology Techniques* **10**, 413–418 (1996).

72. He, J., Chen, S., and Yu, Z. Determination of poly-β-hydroxybutyric acid in *Bacillus thuringiensis* by capillary zone electrophoresis with indirect ultraviolet absorbance detection. *Journal of Chromatography A* **973**, 197–202 (2002).

73. Kwan, R. C. H. et al. Biosensor for rapid determination of 3-hydroxybutyrate using bienzyme system. *Biosensors and Bioelectronics* **21**, 1101–1106 (2006).

7 Future Directions for Breath Sensors

Arunima Panigrahy, Jean-Pierre Delplanque, and Cristina E. Davis

CONTENTS

7.1 INTRODUCTION TO BREATH ANALYSIS

7.1.1 HISTORICAL PERSPECTIVE

Characteristic smells on human breath have been widely used for clinical purposes over the last several millennia (Figure 7.1). Hippocrates was one of the earliest clinicians to identify characteristic odors from the breath of patients that can provide vital clues to their ailments. Since then, physicians have anecdotally noted, for example, that a very sweet smell on the breath could indicate the onset of diabetes, a foul fishy smell may indicate liver disease, and an ammonia-like smell on the breath may

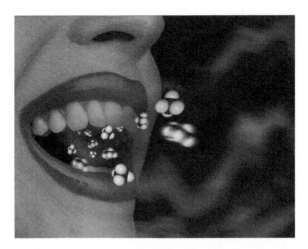

FIGURE 7.1 Compounds in exhaled breath are metabolites related to general human physiological changes. These can be important markers of disease, and these biomarkers can be monitored noninvasively in the breath. (This figure is reprinted with permission from Mukhopadhyay, R., *Anal. Chem.*, 76(15), 273A, 2004.)

indicate kidney problems. Ancient Chinese medical texts also describe body odor as an important diagnostic criterion to be used by a well-trained clinician. Knowing this historical context, it is surprising that few advances were made in the systematic study of exhaled breath odors until relatively late in the twentieth century [1].

Modern forensic studies have shown that the chemical composition of human breath is far more complex than initially believed. Exhaled breath contains a multitude of volatile and nonvolatile compounds—each of which has the potential to be a disease biomarker [2–5]. These breath metabolites are thought to be by-products of the constitutive biochemical processes ongoing within the human body [6,7]. In addition to the biochemical diversity of compounds in the breath, it can be challenging to reproducibly sample this chemical milieu [8]. There are two major methods of breath sampling that have been used in clinical tests, both of which have advantages and disadvantages. Gas phase sampling methods allow instant access to the volatile breath odor components that are exhaled; however, there can be reproducibility issues with these approaches. It is also possible to condense human breath from a vapor into a liquid phase and to simultaneously collect small diameter aerosol particles that are also present in the exhaled gas stream. This method can collect both the volatile and nonvolatile fractions of the breath compounds and thus may provide access to a wider spectrum of potential disease biomarkers. Both of these approaches will be described in greater detail later in this chapter.

Standardization and normalization procedures for breath sample collection, preconcentration, and biochemical analysis have been slow to emerge [2], although some studies on exhaled breath condensate (EBC) sample stability while frozen indicates EBC collection may be one method to enable large-scale clinical trials [9]. While some breath analysis approaches to sampling, analysis, and storage have helped advance the repeatability of processes [10], there are still significant topics

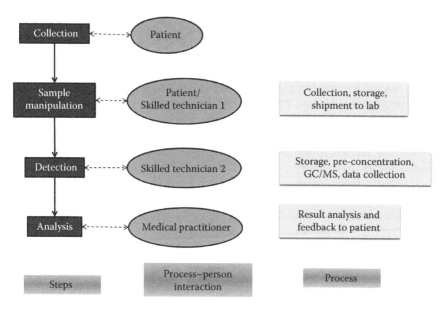

FIGURE 7.2 Current exhaled breath collection process consists of multiple steps that are time consuming and logistically challenging to orchestrate in short time frames.

within this research field that require technological investments in coming years. Figure 7.2 shows various basic steps followed from collection to analysis of exhaled breath and provides an overview of human interactions needed at each step along with various subprocesses involved.

We predict that most significant advances in the field of breath analyzer systems in the future are likely to be enabling technology modules that allow for miniaturization and, thus, the design of portable devices that could be easily used by doctors as clinical diagnostic tools. The main purpose of this chapter is to outline state-of-the-art miniature instrumentation systems available at this time, and to identify areas of technology investment that could lead to substantial advances in the development of clinical medicine tools. Before outlining the requirements for these devices, we will first give a more detailed introduction to the application areas that require these instruments.

7.1.2 CLINICAL USES OF BREATH DIAGNOSTICS FOR DISEASES AND DISORDERS

Breath analysis has the potential to be a powerful noninvasive technique for the diagnosis and monitoring of many different respiratory diseases such as asthma, lung cancer, or other airway inflammation disorders as well as other diseases such as diabetes and nephropathies. Diagnostic tests that utilize urine, blood, or tissue samples have existed since the late nineteenth century. Blood and tissue tests are invasive and patients, especially elderly patients and young children, frequently display some amount of discomfort and uneasiness with sample collection. This can result in patient inhibition or fear of visiting their doctor, thus potentially delaying the detection of early onset of certain diseases. Urine tests—like breath analysis—may yield interesting new diagnostic tools as they provide a minimally invasive route to monitor physiological metabolites. However, relatively few have been routinely available for common diseases so far. Patients are more likely to permit noninvasive testing, but also to accept frequent or regular testing which may allow early detection of potential health problems. These noninvasive methods (especially breath analysis) can enable easier testing for children, neonates, and patients with severe level of diseases on whom invasive tests are otherwise difficult to perform [11].

Extensive reports in the literature have suggested the association of specific exhaled breath biomarkers to various diseases. The number of breath studies reported in the literature has continued to increase each year, and biomarker correlation studies are among those most frequently published. Some examples of these studies are reported here, although these are by no means exhaustive. Studies have shown abnormal breath test findings were consistent with increased catabolism, alkanes, and monomethylated alkanes in lung cancer patients [12]. Exhaled nitric oxide has been considered a very promising marker for prognosis of asthma and obstructive pulmonary diseases (COPD) [13]. Changes in the pH levels in EBC samples have been used to evaluate airway inflammation in children with cystic fibrosis and asthma [14]. Studies have also shown a strong correlation between exhaled methyl nitride with hyperglycemia in childhood type 1 diabetes [15]. Phillips et al. concluded that breath test with multivariate analysis could help in rapid and accurate detection of patients with high risk of active pulmonary tuberculosis [16]. Studies have provided evidence that oxidative stress could identify low risk of grade 3 rejection of heart transplant recipients, thus enabling the reduction of invasive endomyocardial biopsies performed as well as reduction in health care cost [17]. Finally, quite a few literature reports provide excellent summaries of various other potential breath biomarkers that could be used as evaluation tools for chronic diseases [5,18–23].

7.1.3 FORENSIC APPLICATIONS OF BREATH DIAGNOSTICS FOR EXPOSURE TO EXOGENOUS CHEMICALS

Some initial studies in modern breath analysis date back to the 1960s, primarily for use in occupational health monitoring to evaluate exposure to exogenous chemicals [24]. These chemicals can constitute a substantial health risk for those exposed in high volumes or concentration, such as in the case of manufacturing process environment. Many industrial chemicals have high toxic potential and are easily absorbed by the human body through the respiratory system or the skin. Regular and timely monitoring of these occupational exposures via their corresponding metabolic biomarkers could help in early detection and thus significantly reduce the chances of health impact. Breath analysis studies in occupational and environmental exposure not only help to reduce ill health effects, but also help provide critical information to design and frame biological exposure limits (BEL) by providing reference values for "maximum safe permissible exposure quantities" [25].

As an example of how easily breath analyzers could be implemented into our culture, we can look toward the common breath alcohol test already used by law enforcement to determine blood alcohol levels of drunk drivers. With each nation aiming at better securing its borders, breath analysis methods may also be developed to rapidly analyze human breath for counterterrorism and national security applications. For example, breath analysis could be useful to determine whether an individual has recently been exposed to biological or chemical weapons agents [26,27], or potentially if a person has been exposed to unsafe levels of radiation.

7.2 BREATH SAMPLERS AND SENSORS

7.2.1 STATE-OF-THE-ART BREATH SAMPLERS/SENSORS

There are different methods and approaches for collecting exhaled breath depending on what compounds are to be analyzed. Some commercial devices already exist to collect specific breath fractions. It is possible to collect only the gas phase of human breath, which mainly involves sampling nonreactive gas species and volatile organic compounds (VOCs). It is also possible to sample breath compounds by condensing the exhaled breath and collecting both the volatile compounds (that are now in liquid form after condensation) and the nonvolatile compounds (that are carried by droplets some of which collect on the inner wall of the condenser) together with the condensate. The sample collected can then be store for later biochemical analysis. We refer to this fraction of breath as "exhaled breath condensate". Various devices have been designed for the collection of EBC and VOCs, and all biochemical analysis is currently performed off-line in a laboratory setting. In both sampling methods, the breath biomarker analytes are at very low concentrations, and it is technically challenging to identify the compounds of interest.

Examples of commercially available EBC and VOC collection devices include the RTube™ (Respiratory Research, Inc., Charlottesville, VA), the ECoScreen™ (Erich Jaeger GmbH, Hochberg, Germany), the Breath Collection Apparatus (BCA™; Menssana Research, Inc., Fort Lee, NJ), and the Bio-VOC™ (Markes International, Llantrisant, United Kingdom). None of these commercial devices are currently coupled directly with on-site analytical capabilities.

The RTube (Figure 7.3a) is one of the most inexpensive exhaled breath collection devices available. It is primarily used to collect EBC. It is a multi-part system with a disposable mouthpiece through which the subject exhales, a wide-diameter primary collection tube made of injection-molded polypropylene, a one-way valve that routes the breath from the mouthpiece to the collection tube, and an insulated aluminum sleeve. Immediately before the EBC sample is to be collected, the aluminum sleeve, which has been cooled to subzero Celsius temperature, is placed onto the RTube exterior. The patient breathes out through the device for several minutes while the exhaled breath condenses onto the interior of the polypropylene tube. Aerosol droplets generated within the respiratory system are already present in the exhaled breath and some will deposit on the interior of the tube as well. When the EBC collection is complete, the EBC is removed from the tube and can either be analyzed immediately or frozen at −80°C until further biochemical analysis is performed at a later time [28]. This is an inexpensive and effective breath collection method, but it does have limitations. First, the sampling method must take place over the course of several

FIGURE 7.3 (a) RTube™ with cooled aluminum sleeve covered with insulator, (b) person exhaling into the ECoScreen™ (partially visible) via mouth piece, (c) nitric oxide capture device, and (d) analyzer.

minutes, and it relies on an external freezer to precool the aluminum tube before use. In addition to the time factor, the repeatability of the sampling method may vary. Reports in the literature have suggested that the condensation rate of the exhaled breath may change over time as the aluminum tube warms, and depending on the temperature gradient driving the condensation. This, in turn, may affect the measured chemical composition in the EBC sample [29–32]. Nevertheless, this sample collection method has remained popular due to the extremely low cost of the device and the disposable nature of the sample collection tubes.

The ECoScreen (Figure 7.3b) is another type of EBC collection device that employs a more complex condensation design than the RTube. It includes a specially designed mouthpiece which consists of two non-rebreathing valves as well as a saliva trap, preventing accidental contamination of the EBC sample from mouth fluids.

The subject is required to breathe with normal frequency into the mouthpiece for a few minutes while wearing a nose clip, and the sample is collected as a frozen EBC which is deposited into a collection vessel within the instrument held at −20°C. The sampling time required can be up to 10 min to collect approximately 1–2 mL of content. As with the RTube, the EBC sample is then removed from the collection instrument, and can either be analyzed immediately in a laboratory or stored at −80°C until later biochemical analysis [33]. The earlier version of this instrument is relatively bulky (weighs about 37 kg) and non-portable. It consumes a large amount of power (a 460 VA power supply is required). These characteristics make it unsuited for field use, or even for routine use in a medical clinic. The more recently released model of this device, the ECoScreen Turbo™, actively maintains the temperature of the condenser tube using advanced thermoelectric principles. This newer device may provide enhanced reproducibility in sample collection, although there have been no published reports to date documenting this. This newer device is much smaller than the previous version; it weighs about 3.5 kg and requires only about 70 W of power. This is nonetheless only a collection device and still faces the limitation associated with off-site sample analysis.

The BCA is used for VOC collection. The patient is required to breathe into an adsorbent tube for several minutes while wearing a nose clip. The exhaled breath is collected in a sorbent reservoir. Room air is also collected in a separate sorbent reservoir to allow the subtraction of inhaled VOCs from breath VOCs and thus to determine which chemicals are generated from within the body and which chemicals are likely from the surrounding environment [34]. The samples collected are analyzed using gas chromatography (GC) for the VOCs present. The flow through the tube is electronically controlled in order to minimize sampling errors. In order to prevent any condensation of the VOCs, the reservoir is maintained at 40°C. The device consists of a digital controller to which a stainless steel tubular reservoir is connected. The reservoir is held on a tripod for stability. A filter and mouth piece are connected to the reservoir. A breath trap is attached to the reservoir and an ambient trap is connected to the BCA controller. After the entire device is set up, the patient is asked to breathe for some time into the mouth piece without actually collecting the breath. This is done to equilibrate the concentrations of the chemicals in the breath with the sampling space within the reservoir and enabling the patient to become comfortable with the sample collection process. Once this is done, the patient is required to exhale into the mouth piece for several additional minutes as the sample is collected. The sample is then stored and sent to the laboratory for analysis. The BCA is considered to be patient friendly and user friendly. It allows collection virtually anywhere and, when coupled with GC/MS, it has picomolar sensitivity of the breath sample it collects. However, it is a sizable instrument and its setup requires a skilled technician. Furthermore, the sorbents reservoirs are analyzed off-site by gas chromatography/mass spectrometry (GC/MS).

The Bio-VOC is also a low-cost device for the collection and VOC analysis of end-tidal air. It consists of a non-emitting plastic container (Teflon™ bulb) into which the patient breathes through using a cardboard mouthpiece. The Teflon bulb collects the last portion of the exhaled air [35]. Once the exhaled breath is collected into the container, a screw in piston–plunger allows it to be pushed into a sorbent tube, which is subsequently sent to the laboratory for thermal desorption and chemical analysis using GC [36]. The container can be reused one more time on the same day by the same patient with a new mouthpiece or later after disinfection and cleaning. The design is rather simple and low cost, and the device does not require a medically qualified technician to operate. Samples still have to be sent off-site for analysis.

Various research groups are conducting further studies on overcoming the limitations of current EBC and VOC collection methods. There are currently many open questions for these sample collection techniques, such as the following:

- Does ambient temperature during collection and/or analysis affect the pH levels of the samples? In the affirmative, is there a need to interpret results for standardized temperatures?
- What effect does a nonuniform temperature along the condenser have on the amount of EBC collection or the chemical composition of EBC?
- Do storage temperature and duration affect the quality of the EBC? That is, are the results obtained with immediate testing different from those obtained when testing occurs after a reasonable amount of time? What is the effect of storage temperatures?
- Does repeated thawing of a cryo-frozen sample have an impact on sample quality?

These and other unresolved issues in the area of breath sampling directly affect the development of future sampling instrumentation and sensor platform designs.

7.2.2 BIOCHEMICAL ANALYSIS STRATEGIES FOR BREATH SAMPLES

After breath samples are collected, several important steps are needed for biochemical analysis. There is currently no standardization of these steps and many studies use some but not all of these possible post-processing methods. These steps can be broadly categorized as pre-concentration, separation, and detection. The first category comprises a variety of sample manipulation methods used to enhance the concentration of one or more specifically selected breath biomarker analyte(s) so that it is easier to detect than other "less-important" background compounds in the breath. This allows for easier detection, and some techniques that would not detect the compound in the original sample can sometimes be employed on the concentrated aliquots. One method for pre-concentration is the solid-phase micro-extraction (SPME) technique where the VOCs are concentrated on a solid sorbent phase (Figure 7.4). The SPME technique can be applied actively or passively. In the case of active collection, the individual exhales directly on to the solid sorbent fiber which then concentrates breath compounds. The SPME fiber is later desorbed into an analytical instrument for chemical analysis of these concentrated compounds. In the passive method, the individual exhales into a collection container into which the sorbent fiber is later introduced to concentrate compounds. However, this method also has limitations if it is to be widely used within the medical community. Typically, the presence of more than one biomarker or a combination of multiple compounds is needed to correctly diagnose a disease or disorder. In such cases, the medical practitioner may need to

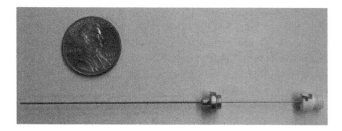

FIGURE 7.4   SPME fiber.

use different absorptive substrates to capture the widest range of different markers. Some patients (especially neonates or elderly sick people) cannot be expected to breathe into/onto a collection device multiple times.

A second type of breath sample manipulation that might be employed is a separation or pre-separation phase where important biomarkers are separated from other "background" breath metabolites. GC and liquid chromatography (LC) are two widely used methods for compound separation. In GC applications, the exhaled breath enters a long, narrow, fused silica capillary column where molecules below their boiling point attach to the stationary phase polymer (predominantly polydimethyl siloxane) coated on the inside of the column. Later the temperature of the column is raised at a controlled rate thus releasing the specific molecules from the stationary phase by boiling. As a result, low boiling point molecules will be released first and the higher boiling point molecules are released later thus separating the compounds [37]. GC is widely used because it can be combined with many different detectors, including thermal conductivity detectors (TCD), flame ionization detectors (FID), mass spectrometry detectors (MS), or differential mass spectrometers (DMS). Figure 7.5 shows an example of one commercially available GC/MS system.

Breath compounds can be detected in concentration ranges as low as parts per billion or even parts per trillion but the required detection methods have significant drawbacks. These are often expensive techniques with relatively high setup cost, essentially non-portable and requiring appreciable analysis time. In recent years, various research groups have been working on micro-fabricated GC columns (μGC)

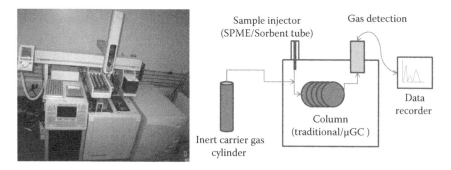

FIGURE 7.5   GC/MS and gas chromatograph schematic.

to incorporate chip-based separation [38–41]. These published works demonstrate several different ways to fabricate the on-chip columns with high aspect ratios that will likely be needed to reduce power consumption since they are frequently fabricated on silicon substrates. These reports also explore ways to reduce fabrication cost and enhance the performance and speed of the device by reducing the distance covered by the gas to obtain similar accuracies. These μGC devices may in fact play an important role in mobile breath sensor platforms, as one possible route to couple microscale breath collection with novel miniature detection modules.

LC or high-performance liquid chromatography (HPLC) differs slightly from GC in the way it operates. LC is mainly used to separate nonvolatile compounds and implements mass transfer between a stationary phase and a mobile phase. Unlike GC, LC uses a liquid mobile phase for separation of various chemical components from a sample mixture. In this method, the EBC is mixed in a solvent and passed through stationary phase columns consisting packed beds of silica particle. The efficiency of chemical separation of this method increases as interactions between mobile phase and stationary phase improve, and this is achieved by keeping the particle size of the stationary phase relatively small. This results in a significant head loss thus requiring high pressure to move the solvent through the column. Hence, pump must be used, unlike in GC systems where the gas moves through the column due to a relatively low-pressure constant flow of a carrier gas. In HPLC, the particle size is about 3–10 μm and requires 7–40 MPa of pressure to maintain a flow rate of 0.5–5.0 mL/min through the columns [42]. Unlike GC where the separation results from boiling point temperature difference between chemicals in the sample mixture, the chemicals in LC are separated based on their polarity as they pass through the stationary phase [37].

The last step in most breath analysis protocols rely on some type of analyte detection strategy. These range from very traditional approaches, like MS [1,43–46], or recently developed analytical methods such as micromachined differential mobility spectrometer (DMS) [7,47–50]. Since the 1970s, GC/MS has been a standard and reliable method used to detect concentrations of compounds in complex mixtures. Many different variations of MS techniques have been developed since then, and many are effective at detecting even very low concentrations of chemicals and proteins. Furthermore, many are compatible with either gas phase or liquid phase samples meaning that they can be used for both types of exhaled breath phases. In the past few years, there has been an effort to miniaturize MS to enable the development of portable, low-power, and low-cost devices for analyte detection while preserving all the features of traditional MS such as high sensitivity, reliability, and self-sustainability [51,52]. Taylor and France provide a comparison of performance and size between a conventional quadrupole mass spectrometer (QMS), a miniature, MEMS-based QMS, and a QMS produced using rapid prototyping [53]. Gao et al. have successfully designed and characterized a handheld rectilinear ion trap mass spectrometer which is the size of a shoebox, battery-powered, consumes less than 70 W and weighs about 10 kg [54]. Ouyang and Cooks provide a very nice comparison of various miniature mass spectrometers. Figure 7.6 shows a comparison of the size of Finnigan's ITMS™, the first commercial ion trap mass spectrometer, with that of a recently designed handheld miniature mass spectrometer, Mini 11™ [51]. The

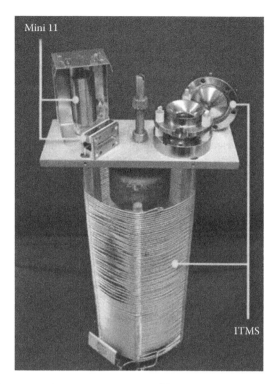

FIGURE 7.6 Size comparison of Finnigan ITMS™ and recently designed Mini 11™. (From Ouyang, Z. and Cooks, R.G., *Annu. Rev. Anal. Chem.*, 2, 187, 2009.)

future trend for these devices is further miniaturization using microfabrication and other techniques. These sensors (and others like them) will be important parts of future mobile breath analysis platforms.

The benefits of breath analysis are widely acknowledged in both the biomedical and national security communities. Microscale, handheld devices that would enable the collection and analysis of breath samples in real time have the potential to revolutionize such applications. Yet, there have been only a few advancements in the design of such devices. The cost of current mass spectrometer devices ranges from the tens of thousands to the hundreds of thousands of dollars. They also require a substantial setup space. Although the weight of the device varies across manufacturers, the devices produced by some of the leading brands weigh about ~40 kg making them practically unmovable. Highly skilled technicians are also required to use these devices and correctly analyze the data obtained. In most cases, semiskilled technicians can be allowed to collect breath samples, but otherwise, the process of pre-concentration, separation, and analysis requires trained individuals to avoid false results. This adds to the overall cost structure of the device as a whole as well as the cost per test. Looking at the given limitations of existing devices paves the way for the need to design a MEMS-based device which could address and overcome all these limitations [6,7,55,56].

7.2.3 Functional Requirements for Breath Sensor Systems

Only a few commercially available breath analyzing devices/sensors are currently used by the medical industry. This is primarily attributable to the scarcity of sensors with an acceptable level of technical accuracy. Furthermore, only a handful of the available devices have been approved by the Food and Drug Administration (FDA). As a result, medical practitioners still prefer to go through the invasive route of testing and analysis of diseases. Examples of devices approved by the FDA for clinical use include the Heartsbreath™ test. This test is a noninvasive and intrinsically risk-free procedure to assess heart transplant rejection that has the potential benefit of reducing the risk for a patient to receive incorrect treatment due to an erroneous heart (endomyocardial) biopsy report [57]. The NIOX™ Breath Nitric Oxide Test System (Aerocrine AB, San Diego, CA) is another device that was approved by the FDA in 2003. The device (as well as its later version NIOX-MINO™) is used to measure the amount of nitric oxide in human breath to evaluate asthma patient response to anti-inflammatory therapy. It adds to established clinical and laboratory assessments methods of asthma including spirometry, peak expiratory flow (PEF) determination, or chest x-rays [58].

In order for medical practitioners and end users to be truly interested in using new generations of exhaled breath analysis devices as the primary method of detecting diseases, the devices will need to possess several important features. Rapid response is one of the requirements. Most current collection methods, like the RTube, require that the samples be collected and then later sent to the lab for analysis which might take anywhere from few hours to few days. Most currently used sensor devices are non-portable and collection devices are portable but bulky and inconvenient to use. The availability of an "all in one" simple, small, and portable collection and analysis device would contribute greatly to the wide adoption of such diagnostic techniques. Commercially available blood glucose level meters are good examples of small and portable devices delivering rapid results, which helps monitor health conditions more effectively. Another important aspect to be considered while designing a device is the influence of physical breath variables such as flow rate, variation in lung capacity, and core body temperature. Work conducted by Landini and Bravard tries to distinguish effects of physical breath variables on current response of an enzymatic breath acetone sensing device and compares the response obtained from controlled breath simulator versus human breath samples [59].

A good device design should be able to distinguish a breath chemical biomarker with high specificity and certainty, and be able to do this across a large chemical concentration range. Current breath sampling procedures are time consuming, labor intensive, and multistage processes which tends to introduce errors at each step [60]. Reliable reproducibility of the data is key to a viable design. Experiments conducted by Leung et al. showed highly reproducible results for EBC pH values for the RTube and ECoScreen individually but they had significantly different results when compared to each other. This indicates that the choice of collection device may be an important contributor to the variability of exhaled breath biomarker measurement [61].

Breath sensors should also be sensitive enough to detect and analyze low concentration levels of biomarkers. This will enable a patient to exhale into the EBC collection device for a relatively short duration and still collect enough sample amounts to conduct a reliable analysis. Current collection devices require about 5–10 min of breathing into the collector to allow sufficient amount of sample collection. For example, according to the RTube manufacturer, about 75–150 μL/min of condensate can be collected for a child and about 100–250 μL/min for an adult [62]. The recommended collection times are 2 min for single analysis and about 7–10 min to store extra samples for future investigation. In some cases, this could be too long duration for the sick, the elderly, or for small children. Also, as the exhaled breath passes through the condenser, not all of saturated vapor condenses, and a significant portion of the breath is likely to be lost. This is due to the design of the device itself, or the surface temperature of the condenser is not low enough to allow the condensation rate to be comparable to the volumetric flow rate of the exhaled breath. A device that could reduce these losses would help in reducing the collection time as well. A comfortable collection time frame of <1 min would be a good goal, although any reduction in collection time would be helpful.

In most cases, multiple exhaled breath biomarkers will be monitored to provide evidence of a certain disease in a patient; hence a comprehensive treatment plan could be designed if the sensor device is able to capture all of those indicators. Hence another key design feature for a new device would be multivalent detection, that is, it should be able to detect multiple compounds either simultaneously or by multiplexing an array of sensors, with each individual unit detecting different compounds of interest. In both cases simultaneous multiple analyses of a variety of compounds would be needed. If separate devices are designed detect each marker, the device management itself could become an issue; however, this can be solved using parallel manufacturing techniques that are readily available. It would require that the patient exhale through a single breath collection device for all different compounds to be analyzed, even if the breath effluent were tracked and routed toward multiple subsequent sensors.

7.2.4 Cost Issues

In order to promote wider use of breath analysis devices in the medical community, these devices need to be reliable, have rapid detection, be highly sensitive, handheld, multivalent, and at the same time be cost effective. For example, a set of 25 RTube along with one aluminum sleeve currently costs more than $1000 averaging to $40 per exhaled breath collection. The cost slightly drops for higher quantity orders [62]. This cost does not include the cost of refrigerator needed to maintain the sleeve at subzero temperatures. The price of ECoScreen device is in the range of $9,000~10,000 not including the separate cost for disposable mouth piece. The Bio-VOC breath sampler costs about $80 per pack. Breath analyzers such as GC/MS are much more expensive and the price range falls around $40,000~$100,000 and 99–100% depending upon the features and sensing capabilities of these devices. In addition to the various prices of collection device, sample storage, sample transportation, and the hourly rate of a skilled technician add to the overall cost of the collection and analysis process.

## 7.3	FACTORS AFFECTING CURRENT BREATH SENSOR SYSTEM PERFORMANCE

It is important to understand the basic principles of fluid flow, thermodynamics, and heat transfer and how they influence the design of a viable exhaled breath collection device. The phrase "exhaled breath" describes a complex flow and heat transfer system: a multicomponent, multiphase jet in a typically still, multicomponent ambient at a different temperature. The addition of a breath-sampling device confines the flow and introduces obstacles. Exhaled breath is a multiphase mixture of gases and droplets that exits the mouth at 37°C [32,63]. The carrier gas contains nitrogen, oxygen, carbon dioxide, water vapor (99%~100% relative humidity), and a multitude of volatile compounds (ethane, pentane, acetone, dimethylsulfide, ammonia, carbon monoxide) [64,18]. The droplets are mostly liquid water and nonvolatile compounds (salts, proteins, mediators) [64–66]. The size of aerosols present in the exhaled breath ranges from 0.3 μm to about 2.5 μm. Ten to twenty percent of the aerosol droplets have a diameter larger than 1 μm [67]. The variation in the size of these aerosols also impacts flow dynamics. In EBC devices, most of the condensation occurs heterogeneously at the cooled wall surface. Convection is the dominant mode of heat transfer between the exhaled breath and the inner walls of the device, but conduction and thermal resistances play an important role in the evaluation of the overall heat balance for the device. In order to collect a useable sample and overcome any losses, patients generally need to breathe into the mouthpiece for 5–10 min. This duration also allows averaging of the breath content collected especially when alveolar breath is sought [68].

The literature is replete with reports of analytical, numerical, and/or experimental investigations of developing flow, with or without heat transfer and condensation in cylindrical ducts of arbitrary cross sections. This is a classical thermo-fluid problem [69–75]. On the contrary, to the best of these authors' knowledge, no reports have been published to date focusing on models and simulations of flow in breath sensors systems.

### 7.3.1	FLOW DYNAMICS

Because flow dynamics is inherent to exhaled breath analysis, the nature of the flow must be determined to inform the design effort. First, the flow regime, laminar, turbulent, or transitional, needs to be identified. Turbulent flow is characterized by macroscopic mixing and therefore desirable for enhanced heat and mass transfer. The flow regime that exists in a pipe of diameter D, in which a fluid (density ρ and viscosity μ) flows with a velocity V is determined by evaluating the Reynolds number: $Re = \rho VD/\mu$. In terms of the volume flow rate, Q (and for a circular duct): $Re = 4\rho Q/(\pi \mu D)$. For Reynolds numbers below 2300 the flow is laminar. It is turbulent for Reynolds numbers above 4000 and transitional in between [76]. The measured volumetric flow rate during tidal breathing for adult is typically on the order of 0.5 L/s [65]. In the case of a device like the RTube, this flow rate results in a Reynolds number on the order of 3000 and, thus, a transitional flow regime. Lower flow rates occur for neonates and children with Reynolds numbers between 1000 and 2000, corresponding to a laminar flow regime. Note that for a given flow rate, reducing the size of the device (decreasing D) will result in an increase of the Reynolds number. The analysis and simulation tools needed to capture such transitional or turbulent flow behavior that are more complex than those available for laminar flow.

The dynamics of the exhaled aerosol also need to be well understood. Droplet trajectories as well as their size and composition histories as they move through the collection tube must be determined. Interactions of droplets with internal obstacles and their deposition rate onto the collector tube surface are essential aspects of the performance of the sampling device. Here again, while the literature abounds with information about the general topic aerosol and spray dynamics [77,78], it is somewhat sparse when it comes to the topic of exhaled aerosols [79] although the recent works of Morawska and coworkers have provided much needed information [80,81].

7.3.2 HEAT TRANSFER AND CONDENSATION

In EBC devices, the entering multiphase mixture condenses as it comes in contact with a cooler surface, going through phase change from saturated vapor to a liquid form. There are two possible condensation regimes: drop-wise condensation or film-wise condensation. Which condensation regime actually occurs in the device greatly depends on the surface properties as well as the rate of heat transfer. The overall rate of condensation will depend on the contact surface temperature, the flow rate and relative humidity of the exhaled breath, and the temperature distribution along the condensation surface. Other factors that may impact the condensation rate and eventually the amount of sample collected include surface material, conduction from the cooling source to the surface of the collector, and condensing surface area [82–84]. A good device design maximizes the amount of condensate collected with the least amount of time required while keeping the size of the collector as small as possible. This can be achieved by maximizing the heat transfer between the cooling source and collection tube surface.

Consider, for example, the collection protocol described in Section 7.2 for the RTube. Multiple instances of heat losses and parasitic heat transfer can be identified. The initial heat flux at the outer surface (surface "4" marked in Figure 7.7) of the insulator that covers the aluminum sleeve as it is brought out from the freezer ($-20°C$) to the ambient ($25°C$) is on the order of 3 kW/m^2. In order to reduce heat addition to the aluminum sleeve from the ambient an insulating layer of polyester is used, since the heat gain from the environment is proportional to the surface area exposed, minimizing that surface (i.e., a smaller device) would limit such parasitic heat transfer. However, by the same token, it is desirable to maximize the internal surface area (surface "1") which is in contact with the exhaled breath. The heat transfer from the low-temperature element to the collector surface also needs to be well designed to maximize heat transfer and, thus, minimize thermal resistance. However, since the prechilled sleeve is designed to slide over the polypropylene collection tube of the RTube it is impossible to have a perfect thermal contact between the tube and the sleeve (surface "2"). Polypropylene is a stable material with limited chemical interaction with the exhaled breath but it has a very low thermal conductivity (three orders of magnitude lower than that of aluminum), thus hindering heat extraction from the exhaled breath.

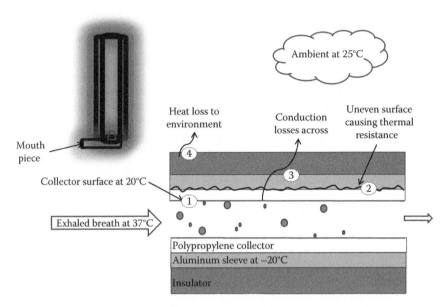

FIGURE 7.7 Schematic of RTube™ collection tube with possible heat transfer zones.

Studies have shown that the chemical composition of compounds present in exhaled breath is likely to be affected by the temperature at which collection occurs or by any changes in temperature during the collection process [32,85–87]. Koczulla et al. found that ambient temperature during collection had negligible effect on the pH of the condensate. However, they also found that the pH changed notice-ably when the analysis of the condensate was made in an ambient environment of 37°C compared to an analysis made at 23°C [88]. Recent experiments showed that acetone concentration was affected when the exhaled breath sample was collected by RTube (which has a time-dependent condensation rate) and compared to samples collected using a device equipped with thermoelectric cooling units which allowed the temperature to be kept constant over time [32]. Other relevant thermal issues include the impact of temperature cycling of the sample because of collection at subzero temperature, then cryo-freezing and eventually thawing for final chemi-cal detection and analysis. Also, the effect of storage temperature on the chemical composition of samples in long-term storage needs to be evaluated. For example, although some studies have indicated that the H_2O_2 levels in EBC do vary substan-tially during long-term storage at low temperature, more research is required to fully assess the impact of storage temperature and duration on other exhaled breath biomarkers [9].

7.3.3 MATERIAL SELECTION

The materials selected for the interior of the breath analysis devices will come into contact with the exhaled breath during sampling. Currently the effect of this mate-rial interfaces is not well understood, and more work is needed to characterize the interactions and possible contamination effects on measured biomarkers [89]. Most current exhaled breath collectors are made of various polymers, glasses, or metals. One study conducted by Vogelberg et al. showed that contamination of exhaled nitrite, a marker for asthma, occurred even after short-term contact of the exhaled breath with the condenser walls which were air dried and stored in sealed container after a disinfection process [90]. This not only brings up the issue of contamination of certain compounds due to exposure to specific polymers but also that of the methodology used to disinfect the condenser. Only a limited number of studies have been performed on exhaled breath contamination by condenser walls but much can be learned from investigations of beverages and food contamination from containers. For example, there is a plethora of data on the contamination effects of liquids with bisphenol A (BPA) from long-term use or exposure at higher temperature of polycarbonate bottles, water bottles, and other food containers [91,92]. The issue of material contamination resulting from long-term storage of EBC also needs to be examined. Studies have also been conducted to understand how metal condensing surfaces could alter EBC study results. Rosias et al. studied the direct effects of condenser coating on EBC chemical content. The study included coating of silicone, glass, aluminum, polypropylene, and Teflon. They found that silicone and glass coating were most effective and provided reliable and consistent EBC biomarker analysis results [30].

Chemical consistency between the sample analyzed and the exhaled breath as it exits the patient's mouth is a key factor in the reliability of a breath analyzer. Consider the example of RTube whose collection protocol is described in Section 7.2. The amount of condensate collected each time a sample is taken will be determined mainly by the cooling rate experienced by the breath as it passes through the collection tube, the amount of condensate lost due to RTube design flaws which may prevent collecting all the exhaled breath into the collection tube with wastage, the value of the ambient temperature, the initial temperature of the aluminum sleeve, the magnitude of heat losses due to thermal resistance, and rate of breathing. These are in addition to other expected human errors. Consequently maintaining sample biochemical consistency and ensuring repeatability of biomarker measurements are very important challenges that must be addressed.

7.3.4 LIMITATIONS OF EXISTING SENSOR SYSTEMS

Breath sensor systems may be categorized in two major classes: condensation devices (e.g., RTube, ECoScreen) and absorption devices (e.g., BCA, Bio-VOC). Condensation devices can be further distinguished based on whether the condenser is passively (e.g., RTube) or actively cooled (e.g., ECoScreen Turbo). All these devices are essentially collectors and samples need to be sent for off-site analysis by GC/MS. Another common feature of these devices is that they require patients to breathe into them for times ranging from 2 to 10 min. Passively cooled condensers rely on the availability of an external freezer to lower the temperature of cooling body before use. Repeatability is also questionable. As mentioned earlier in Section 7.2.1, published reports [29–32] have suggested that the condensation rate of the exhaled breath may change over time as the device warms up and the temperature gradient driving the condensation process changes. This, in turn, can affect the measured chemical composition in the EBC sample [29–32]. Sample collected with condensation devices must be stored at low temperature (−80°C for RTube or ECoScreen) for later biochemical analysis [33]. Size and power requirements are common obstacles to routine use in a medical clinic.

Existing common breath analyzers such as GC/MS or LC/MS also have limitation of size, high cost, and time required for analysis. Various research groups have been working toward miniaturization of these devices as discussed in detail in Section 7.2.2.

7.4 FUTURE TRENDS IN BREATH SENSOR SYSTEM DESIGN

Various research groups have been working on miniaturization of breath sensor systems due to the limitations of current devices as discussed in the previous sections. Most of the recent work has been in the area of miniaturization due to the relative complexity of the analysis systems, the high cost of benchtop instrumentation, and need for off-site chemical analysis of the current systems. As a whole, minimal progress has been made in the area of scaling down the size of the breath collection device for fieldable instruments. There are substantial literature reports on miniaturization of other unrelated medical devices that have structures similar to the exhaled breath condensers. Hence, we expect that further miniaturization of breath sensors in general should be possible. Both the small-scale condenser and microscale chemical analyzer also ultimately need to be integrated into one single device, which still presents technical challenges. The following sections briefly discuss some of the recent work pertaining to the miniaturization of breath sensor systems.

7.4.1 Challenges Associated with Miniature Breath Sensor Systems

There has been vast interest in designing a real-time miniature device that could collect a sample and detect multiple breath biomarkers concurrently. Such devices could be used as a personalized self-monitoring breath sensor. An integrated instrument combining a small-scale sampling device with a micro-sensor which could detect both EBC and breath VOCs would fulfill the medical and commercial need for such a biosensor.

These microscale devices have great potential in terms of portability, ease of use, and turnaround time, but their reusability introduces a few additional constraints that do not exist in current disposable condensate collection devices and need to be evaluated.

For instance, the impact of cycling on thermal behavior and biomarker contamination needs to be well understood. The life of the condenser will also be limited by the cumulative contamination caused by reuse. Although no data is available to assess the effects of residual disinfectant agents on reusable EBC condensers, studies have shown that traces of external contaminants were present in the breath samples collected even after extensive cleaning and soaking of the condenser in water for prolonged periods of time [93]. Lessons can be learnt from contamination issues in respiratory applications due to reuse since abundant literature is available in that field [94–96]. All these concerns also exist for the chemical sensors, with further complications associated with the presence of electronic components.

Another concern associated with reuse of the condenser is another form of chemical contamination, which affects the wetting properties of the condensation surface. This will not only affect the droplet size and shape but also the movement of drops and their coalescence properties. The performance of a device designed to operate on either of the condensation regimes—drop-wise condensation or film-wise condensation—could be significantly altered by contamination. Ample literature is available on wetting, condensation rate, and droplet formation for various surfaces [97–99]. Studies have shown that nanoscale structures can greatly alter the phase-change properties of fluid and hence affect the rate of condensation [100–105].

Small, MEMS-based devices will require smaller samples and because sample wastage is likely to be less prevalent, shorter sampling times may be possible. Recent studies have shown promising results in collecting gaseous and aerosol particles through micro-channel [106] but attention must be paid to the effect that device size reduction may have on flow physics. While it is unlikely that breath sensor system will get small enough for the continuum description of the fluid to be questionable (at least for the foreseeable future), as channel sizes are shrunk from centimeter scale (e.g., RTube) to submillimeter scale, the importance of capillarity will increase and that of body forces and buoyancy forces will wane. Microdevices are characterized by large surface-to-volume ratio, which directly affects heat and mass transfer processes.

7.4.2 CMOS Technologies

Among the many technologies available for incorporation into breath sensors, researchers have recently implemented complementary metal oxide semiconductor (CMOS) approaches into new chemical sensors. However, the number of publications in this area is much less than the work going on in the field of MEMS chemical sensors (discussed in the next section). As the name suggests CMOS is a type of semiconductor chip. It is widely used in electronic industry due to low power consumption as compared to the transistor chip. This makes the CMOS chip very useful for battery-powered devices and an ideal candidate for designing a handheld breath analyzer. Rairigh et al. successfully experimented the use of thiolate-monolayer-protected gold nanoparticles (MPN) coated chemiresistor (CR) sensors with four element integrated array to analyze synthetically test-breath sample containing four known biomarkers for lung cancer [107]. Benkstein et al. in their study have successfully implemented CMOS-based microsensor arrays for EBC detection [108] which were originally designed by various research groups to analyze chemical mixtures [109–111]. Figure 7.8 shows a 4×4 array configuration of microhotplates, used by Benkstein et al. with five different sensing material deposits set at different temperatures to detect various breath biomarkers. This array of sensor elements allows for a potentially large number of discrete breath metabolites to be measured over time.

SnO$_2$ 375°C 8 s	WO$_3$ 1 drop	TiO$_2$ 475°C 30 s	Sb:SnO$_2$ µshell 2 drops
Nb:TiO$_2$ np 2 drops	Sb:SnO$_2$ µshell 2 drops	WO$_3$ 1 drop	SnO$_2$ 375°C 8 s
TiO$_2$ 475°C 30 s	Nb:TiO$_2$ np 2 drops	SnO$_2$ 375°C 8 s	
WO$_3$ 1 drop	TiO$_2$ 475°C 30 s	Sb:SnO$_2$ µshell 2 drops	Nb:TiO$_2$ np 2 drops

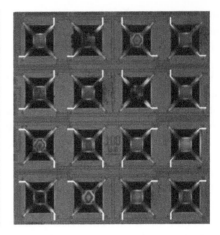

FIGURE 7.8 4×4 array of micro-hotplates with five different sensing compounds deposits set at different temperatures. (From Benkstein, K.D. et al., *IEEE Sens. J.*, 10(1), 137 © [2010] IEEE.)

7.4.3 MEMS Architecture

Some recent studies have assessed the use of MEMS devices for exhaled breath analysis and there appears to be great potential in implementing this architecture in the near future. A successful MEMS-based device should meet various requirements and be a complete system with features such as data processing, data interpretation, data recording, as well as self-calibration [112]. Other physical features required in the design are minimal weight and size to allow it to have handheld functionality, and low power consumption so that it can be easily run on battery power. For example, the ECoScreen Turbo requires 70 W for sample collection and certainly does not fall into the category of low-power device. As mentioned previously, most recent research reports in the area of breath sensors platforms have been in the area of miniaturization of chemical analyzer. Minimal progress has been made for small-scale condensers or other sampling approaches. This section briefly discusses the work being done by various research groups in implementing MEMS architecture for exhaled breath diagnostics.

One research group has recently published a report of a MEMS-based infrared (IR) microsensor design which operates in the mid-IR spectral range of 2200–2250 cm^{-1}. Their sensor detects and quantifies chemicals based on their vibration absorption properties. The prototype device built for the study had the capability of high scanning speed, weighed less than a kilogram, and consumed about 5 W [113]. In addition to this work, Alfeeli et al. presented a MEMS-based µGC which consumed less power, had fast analysis time, and was microscaled. The design was based on micro fabricated columns with high aspect ratio etched on silicon. These columns were then coated by a specific stationary phase to enable separation of VOCs based on their boiling points [114].

In traditional benchtop breath analysis platform designs, pre-concentrators are typically used to absorb VOCs from a complex sample, and these pre-concentrators frequently consist of packed tubes of small absorbent particles. While this design provides a large surface area to absorb and concentrate trace chemicals, it also results in a high pressure drop through the tubes. An alternate design consists of tube walls coated with absorbent material to overcome the issue of pressure losses, and provides a limited absorption surface area. Alfeeli et al. designed micro-pillars for a µGC to increase surface area and chemical pre-concentration efficiency. The group studied different design and structures of micro pillars for maximizing flow and efficiency [115]. Figure 7.9 shows different screen shots of flow analysis for different designs and layout of micro pillars.

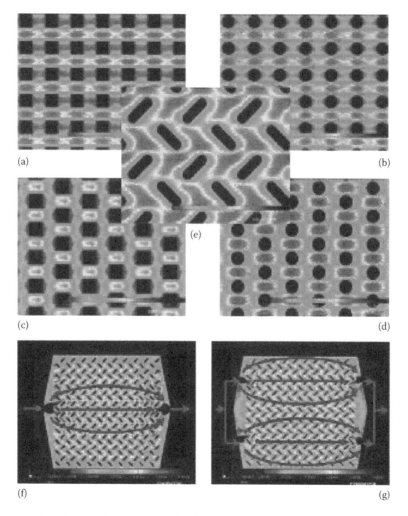

(a)

(b)

(e)

(c)

(d)

(f)

(g)

FIGURE 7.9 Three different micro-pillar shapes aligned differently to analyze flow behavior. (a) Ordered square pillars, (b) ordered circular pillars, (c) staggered square pillars, (d) staggered circular pillars, (e) crisscross pillars, (f) crisscross with single inlet/outlet showing the flow path, (g) crisscross pillars with multiple inlet/outlet showing the flow path. (From Alfeeli, B. and Agah, M., Micro preconcentrator with embedded 3D pillars for breath analysis applications, *2008 IEEE Sensors*, Leece, Italy, pp. 736–739 © [2008] IEEE.)

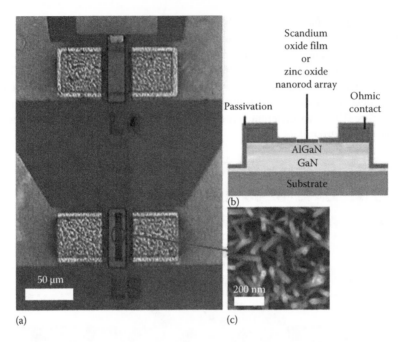

FIGURE 7.10 (a) Integrated pH HEMT sensor (top) and glucose HEMT sensor (bottom). (b) Schematic of pH HEMT sensor. (c) Zinc oxide nanorod SEM image and location. (From Chu, B.H. et al., *IEEE Sens. J.*, 10(1), 64 © [2010] IEEE.)

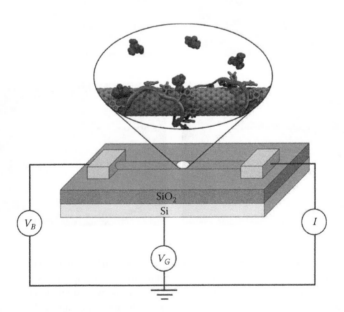

FIGURE 7.11 Single-stranded DNA-coated semiconducting carbon nanotube. (From Johnson, A.T.C. et al., *IEEE Sens. J.*, 10(1), 159 © [2010] IEEE.)

In addition to the previously discussed studies, some recent work had been conducted on combining integrated high electron mobility sensor (HEMT) with pH and glucose sensor on a single chip to detect the presence of exhaled breath biomarkers for diabetes. Figure 7.10 shows the implemented design on a wireless handheld device for remote monitoring [116].

Another interesting study recently reported used DNA-coated nanosensors for breath analysis. The study used an artificial biological olfactory system and implemented it on a solid-state device. A single-stranded DNA-coated semiconducting carbon nanotube was used for odorant detection and subsequent chemical analysis of VOCs in human breath (Figure 7.11) [117].

REFERENCES

1. Pauling L, Robinson AB, Teranish R, Cary P: Quantitative analysis of urine vapor and breath by gas–liquid partition chromatography. *Proceedings of the National Academy of Sciences of the United States of America* 1971, 68(10):2374–2376.
2. Cao WQ, Duan YX: Current status of methods and techniques for breath analysis. *Critical Reviews in Analytical Chemistry* 2007, 37(1):3–13.
3. Manolis A: The diagnostic potential of breath analysis. *Clinical Chemistry* 1983, 29(1):5–15.
4. Whittle CL, Fakharzadeh S, Eades J, Preti G: Human breath odors and their use in diagnosis. *Annals of the New York Academy of Sciences*, Malamud D, Niedbala RS, eds., 2007, 1098:252–266.
5. Buszewski B, Kesy M, Ligor T, Amann A: Human exhaled air analytics: Biomarkers of diseases. *Biomedical Chromatography* 2007, 21(6):553–566.
6. Davis CE, Bogan MJ, Sankaran S, Molina MA, Loyola BR, Zhao WX, Benner WH, Schivo M, Farquar GR, Kenyon NJ et al.: Analysis of volatile and non-volatile biomarkers in human breath using differential mobility spectrometry (DMS). *IEEE Sensors Journal* 2010, 10(1):114–122.
7. Molina MA, Zhao W, Sankaran S, Schivo M, Kenyon NJ, Davis CE: Design-of-experiment optimization of exhaled breath condensate analysis using a miniature differential mobility spectrometer (DMS). *Analytica Chimica Acta* 2008, 628(2):155–161.
8. Di Francesco F, Fuoco R, Trivella MG, Ceccarini A: Breath analysis: Trends in techniques and clinical applications. *Microchemical Journal* 2005, 79(1–2):405–410.
9. Brooks WM, Lash H, Kettle AJ, Epton MJ: Optimising hydrogen peroxide measurement in exhaled breath condensate. *Redox Report* 2006, 11(2):78–84.
10. Vaughan J, Ngamtrakulpanit L, Pajewski TN, Turner R, Nguyen TA, Smith A, Urban P, Hom S, Gaston B, Hunt J: Exhaled breath condensate pH is a robust and reproducible assay of airway acidity. *European Respiratory Journal* 2003, 22(6):889–894.
11. Kharitonov SA, Barnes PJ: Biomarkers of some pulmonary diseases in exhaled breath. *Biomarkers* 2002, 7(1):1–32.
12. Phillips M, Cataneo RN, Cummin ARC, Gagliardi AJ, Gleeson K, Greenberg J, Maxfield RA, Rom WN: Detection of lung cancer with volatile markers in the breath. *Chest* 2003, 123(6):2115–2123.
13. Zeidler MR, Kleerup EC, Tashkin DP: Exhaled nitric oxide in the assessment of asthma. *Current Opinion in Pulmonary Medicine* 2004, 10(1):31–36.
14. Carpagnano GE, Barnes PJ, Francis J, Wilson N, Bush A, Kharitonov SA: Breath condensate pH in children with cystic fibrosis and asthma—A new noninvasive marker of airway inflammation? *Chest* 2004, 125(6):2005–2010.

15. Novak BJ, Blake DR, Meinardi S, Rowland FS, Pontello A, Cooper DM, Galassetti PR: Exhaled methyl nitrate as a noninvasive marker of hyperglycemia in type 1 diabetes. *Proceedings of the National Academy of Sciences of the United States of America* 2007, 104(40):15613–15618.

16. Phillips M, Cataneo RN, Condos R, Erickson GAR, Greenberg J, La Bombardi V, Munawar MI, Tietje O: Volatile biomarkers of pulmonary tuberculosis in the breath. *Tuberculosis* 2007, 87(1):44–52.

17. Phillips M, Boehmer JP, Cataneo RN, Cheema T, Eisen HJ, Fallon JT, Fisher PE, Gass A, Greenberg J, Kobashigawa J et al.: Heart allograft rejection: Detection with breath alkanes in low levels (the HARDBALL study). *Journal of Heart and Lung Transplantation* 2004, 23(6):701–708.

18. Miekisch W, Schubert JK, Noeldge-Schomburg GFE: Diagnostic potential of breath analysis—Focus on volatile organic compounds. *Clinica Chimica Acta* 2004, 347(1–2):25–39.

19. Pleil JD: Role of exhaled breath biomarkers in environmental health science. *Journal of Toxicology and Environmental Health-Part B: Critical Reviews* 2008, 11(8):613–629.

20. Borrill ZL, Roy K, Singh D: Exhaled breath condensate biomarkers in COPD. *European Respiratory Journal* 2008, 32(2):472–486.

21. Comandini A, Rogliani P, Nunziata A, Cazzola M, Curradi G, Saltini C: Biomarkers of lung damage associated with tobacco smoke in induced sputum. *Respiratory Medicine* 2009, 103(11):1592–1613.

22. Chan HP, Lewis C, Thomas PS: Exhaled breath analysis: Novel approach for early detection of lung cancer. *Lung Cancer* 2009, 63(2):164–168.

23. Hunt J: Exhaled breath condensate: An evolving tool for noninvasive evaluation of lung disease. *Journal of Allergy and Clinical Immunology* 2002, 110(1):28–34.

24. Wallace L, Buckley T, Pellizzari E, Gordon S: Breath measurements as volatile organic compound biomarkers. *Environmental Health Perspectives* 1996, 104:861–869.

25. Amorim LCA, Cardeal ZDL: Breath air analysis and its use as a biomarker in biological monitoring of occupational and environmental exposure to chemical agents. *Journal of Chromatography B: Analytical Technologies in the Biomedical and Life Sciences* 2007, 853(1–2):1–9.

26. Frank M, Farquar G, Adams K, Bogan M, Martin A, Benner H, Spadaccini C, Steele P, Sankaran S, Loyola B et al.: Modular sampling and analysis techniques for the real-time analysis of human breath. *IEEE Sensors 2007 Conference*, Atlanta, GA, pp. 10–13, 2007.

27. Shnayderman M, Mansfield B, Yip P, Clark HA, Krebs MD, Cohen SJ, Zeskind JE, Ryan ET, Dorkin HL, Callahan MV et al.: Species-specific bacteria identification using differential mobility spectrometry and bioinformatics pattern recognition. *Analytical Chemistry* 2005, 77(18):5930–5937.

28. Esther CR, Jasin HM, Collins LB, Swenberg JA, Boysen G: A mass spectrometric method to simultaneously measure a biomarker and dilution marker in exhaled breath condensate. *Rapid Communications in Mass Spectrometry* 2008, 22(5):701–705.

29. Corradi M, Rubinstein I, Andreoli R, Manini P, Caglieri A, Poli D, Alinovi R, Mutti A: Aldehydes in exhaled breath condensate of patients with chronic obstructive pulmonary disease. *American Journal of Respiratory and Critical Care Medicine* 2003, 167(10):1380–1386.

30. Rosias PP, Robroeks CM, Niemarkt HJ, Kester AD, Vernooy JH, Suykerbuyk J, Teunissen J, Heynens J, Hendriks HJ, Jobsis Q et al.: Breath condenser coatings affect measurement of biomarkers in exhaled breath condensate. *European Respiratory Journal* 2006, 28(5):1036–1041.

31. Bloemen K, Lissens G, Desager K, Schoeters G: Determinants of variability of protein content, volume and pH of exhaled breath condensate. *Respiratory Medicine* 2007, 101(6):1331–1337.

32. Loyola BR, Bhushan A, Schivo M, Kenyon NJ, Davis CE: Temperature changes in exhaled breath condensate collection devices affect observed acetone concentrations. *Journal of Breath Research* 2008, 2(3):037005 (037007 pp.).
33. Carpagnano GE, Kharitonov SA, Foschino-Barbaro MP, Resta O, Gramiccioni E, Barnes PJ: Increased inflammatory markers in the exhaled breath condensate of cigarette smokers. *European Respiratory Journal* 2003, 21(4):589–593.
34. Phillips M: Method for the collection and assay of volatile organic compounds in breath. *Analytical Biochemistry* 1997, 247(2):272–278.
35. Poli D, Carbognani P, Corradi M, Goldoni M, Acampa O, Balbi B, Bianchi L, Rusca M, Mutti A: Exhaled volatile organic compounds in patients with non-small cell lung cancer: Cross sectional and nested short-term follow-up study. *Respiratory Research* 2005, 6:71.
36. Scheepers PTJ, Heussen GAH: Assessing health risk of toxic substances by analysis of body fluids and exhaled air. *Trends in Analytical Chemistry* 2002, 21(3):XI–XIV.
37. Simon MG, Davis CE: Instrumentation and sensors for human breath analysis. In: *Advances in Biomedical Sensing, Measurements, Instrumentation and Systems*, Subhas, C.M. and Animé, L.-E., eds., Vol. 55, pp.144–165. Berlin, Germany: Springer, 2010.
38. Agah M, Wise KD: Low-mass PECVD oxynitride gas chromatographic columns. *Journal of Microelectromechanical Systems* 2007, 16(4):853–860.
39. Kolesar ES, Reston RR: Review and summary of a silicon micromachined gas chromatography system. *IEEE Transactions on Components Packaging and Manufacturing Technology Part B: Advanced Packaging* 1998, 21(4):324–328.
40. Agah M, Potkay JA, Lambertus G, Sacks R, Wise KD: High-performance temperature-programmed microfabricated gas chromatography columns. *Journal of Microelectromechanical Systems* 2005, 14(5):1039–1050.
41. Bhushan A, Yemane D, Trudell D, Overton EB, Goettert J: Fabrication of micro-gas chromatograph columns for fast chromatography. *Microsystem Technologies: Micro-and Nanosystems—Information Storage and Processing Systems* 2007, 13(3–4):361–368.
42. Harris DC: *Quantitative Chemical Analysis*, 6th edn. New York: W.H. Freeman and Company, 2003.
43. Pleil JD, Lindstrom AB: Exhaled human breath measurement method for assessing exposure to halogenated volatile organic compounds. *Clinical Chemistry* 1997, 43(5):723–730.
44. Mueller W, Schubert J, Benzing A, Geiger K: Method for analysis of exhaled air by microwave energy desorption coupled with gas chromatography flame ionization detection mass spectrometry. *Journal of Chromatography B* 1998, 716(1–2):27–38.
45. Leone AM, Gustafsson LE, Francis PL, Persson MG, Wiklund NP, Moncada S: Nitric-oxide is present in exhaled breath in humans—Direct GC-MS confirmation. *Biochemical and Biophysical Research Communications* 1994, 201(2):883–887.
46. Dyne D, Cocker J, Wilson HK: A novel device for capturing breath samples for solvent analysis. *Science of the Total Environment* 1997, 199(1–2):83–89.
47. Davis CE, Kang JM, Dube CE, Borenstein JT, Nazarov EG, Miller RA, Zapata AM: Spore biomarker detection using a MEMS differential mobility spectrometer. *TRANSDUCERS '03, 12th International Conference on Solid-State Sensors, Actuators and Microsystems Digest of Technical Papers (Cat No03TH8664)*, Boston, MA, vol.1232|1232 vol.(xl + xxxix + 1938), pp. 1233–1238, 2003.
48. Eiceman GA, Krylov EV, Nazarov EG, Miller RA: Separation of ions from explosives in differential mobility spectrometry by vapor-modified drift gas. *Analytical Chemistry* 2004, 76(17):4937–4944.
49. Eiceman GA, Wang M, Prasad S, Schmidt H, Tadjimukhamedov FK, Lavine BK, Mirjankar N: Pattern recognition analysis of differential mobility spectra with classification by chemical family. *Analytica Chimica Acta* 2006, 579(1):1–10.

50. Krebs MD, Zapata AM, Nazarov EG, Miller RA, Costa IS, Sonenshein AL, Davis CE: Detection of biological and chemical agents using differential mobility spectrometry (DMS) technology. *IEEE Sensors Journal* 2005, 5(4):696–703.

51. Ouyang Z, Cooks RG: Miniature mass spectrometers. *Annual Review of Analytical Chemistry* 2009, 2:187–214.

52. Smith JN, Keil A, Likens J, Noll RJ, Cooks RG: Facility monitoring of toxic industrial compounds in air using an automated, fieldable, miniature mass spectrometer. *Analyst* 2010, 135(5):994–1003.

53. Taylor S, France N: Miniature and micro mass spectrometry for nanoscale sensing applications. *Journal of Physics: Conference Series* 2009, 178:012003.

54. Gao L, Song QY, Patterson GE, Cooks RG, Ouyang Z: Handheld rectilinear ion trap mass spectrometer. *Analytical Chemistry* 2006, 78(17):5994–6002.

55. Strand N, Bhushan A, Schivo M, Kenyon NJ, Davis CE: Chemically polymerized polypyrrole for on-chip concentration of volatile breath metabolites. *Sensors and Actuators B: Chemical* 2010, 143(2):516–523.

56. Sankaran S, Weixiang Z, Loyola B, Morgan J, Molina M, Shivo M, Rana R, Kenyon N, Davis C: Microfabricated differential mobility spectrometers for breath analysis. *IEEE Sensors 2007 Conference*, October 28–31, 2007, Atlanta, GA, pp. 16–19, 2007.

57. FDA approval notification for Heartsbreath. http://www.accessdata.fda.gov/cdrh_docs/pdf3/H030004a.pdf, Food and Drug Administration, Silver Spring, Rockville, MD, 2004.

58. FDA approval notification for NIOX breath nitric oxide test system. http://www.access-data.fda.gov/cdrh_docs/pdf2/K021133.pdf, Food and Drug Administration, Silver Spring, Rockville, MD, 2003.

59. Landini BE, Bravard ST: Effect of exhalation variables on the current response of an enzymatic breath acetone sensing device. *IEEE Sensors Journal* 2010, 10(1):19–24.

60. Mills GA, Walker V: Headspace solid-phase microextraction procedures for gas chromatographic analysis of biological fluids and materials. *Journal of Chromatography A* 2000, 902(1):267–287.

61. Leung TF, Li CY, Yung E, Liu EKH, Lam CWK, Wong GWK: Clinical and technical factors affecting pH and other biomarkers in exhaled breath condensate. *Pediatric Pulmonology* 2006, 41(1):87–94.

62. RTube exhaled breath condensate collector technical specifications. http://www.rtube.com/. Edited by Research R, 2009.

63. Effros RM, Hoagland KW, Bosbous M, Castillo D, Foss B, Dunning M, Gare M, Lin W, Sun F: Dilution of respiratory solutes in exhaled condensates. *American Journal of Respiratory and Critical Care Medicine* 2002, 165(5):663–669.

64. Mukhopadhyay R: Dont waste your breath. *Analytical Chemistry* 2004, 76(15):273A–276A.

65. Davis CE, Frank M, Mizaikoff B, Oser H: The future of sensors and instrumentation for human breath analysis. *IEEE Sensors Journal* 2010, 10(1):3–6.

66. Corradi M, Goldoni M, Caglieri A, Folesani G, Poli D, Corti M, Mutti A: Collecting exhaled breath condensate (EBC) with two condensers in series: A promising technique for studying the mechanisms of EBC formation, and the volatility of selected biomarkers. *Journal of Aerosol Medicine and Pulmonary Drug Delivery* 2008, 21(1):35–44.

67. Papineni RS, Rosenthal FS: The size distribution of droplets in the exhaled breath of healthy human subjects. *Journal of Aerosol Medicine-Deposition Clearance and Effects in the Lung* 1996, 10(2):105–116.

68. Strand NC, Davis CE: Analytical methods in exhaled breath diagnostics. *Separation Science* 2009, 1(4):20–28.

69. Moin P, Kim J: Numerical investigation of turbulent channel flow. *Journal of Fluid Mechanics* 1982, 118(May):341–377.

70. Ferziger JH: *Numerical Methods for Engineering Applications*, 2nd edn. New York: John Wiley & Sons, 1998.
71. Burmeister LC: *Convective Heat Transfer*, 2nd edn. New York: John Wiley & Sons, 1993.
72. Whalley PB: *Boiling, Condensation, and Gas-Liquid Flow*. New York: Oxford University Press, 1987.
73. Siegel R, Sparrow EM, Hallman TM: Steady laminar heat transfer in a circular tube with prescribed wall heat flux. *Applied Scientific Research* 1958, 7(5):386–392.
74. Shah RK, London AL: Thermal boundary conditions and some solutions for laminar duct flow forced-convection. *Journal of Heat Transfer: Transactions of the ASME* 1974, 96(2):159–165.
75. Quaresma JNNC, Cotta RM: Exact-solutions for thermally developing tube flow with variable wall heat-flux. *International Communications in Heat and Mass Transfer* 1994, 21(5):729–742.
76. White FM: *Fluid Mechanics*, 4th edn. New York: McGraw-Hill, 1999.
77. Friedlander SK: *Smoke, Dust, and Haze: Fundamentals of Aerosol Dynamics*, 2nd edn. New York: Oxford University Press, 2000.
78. Sirignano WA: *Fluid Dynamics and Transport of Droplets and Sprays*. Cambridge, U.K.: Cambridge University Press, 1999.
79. Papineni RS, Rosenthal FS: The size distribution of droplets in the exhaled breath of healthy human subjects. *Journal of Aerosol Medicine-Deposition Clearance and Effects in the Lung* 1997, 10(2):105–116.
80. Chao CYH, Wan MP, Morawska L, Johnson GR, Ristovski ZD, Hargreaves M, Mengersen K, Corbett S, Li Y, Xie X et al.: Characterization of expiration air jets and droplet size distributions immediately at the mouth opening. *Journal of Aerosol Science* 2009, 40(2):122–133.
81. Morawska L, Johnson GR, Ristovski ZD, Hargreaves M, Mengersen K, Corbett S, Chao CYH, Li Y, Katoshevski D: Size distribution and sites of origin of droplets expelled from the human respiratory tract during expiratory activities. *Journal of Aerosol Science* 2009, 40(3):256–269.
82. Moran MJ, Shapiro HN: *Fundamentals of Engineering Thermodynamics*, 5th edn. Hoboken, NJ: John Wiley & Sons, Inc., 2004.
83. Kays WM, Crawford ME: *Convective Heat and Mass Transfer*, 2nd edn. New York: McGraw-Hill Book Company, 1980.
84. Incropera FP, DeWitt DP: *Fundamentals of Heat and Mass Transfer*, 4th edn. Hoboken, NJ: John Wiley & Sons, Inc; 1996.
85. Czebe K, Barta I, Antus B, Valyon M, Horvath I, Kullmann T: Influence of condensing equipment and temperature on exhaled breath condensate pH, total protein and leukotriene concentrations. *Respiratory Medicine* 2008, 102(5):720–725.
86. Goldoni M, Caglieri A, Andreoli R, Poli D, Manini P, Vettori MV, Corradi M, Mutti A: Influence of condensation temperature on selected exhaled breath parameters. *BMC Pulmonary Medicine* 2005, 5:10.
87. Prieto L, Ferrer A, Palop J, Domenech J, Llusar R, Rojas R: Differences in exhaled breath condensate pH measurements between samples obtained with two commercial devices. *Respiratory Medicine* 2007, 101(8):1715–1720.
88. Koczulla AR, Noeske S, Herr C, Dette F, Pinkenburg O, Schmid S, Jorres RA, Vogelmeier C, Bals R: Ambient temperature impacts on pH of exhaled breath condensate. *Respirology* 2010, 15(1):155–159.
89. Horvath I, Hunt J, Barnes PJ, Breath AETFE: Exhaled breath condensate: Methodological recommendations and unresolved questions. *European Respiratory Journal* 2005, 26(3):523–548.

90. Vogelberg C, Kahlert A, Wurfel C, Marx K, Bohm A, Range U, Neurneister V, Leupold W: Exhaled breath condensate nitrite—Methodological problems of sample collection. *Medical Science Monitor* 2008, 14(8):CR416–CR422.

91. Lopez-Cervantes J, Paseiro-Losada P: Determination of bisphenol A in, and its migration from, PVC stretch film used for food packaging. *Food Additives and Contaminants* 2003, 20(6):596–606.

92. Biedermann-Brem S, Grob K, Fjeldal P: Release of bisphenol A from polycarbonate baby bottles: Mechanisms of formation and investigation of worst case scenarios. *European Food Research and Technology* 2008, 227(4):1053–1060.

93. de Laurentiis G, Paris D, Melck D, Maniscalco M, Marsico S, Corso G, Motta A, Sofia M: Metabonomic analysis of exhaled breath condensate in adults by nuclear magnetic resonance spectroscopy. *European Respiratory Journal* 2008, 32(5):1175–1183.

94. Jhung MA, Sunenshine RH, Noble-Wang J, Coffin SE, St John K, Lewis FM, Jensen B, Peterson A, LiPuma J, Arduino MJ et al.: A national outbreak of *Ralstonia mannitolilytica* associated with use of a contaminated oxygen-delivery device among pediatric patients. *Pediatrics* 2007, 119(6):1061–1068.

95. Harrel SK, Molinari J: Aerosols and splatter in dentistry—A brief review of the literature and infection control implications. *Journal of the American Dental Association* 2004, 135(4):429–437.

96. Garland JS, Uhing MR: Strategies to prevent bacterial and fungal infection in the neonatal intensive care unit. *Clinics in Perinatology* 2009, 36(1):1–13.

97. Degennes PG: Wetting—Statics and dynamics. *Reviews of Modern Physics* 1985, 57(3):827–863.

98. Cazabat AM: How does a droplet spread? *Contemporary Physics* 1987, 28(4):347–364.

99. Blake TD: The physics of moving wetting lines. *Journal of Colloid and Interface Science* 2006, 299:1–13.

100. Ojha M, Chatterjee A, Mont F, Schubert EF, Wayner PC, Jr., Plawsky JL: The role of solid surface structure on dropwise phase change processes. *International Journal of Heat and Mass Transfer* 2010, 53(5–6):910–922.

101. Bayer IS, Megaridis CM: Contact angle dynamics in droplets impacting on flat surfaces with different wetting characteristics. *Journal of Fluid Mechanics* 2006, 558:415–449.

102. Kannan R, Sivakumar D: Drop impact on a solid surface comprising micro groove structure. *Complex Systems* 2008, 982:633–638.

103. Kannan R, Sivakumar D: Impact of liquid drops on a rough surface comprising microgrooves. *Experiments in Fluids* 2008, 44(6):927–938.

104. Grest GS, Heine DR, Webb EB: Liquid nanodroplets spreading on chemically patterned surfaces. *Langmuir* 2006, 22(10):4745–4749.

105. Heine DR, Grest GS, Webb EB: Surface wetting of liquid nanodroplets: Droplet-size effects. *Physical Review Letters* 2005, 95(10):107801.

106. Greenwood J, Daming C, Ye L, Hongrui J: Air to liquid sample collection devices using microfluidic gas/liquid interfaces. *2008 IEEE Sensors*, Leece, Italy, pp. 720–723, 2008.

107. Rairigh DJ, Warnell GA, Xu C, Zellers ET, Mason AJ: CMOS baseline tracking and cancellation instrumentation for nanoparticle-coated chemiresistors. *IEEE Transactions on Biomedical Circuits and Systems* 2009, 3(5):267–276.

108. Benkstein KD, Raman B, Montgomery CB, Martinez CJ, Semancik S: Microsensors in dynamic backgrounds: Toward real-time breath monitoring. *IEEE Sensors Journal* 2010, 10(1):137–144.

109. Semancik S, Cavicchi R: Kinetically controlled chemical sensing using micromachined structures. *Accounts of Chemical Research* 1998, 31(5):279–287.

110. Semancik S, Cavicchi RE, Wheeler MC, Tiffany JE, Poirier GE, Walton RM, Suehle JS, Panchapakesan B, DeVoe DL: Microhotplate platforms for chemical sensor research. *Sensors and Actuators B: Chemical* 2001, 77(1–2):579–591.

111. Semancik S, Cavicchi RE, Gaitan M, Suehle JS: Temperature-controlled, micromachined, arrays for chemical sensor fabrication and operation. U.S. Patent 5,345,213, 1994.

112. Hunter GW, Dweik RA: Applied breath analysis: An overview of the challenges and opportunities in developing and testing sensor technology for human health monitoring in aerospace and clinical applications. *Journal of Breath Research* 2008, 2(3):037020.

113. Kenda A, Kraft M, Wagner C, Lendl B, Wolter A: MEMS-based spectrometric sensor for the measurement of dissolved CO/sub 2. *2008 IEEE Sensors*, Leece, Italy, pp. 724–727, 2008.

114. Alfeeli B, Ali S, Jain V, Montazami R, Heflin J, Agah M: MEMS-based gas chromatography columns with nano-structured stationary phases. *2008 IEEE Sensors*, Leece, Italy, pp. 728–731, 2008.

115. Alfeeli B, Agah M: Micro preconcentrator with embedded 3D pillars for breath analysis applications. *2008 IEEE Sensors*, Leece, Italy, pp. 736–739, 2008.

116. Chu BH, Kang BS, Chang CY, Ren F, Goh A, Sciullo A, Wu WS, Lin JS, Gila BP, Pearton SJ et al.: Wireless detection system for glucose and pH sensing in exhaled breath condensate using AlGaN/GaN high electron mobility transistors. *IEEE Sensors Journal* 2010, 10(1):64–70.

117. Johnson ATC, Khamis SM, Preti G, Kwak J, Gelperin A: DNA-coated nanosensors for breath analysis. *IEEE Sensors Journal* 2010, 10(1):159–166.

8 Solid-State Gas Sensors for Clinical Diagnosis

Giovanni Neri

CONTENTS

8.1   INTRODUCTION

Patel defined "sensor" as a device or system that responds to a physical or chemical quantity to produce a measurable output of that quantity [1]. In the last decades, sensors that detect chemical quantities (chemical sensors) have shown an accelerated impulse of their applications in many fields (industrial emission control, household security, vehicle emission control and environmental monitoring, agricultural, biomedical, etc.). In particular, solid-state semiconducting devices developed for gas sensing purposes proved to be essential in many of aforementioned application fields. Therefore, in the past few years, research for solid-state gas sensors has assumed a relevant role. This is undoubtedly due to the growing market interest, thanks to the very satisfactory performance and reliability that these devices are now able to provide, so, after decades of R&D effort, we see that companies are commercializing millions of devices each year.

Solid-state gas sensors present a high potential for applications where the use of conventional analytical systems such as gas chromatography or optical techniques is prohibitively expensive. Their operation principle is inherently simple. After exposing to the vapor of an analyte, the active sensing material of the sensor interacts with the analyte, which causes the physical property changes of the sensing material. The interaction between the analyte in the surrounding gas phase and the sensor material is transduced as a measurable electrical signal that most often is a change in the conductance, capacitance, or mass of the sensing element. So, the dominating sensor technology in present devices is based on electrochemical, capacitive, and resistive solid-state gas sensors.

The measurement of volatile chemicals in human breath (breath analysis) is currently receiving attention as a technique for the detection of disease which, being noninvasive, is particularly suited to screening for presymptomatic disease in healthy populations. Like blood or urine, composition of expired air gives helpful indications about the metabolic processes occurring in the body. The growing literature on this subject suggests the potential utility of this alternative methodology. The concept of breath analysis is based on the fact that the concentration of a volatile substance in exhaled breath reflects its concentration in blood. The composition of breath provides a complex mirror image of the biochemical processes within the body, which may be correlated to the physiological status, disease progression, or therapeutic progress of a patient. It has found, for example, that breath ethanol and acetone, both of which are associated with glucose metabolism, correlated with changes in serum glucose levels [2]. A breath sensor detects these volatile organic chemicals in the patient's exhaled breath, transducing all this in a measurable property. The reading is immediately translated by the device's software into blood glucose measurements that are expected to correlate closely to glucose readings obtained by conventional blood glucose measuring devices. This could eliminate a patient's need to get a blood glucose reading, can improve diabetics' lives while helping them manage their healthy glucose levels, thereby decreasing the risk of diabetes-related complications and reducing medical costs. The advantages of this simple diagnosis system for healthcare are immediately evident. However, the molecular complexity of breath has rendered the quantitative analysis of breath constituents and metabolites present at ppb or ppm concentration levels an analytical challenge requiring modern instrumental techniques with adequate sensitivity and discriminatory power at a molecular level. In addition, it is extremely important to eliminate potential errors deriving, for examples, from ambient air contamination, interference from other gaseous substances than the marker of interest, smoking, and so on.

Despite all, the science of breath analysis is rapidly expanding, the technology is improving, and several new applications have been developed or are under commercial development. A major breakthrough over the past decade has been the increase in breath-based tests approved by the U.S. Food and Drug Administration (FDA). Amann reported some of breath tests approved by the FDA: ethanol (law enforcement), hydrogen (carbohydrate metabolism), nitric oxide (asthma), carbon monoxide (neonate jaundice), (*Helicobacter pylori* infection), and branched chain hydrocarbons (heart transplant rejection) [3]. It should be also noted that breath analysis is routinely used by anesthetists to monitor the breath of patients under anesthesia for correct dosage and levels of other gases [4,5].

Despite the breath test simplicity, there are several technical obstacles in practice to overcome. First, the breath test must be sufficiently sensitive to detect markers in the sub-ppm concentrations. Second, the breath test must be sufficiently specific to distinguish the marker from hundred of gaseous species. Third, environmental contamination should be minimized or compensating appropriately for the target gases present in ambient air. At last, to ensure reliability and long-term stability, periodic recalibrations are required. A hard work is in progress to overcome these limitations and develop the next generation of breath analyzers based on solid-state gas sensors.

This chapter focuses on the advantages that solid-state gas sensors offer to this field, recalls the basic working principles of sensing, and discusses some examples. Despite the simplicity of these devices, there is also the feeling that many fundamental issues are still poorly understood and that we need breakthroughs in both technology and basic understanding to be able to bring about new advancements in the field. The description of miniaturized, handheld, low-cost, solid-state sensors for clinical diagnosis of important pathologies (diabetes, renal diseases, asthma, etc.) through analysis of the exhaled breath is reported. Their application in the healthcare will be thoroughly reviewed.

8.2   WORKING PRINCIPLES OF SOLID-STATE SENSORS

To be able to use correctly solid-state gas sensors for medical applications, it is necessary first to be familiar with the underlying sensing principles. It is then possible to apply these in a form that addresses the problems at hand.

A lot of recent work focuses on the development of solid-state gas sensors [6–9]. All the overreported applications of chemical sensors involve the use of different dedicated technologies regarding the sensing-transduction mechanism (resistive, electrochemical, capacitive, optical, acoustic, etc.). Here, we refers only to gas sensors involving as transduction mechanism relying on electrical properties change, that is, resistive, electrochemical, and capacitive.

8.2.1　Resistive Gas Sensors

Resistive chemical sensors are attracting significant attention due to their simplicity, low cost, small size, and ability to be integrated in electronic devices [10,11]. The fabrication simplicity of resistive sensors is the main factor contributing to their large use. They are generally constituted of a porous metal oxide or conductive polymer thick/thin film deposited onto a ceramic/polymeric substrate with interdigitated electrodes (Figure 8.1).

Many parameters affect the sensing properties of the sensing layer (thickness, porosity, surface area, etc.); therefore, the deposition process and successive posttreatments should guarantee a reliable deposition of the layer on the substrate. Moreover, to ensure the long-term stability of the sensor, it is also important that the sensing layer does not undergo any significant structural/morphological modification during working.

To be effective for practical use, resistive gas sensors should satisfy the following main requirements: (1) detection of low levels of the gas target in the surrounding environment (sensitivity); (2) discrimination of the gas target from the others simultaneously present (selectivity); and (3) reproducibility of response in the short, medium, and long period (stability). Moreover, in order to meet the commercial requirement, they have to be cheap, low power consuming, easy to use, durable, resistant, and miniaturizable.

Resistive gas sensors based on metal oxide semiconductors (MOS) were put in practical use for the first time by Taguchi at the beginning of the 1970s [12]. The main application of these devices was as alarms to prevent accidents in domestic houses by monitoring the presence of hazardous levels of explosive gases. Due to the extensive range of gas detected by MOS resistive sensors, during the successive four decades till today, many other applications in various different sectors arises [13–18]. For example, in environmental applications, the sophisticated technology of

FIGURE 8.1　Prototype of planar resistive sensor to monitor breath acetone in diabetics. (Swiss Federal Institute of Technology (ZTH), Zurich, Switzerland).

MOS sensors is particularly suited to comply with the newly reinforced environmental regulations, and give a valuable alternative to conventional analytical techniques, accurate but more expensive and time-consuming.

The future of MOS sensors based on nanosized sensing elements looks set to expand into areas that could have been little contemplated so far. This is not merely because of the economic aspects of the technology but also because of other technical characteristics. As MOS sensors are able to detect very low concentrations of gases (in the ppt-ppm range), this make them potential detectors in biomedical applications (e.g., breath test) where the concentrations of gaseous species to be detected, coming from biochemical processes occurring in the human body, are very small.

The most common metal oxides utilized as sensing layer for MOS devices are binary oxides like SnO_2, ZnO, TiO_2, etc., however, also ternary and more complex oxides are applied in practical MOS sensors [19]. The electronic structure, composition, valence state, and the acid–base and redox properties of the metal oxide layer are the main characteristics determining the electrical properties of the sensing layer. Doping the metal oxide layer with metal particles is a common way to enhance their sensing characteristics. A comprehensive review dealing with all aspects of metal doping can be found in the Ref. [20].

The sensing mechanism of MOS sensors relies on reactions which occur between adsorbed oxygen species and the probed gas on the surface of the sensing layer. Details on this subject can be found in many books and reviews [21–24]. In first approximation, oxygen adsorbed on the surface of n-type metal oxide semiconductors plays a key role, trapping free electrons because of its high electron affinity, and forming a potential barrier at the grain boundaries. This potential barrier restricts the flow of electrons, causing the electric resistance to increase. When the sensor is exposed to an atmosphere containing reducing gases, for example, CO, the gas molecules adsorbs on the surface and reacts with active oxygen species, for example, O^-, which liberates free electrons in the bulk, as follows:

$$CO + O^- \rightarrow CO_2 + e^-$$

This lowers the potential barrier allowing electrons to flow more easily, thereby reducing the electrical resistance. With oxidizing gases such as NO_2 and ozone, the adsorption process increases instead the surface resistance. The converse is true for p-type oxides, where electron exchange due to the gas interaction leads either to a reduction (reducing gas) or an increase (oxidizing gas) in electron holes in the valence band.

Conducting polymers, such as polypyrrole, polyaniline, polythiophene, and their derivatives, have been used as the active layers of gas sensors since the early 1980s [25,26]. In comparison with sensors based on metal oxides and operated at high temperatures, the sensors made of conducting polymers have many improved characteristics. They operate at low temperature with high sensitivities and short response time.

Conducting polymers are easy to be synthesized through chemical or electrochemical processes, and have good mechanical properties, which allow a facile fabrication of the active layer in gas sensors. However, intrinsic conducting polymers display rather low ($<10^{-5}$ S cm^{-1}) conductivity. In order to achieve highly conductive polymers, a doping process is necessary. The doping levels of conducting polymers can be easily changed by chemical reactions with many gases at room temperature, and this provides a simple technique to detect the target analyte. Indeed, interactions between gas molecules such as NH_3, NO_2, H_2S and other redox-active gases, and conducting polymer films, can occur through electrons transfer changing the resistance of the polymeric sensing material.

On the basis of the sensing mechanism illustrated earlier, it follows also that both MOS and polymeric resistive sensors are generally not selective. They are able to sense a large number of gases with the same chemical structure and/or properties. Then the selectivity of these devices remains one of the main problems to solve for many applications.

8.2.2 Electrochemical Gas Sensors

Electrochemical gas sensors are able to detect the majority of common gaseous species and vapor, including CO, H_2S, ethanol, acetone, ammonia, SO_2 etc. in a wide variety of applications [27,28]. These sensors are compact, require very little power, exhibit excellent linearity and repeatability, and generally have a long life span, typically 1–3 years. Response times, denoted as τ_{90}, that is, time to reach 90% of the final response, are typically 30–60s and minimum detection limits range from 0.02 to 50 ppm depending upon target gas. Commercial electrochemical sensors display some common features, that is, they are composed of three active gas diffusion electrodes immersed in a common electrolyte solution, for efficient conduction of ions between the working and counter electrodes (Figure 8.2). Gas enters the cell through an external diffusion barrier that is porous to gas but impermeable to liquid.

Depending on the specific cell the target gas is either oxidized or reduced at the surface of the working electrode. For example, O_2 electrochemical sensors intended for the measurement of O_2 partial pressure, with a gold working electrode and a lead counter-electrode, work according to the following reaction scheme:

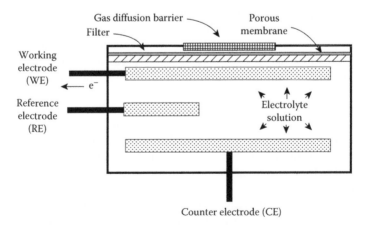

FIGURE 8.2 Electrochemical gas sensor architecture.

- Working electrode:

$$O_2 + 2H_2O + 4e^- \rightarrow 4OH^-$$

- Counter electrode:

$$2Pb \rightarrow 2Pb^{2+} + 4e^-$$

This alters the potential of the working electrode relative to the reference electrode. The associated electronic driver circuit connected to the sensor minimizes this potential difference by passing current between the working and counter electrodes, the measured current being proportional to the O_2 partial pressure.

The aspect of selectivity to the target gas is also of outmost importance. Specificity to the target gas is achieved either by optimization of the electrochemistry, that is, choice of catalyst and electrolyte, or by incorporating filters within the cell which physically absorb or chemically react with certain interferent gas molecules in order to increase target gas specificity. To further reduce cross-interference of other analytes on the sensor signal, the use of highly selective biological components (e.g., enzymes or antigen/antibody pairs) immobilized on the membrane, for various analytes ranging from ethanol and formaldehyde to trimethylamine, methylmercaptan, and acetaldehyde, has been reported [29].

The presence of acid electrolytes within electrochemical sensors results in strong sensitivity to environmental conditions of both temperature and humidity. In order to avoid this, the latest electrochemical technology utilizes solid electrolytes, which offer a number of benefits over acid electrolytes [30]. Solid electrolytes (e.g., Nafion or doped zirconia) are easier to handle with no housing or reservoir required. This means that the sensor provides cost advantages for instrumentation. The reading is fast with immediate adjustment to humidity and quick adjustment to temperature. Production is fully automated ensuring quality and stability. Lifetime is in excess of 5 years. They usually operate at elevated temperatures because most solid electrolytes have low conductivity at room temperature. A higher temperature of operation is an advantage, because electrode reactions proceed much faster.

Depending on the mode of operation, electrochemical sensors are divided into potentiometric and amperometric. In the potentiometric mode, the measured signal is an electromotive force, while in the amperometric mode an electric current is recorded.

In amperometric gas sensing devices, the reaction of a gaseous analyte at an electrode generates a current which is then measured commonly at a fixed applied potential. A linear relationship between current and concentration is observed. A detailed review on amperometric gas sensor devices, encompassing solid polymer electrolyte sensors for the simultaneous detection of NO and O_2, is reported in Ref. [31].

8.2.3 CAPACITIVE GAS SENSORS

Capacitive sensors enable low power consumption, high sensitivity, high selectivity, quick gas reaction rate, simplification of its fabrication process due to its simple structure, and miniaturization, and particularly enable long-term stability against an

FIGURE 8.3 Examples of capacitive sensor design. Capacitive humidity sensor shown on the right is a product of IST, Las Vegas, NV.

external environment and high integration density. In addition, amplification of electrical capacity can be easily realized by an oscillator circuit and price can be lowered due to a simple signal processing circuit.

A capacitive sensor works like a plate capacitor. The lower electrode is deposited on a carrier substrate, for example, a ceramic material. A thin polymer layer acts as the dielectric, and on top of this is the upper plate, which acts as the second electrode but which also allows gas vapor to pass through it, into the polymer. The vapor molecules enter or leave the polymer until the vapor content is in equilibrium with the ambient air or gas. The dielectric strength of the polymer is proportional to the vapor content. In turn the dielectric strength affects the capacitance, which is measured. Examples of capacitive sensor are shown in Figure 8.3.

Humidity sensors are the most known capacitive gas sensors and are used in many fields [32,33]. Besides automotive applications for air quality control inside the car, other typical application areas are building instrumentation, meteorological instruments, or air regulation for sanitary rooms. Capacitive humidity sensors display excellent linearity, low hysteresis, fast response times, and high chemical resistance. They are capable of measuring 0%–100% relative humidity and operating at temperature ranges of −80°C to +200°C.

Many studies on polymer or ceramic mixtures of $BaTiO_3$ and metal oxides such as CaO, La_2O_3, Nd_2O_3, and ceramic mixtures of different metal oxides such as CaO–In_2O_3 are under way for obtain CO_2 capacitive sensors [34,35]. The capacity of such complex oxide materials is changed by adsorption of gases and oxidation/reduction reactions occurring on their surface, so allowing detection.

8.3 CLINICAL DIAGNOSIS THROUGH BREATH ANALYSIS

In modern medicine the role of prevention is acquiring an increasing interest, due to the recognition that investing on this issue will be the only mean to afford economical sustainability in the years to come. An effective prevention relies on the periodic monitoring of some substances (markers) predictive of specific pathologies [36]. Physical, biochemical, and molecular biological methods for medical monitoring and diagnostics have therefore rapidly developed in recent decades.

The major developments in medical monitoring technologies and diagnostic methods have focused on blood and urine analysis for clinical diagnostics. Although enzymatic assay kits have simplified the analysis of these markers in the blood, most methods rely on laboratory instrumentation and trained personnel, requiring a constant presence in people's everyday life. At the present time, more rapid blood tests are currently marketed for use in clinical laboratories and physicians' offices and patient homes. These analytical tools suffer from two common shortcomings: (1) the test strips/test cards are expensive and (2) blood sampling, regardless of the volume, is invasive. A minimally invasive, reliable monitoring of markers disease is therefore strongly demanding.

Breath analysis is foreseen to become a noninvasive tool to diagnose and monitor metabolic modifications induced by several diseases, thanks to the fast exchange of endogenous volatile compounds between pulmonary blood and air in the alveoli of the lung [37]. However, diagnostics based on breath analysis are much less developed and not yet widely used in clinical practice [38]. Although a few types of breath tests have been successfully used as diagnostic tools in clinical analyses, such as the [13/14C] urea breath test (UBT) in the diagnosis of *H. pylori* infection [39] and the NO breath test in the diagnosis of airway inflammation [40], breath analysis could have many more potential applications in the clinical diagnosis of disease and monitoring of exposure to environmental pollutants.

Examples of such compounds strictly correlated to various diseases include nitric oxide (NO) for asthma, acetone for diabetes, ammonia for renal disease, etc. [41,42]. A nonexhaustive list of these compounds, their metabolic origin, and the correlated disease is reported in Table 8.1.

Breath testing devices equipped with solid-state sensors can offer a user-friendly, noninvasive diagnostic tool for these pathologies. This kind of analyzers will have the possibility to bring breath analysis at clinics and point-of-care, which is unlikely to happen with more conventional analytical techniques like GC-MS which are generally expensive and time consuming and therefore not suitable for routine analysis.

TABLE 8.1
List of Some Gaseous Compounds Found in the Breath, Their Metabolic Origin and the Correlated Disease

Molecule	Origin	Disease
Acetone	Decarboxilation of acetoacetate	Diabetes, ketosis
Ethyl-mercaptane	Transamination metabolism	Cirrhosis
Ethane, *n*-pentane	Lipid peroxidation	Oxidative stress
Isoprene	Cholesterol biosynthesis	Oxidative stress
Carbonyl sulfide	Transamination metabolism	Impaired liver function, lung cancer
NO	Airways inflammation	Asthma
Carbon monoxide	CO intoxication, smoking	CO intoxication
Ethanol	Alcohol ingestion	Drunkenness
Ammonia	Protein metabolism	Uremia, kidney failure

8.3.1 Breath Sampling

In order to perform the breath test, the expired breath coming from the patient has to be first collected. The breath composition is changing during the respiration and two phases can be identified. During the first one, including the air contained in the upper airways which experiences no gas exchange with blood (dead space), endogenous volatile organic compounds are absent. Most of the chemical information on blood composition is instead contained in the alveolar air, which contains also the maximum concentration of CO_2. The monitoring of the CO_2 concentration in breathed out air allows distinguishing between the two phases (see Figure 8.4).

Alveolar breath sampling thus allows collecting samples with a higher analyte concentration, improves the data reproducibility, and reduces possible interferences from compounds present in the ambient air. Concentrations of endogenous volatile substances in alveolar air are two to three times higher than those found in mixed expiratory samples, because there is no dilution by dead-space gas.

Sampling is then an essential part of the breath analysis. A breath analyzer equipped with metal oxide semiconductor sensors has been specifically designed for the automatic selection of the alveolar air in breath discarding the dead-space air [43]. In addition, the breath sampler helps to standardize the breath collection process by monitoring and maintaining the pressure of exhaled breath within a certain acceptable range. A decision unity discriminates the alveolar air from dead-space air in tidal breathing on the basis of CO_2 concentration and flow of the exhaled air (Figure 8.5).

Alveolar air collected can be directly delivered to the sensor chamber or stored in a bag before measurement. In the first case, to realize breath-to-breath sampling, additional effort is necessary to provide breath samples of well defined and reproducible composition. The effect of breath flow rate, vapor temperature, and flow duration on the linearity and variation in the response of a breath solid-state sensor should be carefully examined. Variations in sensor response due to physical breathing parameters can be caused by effects such as hydrodynamics, heating, or reaction timing. Such effects are especially important for systems directly measuring breath, compared to controlled sampling from a breath sample container. In this latter case, various collection devices such as Tedlar or Teflon bags were used. Other adsorption techniques making use, for example, of membranes or solid phase microextraction (SPME) have not been widely used.

However, so far, the collection of breath samples has not received comparable attention as other clinical samples such as blood, urine, tissue, or other body fluids.

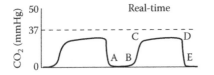

FIGURE 8.4 Graphical representation of CO_2 concentration against time during an expiration.

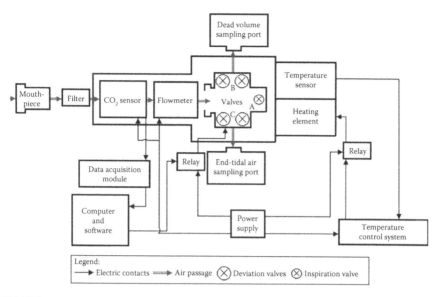

FIGURE 8.5 Schematic representation of a sampling apparatus of alveolar air. (From Di Francesco, F. et al., *J. Breath Res.*, 2, 037009, 2008.)

The main obstacle has been the lack of established sampling and measurement protocols. In order to overcome the existence of potential variability in sampling and ensure a reliable breath analysis, an internal calibration is accomplished by simultaneous measurement of CO_2. The use of breath CO_2 as an internal standard is an effective method to solve the difficulties associated with variations in the target analyte concentrations in a sample, attributed to mass losses and different breathing patterns of different subjects [44].

8.3.2 BREATH COMPOSITION

The bulk matrix of breath is a mixture of nitrogen, oxygen, CO_2, H_2O, and inert gases. Table 8.2 shows the differences between the inhaled air, having the same composition as normal air, and exhaled air. In the latter, there is less oxygen and more carbon dioxide; it is also saturated with water vapor.

TABLE 8.2
Composition of Inhaled and Exhaled Air

Nitrogen, N_2	78	78
Oxygen, O_2	21	17
Carbon dioxide, CO_2	0.04	4
Inert gases, Ar, …	1	1
Water vapor, H_2O	Little	Saturation

The remaining small fraction consists of more than 1000 trace volatile organic compounds (VOCs) and other gases (NH$_3$, H$_2$S, H$_2$, etc.) with concentrations in the range of parts per million (ppm) to parts per trillion (ppt) by volume [45–47].

In terms of their origin, volatile substances may be generated in the body (endogenous) or may be absorbed as contaminants from the environment (exogenous). The composition of VOCs in breath varies widely from person to person, both qualitatively and quantitatively. The bulk matrix and trace VOCs in breath exchange between the blood and alveolar air at the blood–gas interface in the lung. One exception is NO, which is released into the airway in the case of airway inflammation.

8.3.3 BIOMARKERS

The main purpose of a breath analysis sensor is to provide longtime continuous monitoring of physiological gaseous biomarkers to see if they exceed normal range in order to monitor patient status and send alarm for abnormal status. Phillips et al. [45] reports a selected list of the nearly 3500 individual volatile organic compounds (VOCs) detected in exhaled breath. Some of these are reported in the following.

8.3.3.1 Ethanol

Ethanol is found in the breath coming from both physiological and external sources. It was found that endogenous breath ethanol can be a useful biomarker in patients with nonalcoholic fatty liver disease. In particular, breath ethanol can be associated with hepatic steatosis [48]. Exogenous breath ethanol is instead the target marker for alcohol abuse. Indeed, after alcohol beverage intake, an almost instantaneous equilibrium is established between ethanol in the blood and in alveolar air, making feasible the evaluation of the legal limit of blood alcohol through the exhaled air.

8.3.3.2 Acetone

Acetone is a chemical product of fat catabolism in the human body. Breath acetone concentration has been found to be a useful indicator of fat metabolism and dietary compliance, and has been correlated to pound weight loss due to fat [49]. Acetone is related to type 1 diabetes, and its concentration increases from 300 to 900 ppb for healthy humans to more than 1800 ppb for diabetic patients. The acetone concentration found in the breath of diabetes mellitus patients was significantly higher than those of all other persons (in the range of 1.25–2.4 ppm). It can be concluded that there is a clear threshold of ~1.25 ppm of acetone in breath. Above this threshold it can be assumed with high probability that the patient is diseased with diabetes mellitus [50].

8.3.3.3 Carbon Monoxide

Carbon monoxide (CO) is promptly detectable in the exhaled breath. However, exhaled CO (eCO) levels often represent the sum of the endogenous production of the gas and airway contamination from environmental exposure. CO concentrations in the exhaled breath of nonsmokers and smoker peoples are in the ranges of 0.4–0.8

and 2–20 ppm, respectively [51]. Carbon monoxide monitoring in the breath is not only used in smoking cessation clinics for smoking intervention testing but may also be used in the emergency room as a test for CO poisoning. Exhaled CO is suggested also as a potential biomarker in pulmonary and systemic inflammatory diseases, although further studies are necessary to identify its role in human diagnostics.

8.3.3.4 Ammonia and Amine

Ammonia gas is a toxic air pollutant. According to regulations of the Occupational Safety and Health Administration (OSHA), the permissible exposure limit is set at 35 ppm (TWA) over a 15 min period [52]. Thus analyzers, which are able to detect breath ammonia, could be useful as efficient devices to monitor personal exposure of workers undertaking any risk of contact with ammonia.

Moreover, breath ammonia and lower aliphatic amines are important biomarkers in renal disorders [53,54]. Therefore, great interest is devoted to ammonia breath analysis not only as screening and diagnostic tools for renal diseases but also as a noninvasive method to check the efficiency of hemodialysis therapy by real-time continuous monitoring of the ammonia concentration in exhaled breath of patients under treatment [54]. Ammonia is also a by-product of the *H. pylori* bacterium which causes an infection of the stomach. Next to that, there is also an association with stomach cancer and liver diseases.

8.3.3.5 Nitric Oxide

Exhaled nitric oxide (eNO) is one of the most important biological mediators as it takes part in many physiological processes in a human organism. It plays a role in airway inflammatory diseases, respiratory tract infections, neurotransmission, vascular regulation, and host defense [55]. Clinical measurements of exhaled nitric oxide have been shown to be useful for noninvasively diagnosing inflammatory-related diseases [56]. Endogenous nitric oxide is produced from the amino acid L-arginine by the enzyme NO synthase.

Nitric oxide serves a variety of pulmonary functions such as vascular and airway smooth muscle tone and is a mediator of the inflammatory response in the airway. This fact enables to perform quick diagnostics and monitor development of many serious diseases such as bronchial asthma [57]. Persons suffering from asthma are able to monitor the intensity of their condition and to predict the likelihood of an asthmatic attack by monitoring the level of nitric oxide in their exhaled breath. The great advantage is that it is noninvasive and can be performed repeatedly in adult and pediatric patients regardless of airflow obstruction. Future research needs to look at the development of methods for eNO sampling in the infant and very young child. Moreover, eNO may offer us a mechanism for close monitoring of steroid usage in order to prevent overdose and related growth concerns as well as monitor compliance.

8.3.3.6 Carbonyl Sulfide

Carbonyl sulfide (COS) has been identified as a biomarker of liver-related diseases [58]. Elevated concentrations of COS have been observed in patients who have undergone lung transplantation and have had acute allograft rejection [59]. The production of exhaled carbonyl sulfide is attributed to oxidative metabolic processes of carbon disulfide and the incomplete metabolism of methionine. Exposure to COS can cause adverse effects on the nervous system (neurotoxicity), and can induce confusion, fatigue, irritability, and other behavioral changes.

8.3.3.7 Aliphatic Hydrocarbons

The straight chain aliphatic hydrocarbons ethane and pentane have been advocated as noninvasive markers of free radical–induced lipid peroxidation [60]. Their determination in exhaled breath enables accurate assessment of oxidative stress both in vitro and in vivo. Detection of a series of aliphatic hydrocarbons biomarkers in exhaled breath has realistic potential for early disease diagnosis of cancer [61].

8.3.3.8 Inorganic Gases (H_2S, H_2, H_2O_2)

The measurements of the level of H_2S in the nose-exhaled breath and the closed mouth indicate that it is largely produced in the oral cavity [62]. Breath concentration of H_2S is relatively high at typically 20–70 ppb. Many studies pave the way for the accurate analysis of this and related sulfur compounds in halitosis and potentially for probing the diseased state, especially liver disease, by breath analysis [63].

Hydrogen breath test represent a valid and noninvasive diagnostic tool in many gastroenterological conditions [64]. The test is based on the concept that part of the gas produced by colonic bacterial fermentation diffuses into the blood and is found in the breath, where it can be quantified easily. Indeed, in fasting subjects, H_2 production is normally low, but, after ingestion of fermentable substrates, primarily carbohydrates, anaerobic bacteria release appreciable amounts of H_2.

Hydrogen peroxide (H_2O_2) in exhaled air condensate is a potential marker of airway inflammation. An increased content of H_2O_2 has been described in exhaled air of patients with various inflammatory lung disorders [65]. Also uremic patients, regularly on hemodialysis, were found to exhale many times more H_2O_2 than healthy persons [66].

8.4 GAS SENSORS FOR CLINICAL DIAGNOSIS

Due to the particular application considered, gas sensors for breath monitoring need to have special characteristics to meet medical requirements. These are specified in the following points:

1. High sensitivity, able to detect very low gas concentrations
2. High signal-to-noise rate to pick up useful information in the interference and noise background
3. Good accuracy to ensure accuracy and reliability
4. Fast response speed
5. Good stability to keep low shifting for longtime detection and stable output

The sensing features of a semiconducting metal oxide sensor having the characteristics for a possible use as CO breath analyzer are identified in Figure 8.6. The sensor shows a good sensitivity to low carbon monoxide concentrations, allowing the

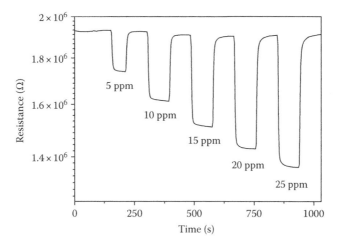

FIGURE 8.6 Transient response of an MOS sensor exposed to low concentrations of CO in air.

detection of CO at sub-ppm levels with response/recovery times of the order of few seconds and maintaining a stable baseline.

However, up to now, low time resolution is the main obstacle to overcome, and prevents a broader use in breath analysis of many sensors [67]. Buffering of sampling by means of the heated loop incorporated into the automatic device may offer a solution. Instead of analyzing substance concentrations in a single breath, averaging of a small number of breaths is applied. In this way, errors due to breath-to-breath variation of substances concentration are reduced and effects of inspired concentrations are minimized.

Breath gas sensors can be classified into different types according to working principle, detection types, and sensing material. Here, a classification based on biomarkers to be detect has been followed. This allows, for each biomarker, to investigate and compare the characteristics and performance of different sensor technologies. However, it is out of the scope of this review to cover the full range of breath sensors, therefore, only some representative examples will be given.

8.4.1 OXYGEN, CARBON DIOXIDE, AND HUMIDITY SENSORS

In the exhaled air, nitrogen, oxygen, water vapor, and CO_2 are the major components. The monitoring of these components is essential to give important information that completes those coming from specific biomarkers. Consequently, a wide number of solid-state sensors have been studied and developed.

Humidity sensors play an important role both in applications where the humidity is an interfering agent and for direct use in monitoring systems, for example, for the assessment of sleep apnea syndrome [68,69].

The detection of oxygen in breath is of central importance to investigations of metabolism and respiration in clinical monitoring applications [70]. For this scope, Drager launched into the market a fast O_2 (response <0.5 s) sensor as a means of measuring and monitoring with breath-by-breath resolution the inspiratory and expiratory oxygen concentrations, and a lifetime of more than 6 months.

Carbon dioxide sensors have numerous biomedical applications. CO_2 is monitored during assisted ventilation, during intubation to ensure correct positioning, for assessment of end tidal CO_2, for respiratory patient monitoring, evaluation of lung function, respiratory therapy control, and in diagnosing and monitoring airway status and pulmonary function [71]. In the measurement of the variation profile of CO_2 concentration in the breath, sometimes referred to as capnography, prevailing technology relies on bulky and expensive nondispersive infrared absorption (NDIR) sensors to determine CO_2 concentration. The high cost, complexity, weight, and other limitations of this technology restrict the use of capnography to high value, controlled environments, such as surgical wards. This limits the medical use of capnography. Thus, lower cost, simplified and integratable devices for the monitoring of CO_2 will greatly improve patients care.

8.4.2 ACETONE SENSORS

Breath acetone concentration is higher in diabetics than in healthy peoples and has been found to correlate with plasma-ketones and β-hydroxybutyrate concentration in venous blood [72]. The increase of ketone bodies in the blood results in ketoacidosis, a severe clinical condition in diabetics. Therefore, breath acetone is a suitable marker to monitor frequently and in a noninvasive way diabetics at risk for ketoacidosis [73,74]. An MOS sensor based on indium oxide for breath acetone detection has been found promising for the control of therapeutic ketogenic diets [75]. Pratsinis' team built an extremely sensitive acetone detector based on a thin film of semiconducting, mixed ceramic nanoparticles as sensing element between a set of gold electrodes. It is sensitive enough to detect acetone at 20 parts per billion, a concentration that is 90 times lower than the level at which it can be found in the breath of diabetic patients [76].

An enzymatic electrochemical device was developed that is capable of measuring breath acetone concentrations from baseline levels across the range expected for healthy individuals on a moderate diet or exercise regimen [77]. The utility of the device shown in Figure 8.7, developed by Kemeta (Phoenix, AZ), for the measure of the rate of fat metabolism during a moderate diet regimen, was demonstrated.

Further studies are however necessary to evaluate the response of these sensor categories to acetone in the presence of interfering gases (CO, ethanol, CO_2, humidity, etc.) present in the breath, with aim to eliminate false positive results.

8.4.3 CARBON MONOXIDE SENSORS

CO is easily detectable in breath with portable electrochemical and semiconducting sensors in the parts per million ranges. Due to its high toxicity, CO sensors are mainly developed for monitoring human exposure to this gas [78]. At lower levels of exposure, CO causes mild effects including headaches, dizziness, disorientation, nausea, and fatigue. At higher concentrations, impaired vision and coordination can occur. Acute effects are due to the formation of carboxyhemoglobin in the blood, which inhibits oxygen intake. At higher concentrations, CO exposure can be fatal.

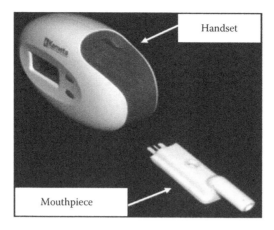

FIGURE 8.7   Enzymatic electrochemical device for measuring breath acetone concentration.

When a patient is admitted to the emergency room in a dazed, incoherent, or unconscious state, a carbon monoxide monitor could be used to perform an immediate
noninvasive CO test. The carbon monoxide breath test will give the same results as
a carboxyhemoglobin blood test.

8.4.4   ETHANOL SENSORS

Breath-ethanol instruments are widely used for testing alcohol-impaired drivers [79].
These breath analyzers are designed to detect ethanol in the breath of drivers in order
to reduce the number of road accidents caused by excessive alcohol consumption.
Infrared technology, electrochemical, fuel cell, or semiconducting instruments were
developed in this context. Among these, semiconductor oxide sensors are claimed
to offer many benefits, including low cost, low power consumption, and small size,
although they need frequent calibrations.

These detection devices (see Figure 8.8) are generally accurate and reliable
enough to establish "probable cause" to reasonably suspect alcohol consumption,
but at present, due to the inherent inability to distinguish interferents, not reliable
enough for evidential testing.

8.4.5   AMMONIA SENSORS

Ammonia is a metabolite present in the exhaled breath, and when found in high
levels it is associated with some kidney diseases due to breakdown of amino acids
or *H. pylori* bacteria stomach infection [80]. There is therefore a considerable interest in breath analyzers that can be applied to measure ammonia levels in exhaled
air for clinical diagnosis. Today, solid-state detectors for ammonia detection in the
breath are not available commercially. An extensive efforts is in progress with aim to
develop miniaturized ammonia sensors with the required features for clinical applications [81–83].

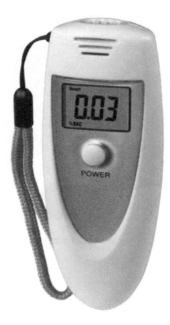

FIGURE 8.8 Commercial digital semiconductor breath alcohol sensor.

8.4.6 NITRIC OXIDE SENSORS

After the first identification of endogenous nitric oxide present in the exhaled air [84], over the past several years, numerous studies have been done describing increased exhaled nitric oxide in asthmatics as compared to healthy subjects [85,86]. It is well known that glucocorticoids reduce eNO levels in asthmatic patients. Therefore, measuring exhaled nitric oxide may be useful in assessing airway inflammation and monitoring the effectiveness and compliance of inhaled steroid medication in the asthmatic patient.

There are several analyzers available commercially based on chemiluminescence that have the capability to measure eNO directly in line to the analyzer or indirectly by obtaining an exhaled air sample to be later analyzed at a more convenient time. However, they are very inexpensive and suitable only for laboratory use.

City Technology has developed a new nitric oxide sensor that provides unparalleled sensitivity, repeatability, stability, and speed of response [87]. It will also be deployed in instruments measuring NO in the breath of patients suffering from chronic obstructive pulmonary disease, where the sensor's performance with differing breath conditions obviates the need for large and expensive sample conditioning.

The sophisticated design of the sensor, which does not require the sample of exhaled breath to be adjusted to a particular temperature and pressure to achieve a reliable reading, will enable lower cost instruments to be developed, allowing their use to be extended into an increased number of more mainstream establishments.

The device itself requires only 10 s exposure to achieve a reliable reading of the NO content of the exhaled breath. It resolves down to 10 ppb, at least 20 times better than any other sensor available on the market today, with excellent stability in the presence of transient humidity changes, further reducing the complexity required of the instrument.

8.4.7 COS SENSORS

A significant increase in exhaled carbonyl sulfide levels (higher than 0.5 ppm) was observed in subjects suffering of acute rejection after a lung transplant [59]. Measurement of expired COS in breath is inherently noninvasive and in some cases could obviate the need for more invasive surgical procedures such as lung biopsies. Due to the very low concentration of this marker to be detected in the breath, the development of COS sensor is a challenge. Neri et al. reported an In_2O_3-based semiconductor sensor showing a high response toward COS. The surface reaction of COS with adsorbed oxygen species to form adsorbed sulfate, which desorbed as sulfur dioxide, was responsible of the electrical resistance variations observed [88].

8.4.8 HYDROGEN SENSORS

Sensors for hydrogen breath test are used as a clinical medical diagnosis for people with common food intolerances. The test is simple and noninvasive and is performed after a short period of fasting (typically 8–12 h). Ingestion of simple and complex carbohydrates, such as potato starch, results in an elevation of breath H_2 levels. An electrochemical device for the measurement of hydrogen in end-expired air is described by Bartlett et al. [89].

The use of hydrogen breath analyzer allows a simplified method of breath collection and rapid estimation of hydrogen concentration, providing a reliable, convenient, and well-tolerated means of detecting sugar absorption compared with the conventional gas chromatographic method.

8.5 NEW TRENDS AND APPLICATIONS

In recent years, according to the requirement of home-care clinical application, medical technology has changed dramatically from traditional larger-size and poor-performance apparatus to new handheld devices. In this regard, the introduction of solid-state gas sensors as detectors in the breath analyzers can open new opportunities to design of novel diagnostic tools based on the chemical information contained in the exhaled breath, in a more simple and convenient way.

However, despite the enormous potentials compared to laboratory-based analytical techniques, numerous problems still remained to be solved. Main limitations could be attributed to poor reproducibility between sensors and selectivity in complex matrices. In fact, with the exception of few cases they suffer of poor selectivity. These devices do not operate any separation into components of the measured samples resulting not sensitive to a limited, and prefixed, number and kind of molecules, but rather, able to sense a large, and sometimes not known, amount of different

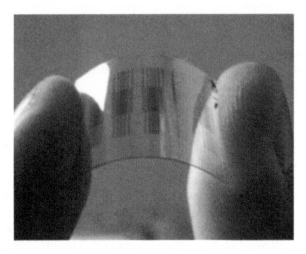

FIGURE 8.9   Sensor circuit printed on flexible substrate. (From Stanford University, Palo Alto, CA.)

volatile compounds. Arrays of nonselective sensors, resembling the functioning of the human olfaction, can be used to overcome this limitation [90,91].

New directions about the developing of gas sensors with novel features lead to the use of flexible devices having very lightweight, robustness, and low cost (see Figure 8.9). Flexibility is fundamental in a number of applications, where the sensor system has to be either conformed to curved surfaces or is subjected to repeated bending. For example, flexible sensors could be used in food applications where it is very important to detect the organic compounds of interest around food package or directly on food surface. In the same way, in medical applications, flexible sensor systems can be utilized either as smart plasters able to detect skin pathologies or for disposable sensors for blood analysis [92].

Resistive and capacitive wearable gas sensors are currently under development on plastic substrates [93]. Besides the performances that can be achieved, the reliability of the devices under operation or mechanical constraint and their packaging, special efforts are also dedicated to realize low-cost and ultra-low-power devices that meet the requirements of wireless applications. In the future, sensors will be wireless and will provide real-time process data monitoring [94].

Nanotechnology is expected to have a substantial impact on sensor technology in the near future, because of an extensive miniaturization of the devices. The potential impact of novel nanosensors and their applications on disease diagnosis is foreseen to change health care in a fundamental way. For example, a CO_2 sensor has been developed by Nanomix Inc. (Emeryville, CA) using polyethylene-imine-coated carbon nanotubes, as an indicator of patient status during administration of anesthetic drugs. The tiny, low-power sensor will be the first disposable electronic capnography sensor and has the potential to extend the reach of quantitative respiratory monitoring beyond the operating room and into ambulatory and emergency settings as well as doctors' offices. Conducting polymer nanojunctions (Figure 8.10) have been used to develop nanosensors for breath ammonia [95].

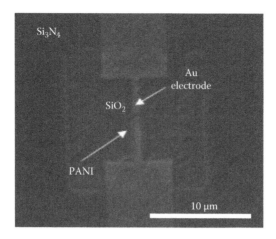

FIGURE 8.10   Micrometer structure of the NH_3 breath sensor based on conducting polymer nanojunctions. (From Aguilar, A.D. et al., *IEEE Sens. J.*, 8, 269, 2008.)

Recent years have seen integrated circuits offering novel solutions for various biomedical applications. As a result there is a new drive for these technologies to be applied in healthcare, to provide cheap, disposable, low power, and intelligent systems. This includes miniaturization of sensors, batch fabrication and incorporation of processing and signal conditioning circuitry to improve signal-to-noise ratio, and implementation of intelligent algorithms. CMOS-based integrated circuits are getting smaller and lower in cost as a result of the decreasing feature size and the economies of scale of semiconductor processes (Figure 8.11). Implementing chemical sensors in CMOS is therefore a widely pursued research area.

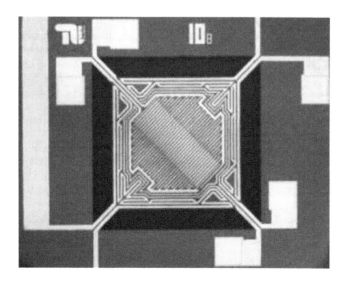

FIGURE 8.11   Micro-hotplate based on 600 nm thick SiC membrane for MOS gas sensor.

8.6　CONCLUSIONS

In recent years, the growing needs of society to improve on healthcare and quality of life have seen high-performance sensors offering novel solutions for various biomedical applications. As a result there is a new drive for these technologies to be applied in healthcare, to provide cheap, disposable, low power, and intelligent systems to provide diagnosis and treatment of medical conditions.

However, at present time, the practical application of specific sensors for use in the medical diagnosis is still a challenge. During past years, the field of breath sensor has been growing due to the increasing trend of different kinds of sensors. Nowadays, due to the appearance of new technologies for sensor integration, breath sensors development encompasses all kind of sensors with its corresponding signal conditioning and/or processing, while putting emphasis on integrated circuits and system design issues. So, although today only few commercial breath devices are available, with respect to rapid, sensitive, and low-cost monitoring of a variety of gaseous biomarkers in the breath, new microelectronics-based breath sensors show promising potential, and their market is expected to grow rapidly in the near future.

REFERENCES

1. P. D. Patel, (Bio)sensors for measurement of analytes implicated in food safety: A review, *Trends Anal. Chem.* 21, 96–115 (2002).
2. P. R. Galassetti, B. Novak, D. Nemet, C. R.-Gottron, D. M. Cooper, S. Meinardi, R. Newcomb, F. Zaldivar, D. R. Blake, Breath ethanol and acetone as indicators of serum glucose levels: An initial report, *Diabetes Technol. Ther.* 7, 115–123 (2005).
3. A. Amann, P. Spanel, D. Smith, Breath analysis: The approach towards clinical applications, *Mini Rev. Med. Chem.* 7, 115–129 (2007).
4. A. Takita, K. Masui, T. Kazama, On-line monitoring of end-tidal propofol concentration in anesthetized patients, *Anesthesiology* 106, 659–664 (2007).
5. C. Hornuss, S. Praun, J. Villinger, A. Dornauer, P. Moehnle, M. Dolch, E. Weninger A. Chouker, C. Feil, J. Briegel, M. Thiel M, G. Schelling, Real-time monitoring of propofol in expired air in humans undergoing total intravenous anesthesia, *Anesthesiology* 106, 665–674 (2007).
6. D.-D. Lee, D.-S. Lee, Environmental gas sensors, *IEEE Sens.* 1, 214–224 (2001).
7. N. Yamazoe, Toward innovations of gas sensor technology, *Sens. Actuators B* 108, 2–14 (2005).
8. D. M. Wilson, S. Hoyt, J. Janata, K. Booksh, L. Obando, Chemical sensors for portable, handheld field instruments, *IEEE Sens. J.* 1, 256–274 (2001).
9. K. Arshak, E. Moore, G. M. Lyons, J. Harris, S. Clifford, A review of gas sensors employed in electronic nose applications, *Sens. Rev.* 24, 181–198 (2004).
10. T. Seiyama, Chemical sensors—Current status and future outlook, in *Chemical Sensor Technology*, T. Seiyama (ed.), Elsevier, Amsterdam, the Netherlands, Vol. 1, pp. 1–13 (1988).
11. B. Hoffheins, Solid state, resistive gas sensors, in *Handbook of Chemical and Biological Sensors*, R. F. Taylor, J. S. Schultz (eds.), Institute of Physics, Philadelphia, PA, pp. 371–398 (1996).
12. N. Taguchi, Gas detecting device, U.S. Patent 3,631,436, December 1971.
13. O. K. Varghese, C. A. Grimes, Metal oxide nanoarchitectures for environmental sensing, *J. Nanosci. Nanotechnol.* 3, 277–293 (2003).

14. C. O. Park, S. A. Akbar, Ceramics for chemical sensing, *J. Mater. Sci.* 38, 4611–4637 (2003).
15. J. Frank, H. Meixner, Sensor system for indoor air monitoring using semiconducting metal oxides and IR-absorption, *Sens. Actuators B* 78, 298–302 (2001).
16. S. Ehrmann, J. Jüngst, J. Goschnick, Automated cooking and frying control using a gas sensor microarray, *Sens. Actuators B* 66, 43–45 (2000).
17. A. Vergara, J. Luis Ramírez, E. Llobet, Reducing power consumption via a discontinuous operation of temperature-modulated micro-hotplate gas sensors: Application to the logistics chain of fruit, *Sens. Actuators B* 129, 311–318 (2008).
18. J. Goschnick, An electronic nose for intelligent consumer products based on a gas analytical gradient microarray, *Microelectron. Eng.* 693, 57–58 (2001).
19. A. Gurlo, N. Bârsan, U. Weimar, Gas sensors based on semiconducting metal oxides, in *Metal Oxides: Chemistry and Applications*, J. L. G. Fierro (ed.), CRC Press, Boca Raton, FL, pp. 683–738 (2006).
20. G. Neri, Metal doping in semiconductor metal oxide gas sensors, in *Encyclopedia of Sensors*, C. A. Grimes, E. C. Dickey, M. V. Pishko (eds.), American Scientific Publishers, Valencia, CA, Vol. 6, pp. 1–14 (2006).
21. A. M. Azad, S. A. Akbar, S. G. Mhaisalkar, L. D. Birkefeld, K. S. Goto, Solid state gas sensors: A review, *J. Electrochem. Soc.* 139, 3690–3704 (1992).
22. N. Barsan, M. Schweizer-Berberich, W. Göpel, Fundamental and practical aspects in the design of nanoscaled SnO_2 gas sensors: A status report, *Fresenius. J. Anal. Chem.* 365, 287–304 (1999).
23. G. Sberveglieri, *Gas Sensors*, Kluwer Academic Publishing, Berlin, Germany (1992).
24. D. E. Williams, Conduction and gas response of semiconductor gas sensors, in *Solid State Gas Sensors*, P. T. Moseley, B. C. Tofield (eds.), Adam Hilger, Bristol, U.K. pp. 71–123 (1987).
25. B. Adhikari, S. Majumdar, Polymers in sensor applications, *Prog. Polym. Sci.* 29, 699–766 (2004).
26. U. Lange, N. V. Roznyatouskaya, V. M. Mirsky, Conducting polymers in chemical sensors and arrays, *Anal. Chim. Acta* 614, 1–26 (2008).
27. U. Guth, W. Vonau, J. Zosel, Recent developments in electrochemical sensor application and technology—A review, *Meas. Sci. Technol.* 20, 042002 (2009).
28. R. Knake, P. Jacquinot, A. W. E. Hodgson, P. C. Hauser, Amperometric sensing in the gas-phase, *Anal. Chim. Acta* 549, 1–9 (2005).
29. S. Achmann, M. Hämmerle, R. Moos, Amperometric enzyme-based biosensor for direct detection of formaldehyde in the gas phase: Dependence on electrolyte composition, *Electroanalysis* 20, 410–417 (2008).
30. C. O. Park, J. W. Fergus, N. Miura, J. Park, A. Choi, Solid-state electrochemical gas sensors, *Ionics* 15, 261–284 (2009).
31. J. R Stetter, J. Li, Amperometric gas sensors—A review, *Chem. Rev.* 108, 352–366 (2008).
32. W. Chen, X. Liu, C. Suo, Z. Zhang, Z. Zhao, A capacitive humidity sensor based on multi-wall carbon nanotubes (MWCNTs), *Sensors* 9, 7431–7444 (2009).
33. E. Zampetti, S. Pantalei, A. Pecora, A. Valletta, L. Maiolo, A. Minotti, A. Macagnano, G. Fortunato, A. Bearzotti, Design and optimization of an ultra thin flexible capacitive humidity sensor, *Sens. Actuators B* 143, 302–307 (2009).
34. T. Isihara, K. Kometani, Y. Mizuhara, Y. Takita, Capacitive type gas sensor for the selective detection of carbon dioxide, *Sens. Actuators B* 13–14, 470–472 (1993).
35. P. L. Kebabian, A. Freedman, Fluoropolymer-based capacitive carbon dioxide sensor, *Meas. Sci. Technol.* 17, 703–710 (2006).
36. W. Cao, Y. Duan, Breath analysis: Potential for clinical diagnosis and exposure assessment, *Clin. Chem.* 52, 800–811 (2006).

37. W. Miekisch, J. K. Schubert, G. F. E. N.-Schomburg, Diagnostic potential of breath analysis; focus on volatile organic compounds, *Clin. Chim. Acta* 347, 25–39 (2004).

38. A. Amann, A. Schmid, S. S.-Burgi, S. Telser, H. Hinterhuber, Breath analysis for medical diagnosis and therapeutic monitoring, *Spectrosc. Eur.* 17, 18–20 (2005).

39. J. P. Gisbert, J. M. Pajares, ^{13}C-urea breath test in the diagnosis of *Helicobacter pylori* infection—A critical review, *Aliment. Pharmacol. Ther.* 20, 1001–1017 (2004).

40. S. M. Stick, Non-invasive monitoring of airway inflammation, *Med. J. Aust.* 177, S59–S60 (2002).

41. B. Buszewski, M. Kęsy, T. Ligor, A. Amann, Human exhaled air analytics: Biomarkers of diseases, *Biomed. Chromatogr.* 21, 553–566 (2007).

42. N. Marczin, S. A. Kharitonov, M. H. Yacoub, P. J. Barnes (eds.), *Disease Markers in Exhaled Breath*, Marcel Dekker, New York (2002).

43. F. Di Francesco, C. Loccioni, M. Fioravanti, A. Russo, G. Pioggia, M. Ferro, I. Roehrer, S. Tabucchi, M. Onor, Implementation of Fowler's method for end-tidal air sampling, *J. Breath Res.* 2, 037009 (2008).

44. W. Ma, X. Liu, J. Pawliszyn, Analysis of human breath with micro extraction techniques and continuous monitoring of carbon dioxide concentration, *Anal. Bioanal. Chem.* 385, 1398–1408 (2006).

45. M. Phillips, J. Herrera, S. Krishnan, M. Zain, J. Greenberg, R. Cataneo, Variation in volatile organic compounds in the breath of normal humans, *J. Chromatogr. B* 729, 75–88 (1999).

46. J. K. Schubert, W. Miekisch, K. Geiger, G. F. N.-Schomburg, Breath analysis in critically ill patients: Potential and limitations, *Expert Rev. Mol. Diagn.* 4, 619–629 (2004).

47. R. Mukhopadhyay, Don't waste your breath. Researchers are developing breath tests for diagnosing diseases, but how well do they work? *Anal. Chem.* 76, 273A–276A (2004).

48. S. F. Solga, A. Alkhuraishe, K. Cope, A. Tabesh, J. M. Clark, M. Torbenson, P. Schwartz, T. Magnuson, A. M. Diehl, T. H. Risby, Breath biomarkers and non-alcoholic fatty liver disease: Preliminary observations, *Biomarkers* 11, 174–183 (2006).

49. S. K. Kundu, J. A. Bruzek, R. Nair, A. M. Judilla, Breath acetone analyzer: Diagnostic tool to monitor dietary fat loss, *Clin. Chem.* 39, 87–92 (1993).

50. A. P. Briggs, The management of diabetes as controlled by tests of acetone in expired air, *J. Lab. Clin. Med.* 25, 603–609 (1940).

51. T. Dolinay, A. M. K. Choi, S. W. Ryter, Exhaled carbon monoxide: Mechanisms and clinical applications, *Exhaled Biomarkers* 9, 82–95 (2010).

52. Health and Safety Executive, Guidance Note EH40/93, Occupational Exposure Limits 1993, HMSO, London (1993).

53. D. Smith, P. Spanel, The challenge of breath analysis for clinical diagnosis and therapeutic monitoring, *Analyst* 132, 390–396 (2007).

54. L. R. Narasimhan, W. Goodman, C. Kumar, N. Patel, Correlation of breath ammonia with blood urea nitrogen and creatinine during hemodialysis, *Proc. Natl Acad. Sci. U. S. A.* 98, 4617–4621 (2001).

55. M. Bernaraggi, G. Cremona, Measurement of exhaled nitric oxide in humans and animals, *Pulm. Pharmacol. Ther.* 12, 331–352 (1999).

56. S. A. Kharitonov, P. J. Barnes, Clinical aspects of exhaled nitric oxide, *Eur. Respir. J.* 16, 781–792 (2000).

57. P. J. Barnes, Nitric oxide and airway disease, *Ann. Med.* 27, 389–393 (1995).

58. S. S. Sehnert, L. Jiang, J. F. Burdick, T. H. Risby, Breath biomarkers for detection of human liver diseases: Preliminary study, *Biomarkers* 7, 174–187 (2002).

59. S. M. Studer, J. B. Orens, I. Rosas, J. A. Krishnan, K. A. Cope, S. Yang, J. V. Conte, P. B. Becker, T. H. Risby, Patterns and significance of exhaled-breath biomarkers in lung transplant recipients with acute allograft rejection, *J. Heart Lung Transplant.* 20, 1158–1166 (2001).

60. C. M. Kneepkens, G. Lepage, C. C. Roy, The potential of the hydrocarbon breath test as a measure of lipid peroxidation, *Free Radic. Biol. Med.* 17, 127–160 (1994).

61. A. Amann, M. Ligor, T. Ligor, A. Bajtarevic, C. Ager, M. Pienz, H. Denz, M. Fiegl, W. Hilbe, W. Weiss, P. Lukas, H. Jamnig, M. Hackl, A. Haidenberger, A. Sponring, W. Filipiak, W. Miekisch, J. Schubert, J. Troppmair, B. Buszewski, Analysis of exhaled breath for screening of lung cancer patients, *Mag. Eur. Med. Oncol.* 3, 106–112 (2010).

62. A. Pysanenko, P. Spanel, D. Smith, A study of sulfur-containing compounds in mouth- and nose-exhaled breath and in the oral cavity using selected ion flow tube mass spectrometry, *J. Breath Res.* 2, 046004 (2008).

63. L. Li, P. K. Moore, Putative biological roles of hydrogen sulfide in health and disease: A breath of not so fresh air? *Trends Pharmacol. Sci.* 29, 84–90 (2008).

64. M. Di Stefano, A. Missanelli, E. Miceli, A. Strocchi, A. G. R. Corazza, Hydrogen breath test in the diagnosis of lactose malabsorption: Accuracy of new versus conventional criteria, *J. Lab. Clin. Med.* 144, 313–318 (2004).

65. I. Horvath, L. E. Donnelly, A. Kiss, S. A. Kharitonov, S. Lim, K. F. Chung, P. J. Barnes, Combined use of exhaled hydrogen peroxide and nitric oxide in monitoring asthma, *Am. J. Respir. Crit. Care Med.* 158, 1042–1046 (1998).

66. J. Rysz, M. Kasielski, J. Apanasiewicz, M. Krol, A. Woznicki, M. Luciak, D. Nowak, Increased hydrogen peroxide in the exhaled breath of uraemic patients unaffected by haemodialysis, *Nephrol. Dial. Transplant.* 19, 158–163 (2004).

67. W. Miekisch, J. K. Schubert, From highly sophisticated analytical techniques to life-saving diagnostics: Technical developments in breath analysis, *Trac-Trends Anal. Chem.* 25, 665–673 (2006).

68. J. B. Grotberg, Respiratory fluid mechanics and transport processes, *Annu. Rev. Biomed. Eng.* 3, 421–457 (2001).

69. A. Tetelin, C. Pellet, C. Laville, G. N'Kaou, Fast response humidity sensors for a medical microsystem, *Sens. Actuators B* 91, 211–218 (2003).

70. C. S. Burke, J. P. Moore, D. Wencel, B. D. Mac-Craith, Development of a compact optical sensor for real-time, breath-by-breath detection of oxygen, *J. Breath Res.* 2, 37012 (2008).

71. D. R. Hampton, B. S. Krauss, Respiratory analysis with capnography, U.S. Patent 6,648,833, November 2003.

72. M. Phillips, J. Greenberg, Detection of endogenous acetone in normal human breath, *J. Chromatogr.* 422, 235–238 (1987).

73. L. Wang, K. Kalyanasundaram, M. Stanacevic, P. Gouma, Nanosensor device for breath acetone detection, *Sens. Lett.* 8, 709–712 (2010).

74. O. E. Owen, V. E. Trapp, C. L. Skutches, M. A. Mozzoli, R. D. Hoeldtke, G. Boden, G. A. Reichard, Acetone metabolism during diabetic ketoacidosis, *Diabetes* 31, 242–248 (1982).

75. G. Neri, A. Bonavita, G. Micali, N. Donato, Design and development of a breath acetone MOS sensor for ketogenic diets control, *IEEE Sens. J.* 10, 131–136 (2010).

76. M. Righettoni, A. Tricoli, S. E. Pratsinis, Si-WO₃ sensors for highly selective detection of acetone for easy diagnosis of diabetes by breath analysis, *Anal. Chem.* 82, 3581–3587 (2010).

77. B. E. Landini, S. T. Bravard, Breath acetone concentration measured using a palm-size enzymatic sensor system, *IEEE Sens. J.* 9, 1802–1807 (2009).

78. P. McGeehin, Self-diagnostic gas sensors which differentiate carbon monoxide from interference gases for residential applications, *Sens. Rev.* 16, 37–39 (1996).

79. K. Sakakibara, T. Taguchi, A. Nakashima, T. Wakita, S. Yabu, B. Atsumi, Development of a new breath alcohol detector without mouthpiece to prevent alcohol-impaired driving, *IEEE International Conference on Vehicular Electronics and Safety*, Columbus, OH, pp. 299–302 (2008).

80. S. Davies, P. Spanel, D. Smith, Quantitative analysis of ammonia on the breath of patients in end-stage renal failure, *Kidney Intern.* 52, 223–228 (1997).

81. G. Neri, G. Micali, A. Bonavita, S. Ipsale, G. Rizzo, M. Niederberger, N. Pinna, Tungsten oxide nanowires-based ammonia gas sensors, *Sens. Lett.* 6, 590–595 (2008).

82. P. Gouma, K. Kalyanasundaram, X. Yun, M. Stanacevic, L. Wang, Chemical sensor and breath analyzer for ammonia detection in exhaled human breath, *IEEE Sens. J.*, Special Issue on Breath Analysis, 10, 49–53 (2010).

83. B. P. J. de Lacy Costello, R. J. Ewen, N. M. Ratcliffe, A sensor system for monitoring the simple gases hydrogen, carbon monoxide, hydrogen sulfide, ammonia and ethanol in exhaled breath, *J. Breath Res.* 2, 037011 (2008).

84. L. E. Gustaffson, A. M. Leone, M. G. Persson, N. P. Wikuld, S. Moncada, Endogenous nitric oxide is present in the exhaled air of humans, *Biochem Biophys. Res. Commun.* 181, 852–857 (1991).

85. K. Alving, E. Weitzberg, J. M. Lundberg, Increased amount of nitric oxide in exhaled air of asthmatics, *Eur. Respir. J.* 6, 1268–1270 (1993).

86. S. A. Kharitonov, D. Yates, R. A. Robbins, R. L.-Sinclair, E. Shinebourne, P. J. Barnes, Increased nitric oxide in exhaled air of asthmatic patients, *Lancet* 343, 133–135 (1994).

87. City Technology Ltd., www.citytech.com (2011).

88. G. Neri, A. Bonavita, S. Ipsale, G. Micali, G. Rizzo, N. Donato, Carbonyl sulphide (COS) monitoring on MOS sensors for biomedical applications, *IEEE Trans. Ind. Electron.* 2776–2781 (2007).

89. K. Bartlett, J. V. Dobson, E. Eastham, A new method for the detection of hydrogen in breath and its application to acquired and inborn sugar malabsorption, *Clin. Chim. Acta* 108, 189–194 (1980).

90. A. Mantini, C. Di Natale, A. Macagnano, R. Paolesse, A. Finazzi-Agrò, A. D'Amico, Biomedical applications of an electronic nose, *Crit. Rev. Biomed. Eng.* 28, 481–485 (2000).

91. F. Röck, N. Barsan, U. Weimar, Electronic nose: Current status and future trends, *Chem. Rev.* 108, 705–725 (2008).

92. F. Axisa, P. M. Schmitt, C. Gehin, G. Delhomme, E. McAdams, A. Dittmar, Flexible technologies and smart clothing for citizen medicine, home healthcare and disease prevention, *IEEE Trans. Inf. Technol. Biomed.* 9, 325–336 (2005).

93. Y. Miyoshi, K. Miyajima, H. Saito, H. Kudo, T. Takeuchi, I. Karube, K. Mitsubayashi, Flexible humidity sensor in a sandwich configuration with a hydrophilic porous membrane, *Sens. Actuators B* 142, 28–32 (2009).

94. H. C. Byung, B. S. Kang, C. Y. Chang, F. Ren, A. Goh, A. Sciullo, W. Wu, J. Lin, B. P. Gila, S. J. Pearton, J. W. Johnson, E. L. Piner, K. J. Linthicum, Wireless detection system for glucose and pH sensing in exhaled breath condensate using AlGaN/GaN high electron mobility transistors, *IEEE Sens. J.* 10, 64–70 (2010).

95. A. D. Aguilar, E. S. Forzani, L. A. Nagahara, I. Amlani, R. Tsui, N. J. Tao, A breath ammonia sensor based on conducting polymer nanojunctions, *IEEE Sens. J.* 8, 269–273 (2008).

Part II

Sensors for Medical Applications

9 Biosensors and Human Behavior Measurement

A State-of-the-Art Review

Yingzi Lin and David Schmidt

CONTENTS

9.1 INTRODUCTION: SENSORS AND BIOSENSORS

All sensors are implemented to provide information that was previously inaccessible. This information, along with the knowledge of the relationship between the sensor and the measured object, provides information regarding the condition of the measured object.

Sensors are implemented practically everywhere to measure the condition of objects. Temperature and pressure sensors in a steam plant provide important information that reflects the process quality. Small accelerometers on a train are capable of detecting when the wheels begin to deform. There are sensors all over cards to provide an awareness of the vehicle's condition to the driver. What about monitoring the human condition? Will there be a situation where the information regarding our own condition will be as easily accessible as the temperature of a car engine? Is this even a goal being worked toward? A look at the development in sensor technology proves that we are already well on our way. The following chapter will discuss the main design constraints of biosensors along some different biosensor applications including specific examples showing how design challenges are met. Additionally, developing technologies that promise exciting advancements in the biosensor field will be introduced.

Car is a perfect platform to show how quickly sensor development is improving over time. Today it is standard practice for cars to provide vehicle information such as fuel levels, speed, rpm, battery life, and engine temperature. Sensors are starting to measure more factors than just those related to engine condition. There are now cameras and proximity sensors to help avoid collisions when backing up or parking. Efforts like Ref. [2] are developing highway collision avoidance trajectory decision algorithms. This requires sensors that can measure vehicle trajectory on the highway as well as environmental conditions including surrounding car locations. So if sensors are becoming capable of measuring more and more factors in a car, should not the same be happening for humans? One major aspect to consider is that the design of cars can be altered to fit new sensors. This is an advantage designing for the technological has over designing for the biological. The option to alter biological organisms to fit sensors simply does not exist. This is one of the greatest challenges faced when designing biosensors. There are a plethora of examples to prove that this limitation has not stopped progress in the biosensor field. A more precise definition of this challenge, provided later in the chapter, will clarify some of the main factors considered in biosensor design.

One of the first and still the largest areas using biosensors is the medical field. The demand to save lives inspires the development of sensors that can measure biosignals from vital signs both in surgery and during transport to high-powered sensors that can practically create a pinpoint image of any location in the human body [3] (Figure 9.1). The use of these devices is not limited to the medical field however. Biosensors are used in many laboratories to monitor much more than just vital signs.

9.2 DESIGN CONSIDERATIONS

9.2.1 Invasive Sensing and Biocompatibility

A device or procedure that requires penetration into the body is referred to as "invasive." A different set of constraints must be considered when designing sensors for use on a human compared to those in a car. One main difference is that many more difficulties are introduced when trying to design a sensor to fit inside a human. For a car, this idea is simple. Just create a gap and mounting somewhere in the car and run wires through. Cars already have a battery and onboard computer, so additional devices for power and signal analyses are not needed. There is no power source or

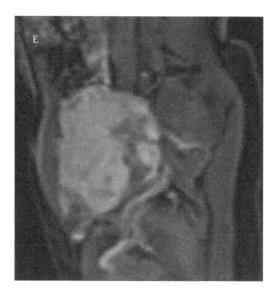

FIGURE 9.1   MRI image of tumor in the pelvic region from Vu et al. [3].

computer to effectively tap into in a person, so additional sources or signal transmission methods would also have to be added. Regardless of these issues, foreign materials in the human body are often rejected and attacked by the immune system. This can lead to dangerous complications for the device and the individual. The complications that go along with designing invasive sensors can make the process very difficult. Some biosensors justify a requirement for being invasive, and the design teams are willing to tackle these difficult challenges. A design by Guiseppi-Elie et al. [4] is a prime example of not only a necessity for a sensor that measures signals in vivo (within the living) but also the magnitude of design challenges that need to be overcome to create a successfully biocompatible sensor. Guiseppi-Elie et al. [4] designed a sensor that can measure glucose and lactate levels from the bloodstream. Two challenges are addressed in detail here—packaging and electrical configuration methods for biocompatibility and filtering methods for signal noise reduction. Packaging for biocompatibility must encapsulate the device in a biocompatible material (silicone tubing was used by Guiseppi-Elie et al.) while maintaining functionality. Sending a current through an in vivo device must be done in a manner to avoid unintentional discharge or unnecessary current heating effects. This is addressed by Guiseppi-Elie et al. with large leads and biocompatible wire sheathing. The blood stream is the body's main method for transporting nutrients and wastes to and from every organ in the body. This can make cognition of specific compounds within the blood stream very difficult. Rather than creating a source of noise, additional compounds in the blood stream can literally clog a chemical sensor. The series of filtration layers developed by Guiseppi-Elie et al. [4] to prevent compounds from creating false signals or clogging the sensor is an excellent example of the complex approaches required to successfully perform in vivo measurements.

9.2.2   NONINVASIVE SENSING

Biosensors are not commonly created for permanent use in a single subject, but rather for many subjects for a variety of tests. This need to repeatedly apply and remove the sensor makes non-invasive sensors desirable. These noninvasive sensors must find ways to effectively collect data from biosignals detectable outside or on the exterior of the body. While designing noninvasive sensors may introduce its own set of challenges, they are usually nowhere near as difficult as those that must be tackled by invasive biosensors. Noninvasive biosensors collect signals either from a distance or, more commonly, are fixed directly on a human subject for data collection. This type of indirect sensing introduces the possibility for many undesired signals to be detected during data collection. The noise created by these undesired or "false" signals is a problem that plagues the non-medical sensor field. In order for accurate data to be interpreted from these signals, the known source or a mathematical definition of these false signals must be defined for effective filtering. Sensors fixed on the subject for data collection introduce another major design constraint that must be considered in the biosensor field—obtrusiveness.

9.2.3   SENSOR OBTRUSIVENESS

Sensors in the medical field have the advantage of collecting data from subjects that are immobile—and sometimes unconscious. This allows medical biosensors to be designed without concern for subject mobility or awareness. Sensors fixed on a human subject in a lab study or in real life do not have this luxury. These sensors will most effectively collect relevant data if they do so while creating little to no distraction or interference to the subject being measured. Sensors that effectively do this are known as unobtrusive biosensors. Sensors collecting data relative to the subject's cognitive state could create false signals if the measured subject is consciously aware of the sensor. Similarly, sensors measuring biosignals while a subject is performing a task should in no way inhibit the subject's ability to perform the task. This requires sensors fixed on the subject to be small and lightweight. Sensor location is another important factor. The sensor must be able to collect relevant data while still remaining out of the subject's way. The best unobtrusive areas on the body for sensor placement are those with no articulations. Put simply, unobtrusive areas on the body to place sensors are those between joints. Mizuno et al. [5] provide an example of sensors designed to unobtrusively collect data for human behavior measurement (see Figure 9.2).

9.2.4   SIGNAL NOISE

Signal noise can be an unavoidable result in unobtrusive biosensors and, if not considered appropriately, can plague study results. There are often a variety of sensors that can detect specific biosignals, and a comparison of these sensors proves that there is often a trade-off found between noise and obtrusiveness. One clear example can be seen from the comparison of two different respiration sensors. A capnograph can be used to measure levels of CO_2 exhaled from the body. This type of sensor

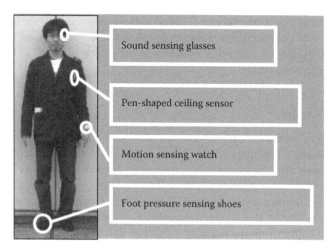

FIGURE 9.2 Location of unobtrusive sensors used in Mizuno et al. [5].

can be very accurate. Not only can the CO_2 levels provide a quantitative analysis for respiration but also directly measuring CO_2 can provide a clean signal. Noise could only be created through the form of other CO_2 sources. This type of noise could come from other nearby people. To account for this, the sensor commonly collects samples directly from the mouth or nose of a subject. In Ref. [6], data are transported to a capnograph by creating a constant airflow in tubes that run from the nostrils to the sensor. This type of accuracy does not come without sacrifice, however. While tubes running form the nostrils might not hinder a wide range of movements, tubes running across your face can be distracting. This method also dictates how the subject must breathe. The constant flow from nostril to sensor in the tubes means that the subject must inhale through the mouth. Additionally, the subject must exhale through the nose for CO_2 levels to be detected. This may not require a strenuous effort, but remembering to do so can be distracting to the subject. Another method for measuring respiration involves wrapping a flexible sensor around the subject's torso, measuring expansion and contraction of the subject's thoracic and abdominal cavities as a result of respiration. This sensing method is much less obtrusive to the user. The single strap can be embedded into or worn under clothing, making it hardly noticeable to the subject or anyone else. The use of wires, compared to tubes, introduces the possibility of wireless signal transmission. This practically eliminates mobility restrictions larger sensors require. Again, these advantages are not without trade-offs. Any movement experienced by the torso can be picked up as signal noise. This can include bending the torso to reach up, down or side to side, along with talking and laughter. While talking and laughter do have an influence over the rates of inhalation and exhalation, they do not actually alter the rate of respiration. While these "false breaths" are a source of noise not detected by the capnograph, performing a visual or kinematic analysis of the subject's torso could provide enough information to filter these out. Both sensor types clearly have their own advantages and disadvantages. Sensor effectiveness depends heavily on the activity of the subject during data collection. If the subject is sitting still and focusing on one specific area, the accuracy of the capnograph may be preferred. If the subject is going to be mobile or is required to interact with other individuals, the unobtrusiveness of the flex sensor could be the best choice.

9.2.5 ARTIFACTING

Accelerometers, inertial sensors, and other motion capturing sensors can be used to collect subject's kinematic data. While these are not particularly measuring biosignals, these are considered biosensors when the kinematic data are used for biomechanical analysis. Winter [7] describes an obstacle these types of sensors must overcome, known as artifacting. This is when kinematic data are recorded from the movement of the sensor independent of the subject, which is usually a result of momentum from the mass of the sensor that causes additional movement from the sensor.

9.3 BIOSENSOR APPLICATIONS

9.3.1 HUMAN ACTIVITY RECOGNITION

9.3.1.1 Sensor Location

What if wearable sensors were able to detect not only where you are, but what you are doing? There are studies dedicated to trying to use wearable sensors for human behavior recognition. Unobtrusiveness is a major factor to consider for this application. While other studies may focus on a specific action, studies in this field must be prepared for the subject to perform any variety of actions, requiring sensors to keep all the subject's degrees of freedom clear. Mizuno et al. [5] (see Figure 9.3) approached this challenge by incorporating sensors into the objects that people already wear. Embedding pressure sensors into shoes and accelerometers and gyro sensors into a watch are surefire methods to avoid obtrusive measurement. While carrying a pen in the front pocket of a button down shirt is not a fashion statement everyone would want to make, it is still an unobtrusive way to implement a ceiling

FIGURE 9.3 Pressure sensing shoes used by Mizuno et al. [5].

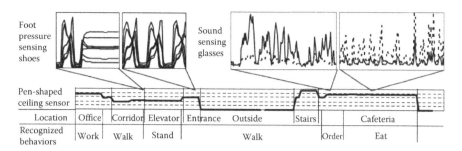

FIGURE 9.4 Behavior prediction from signal analysis from Mizuno et al. [5].

sensor. Finally, while it may be difficult to ignore, a microphone secured to a glass frame will not inhibit any motion.

9.3.1.2 Sensor Communication

Unobtrusively securing the sensors on the body is an accomplishment, but meaningless without a successful means of signal transmission. Mizuno et al. [5] (see Figure 9.4) solve this problem by communicating wirelessly to a network over Bluetooth. A disadvantage this method creates is that it can only work in an area where a Bluetooth station is set up for the sensor system. On the other hand, using Bluetooth systems creates an advantage as well. Knowledge of signal strength can be related to the distance of the sensor system from the Bluetooth station. This information, along with network locations, can be used to estimate the subject's location.

9.3.1.3 Signal Perception

Properly identifying activities from signals requires knowledge of that activity. For example, recognizing whether or not an individual is walking by looking at pressure signals from both shoes may seem easy, but programming a computer to accurately identify this could be difficult. Also, collecting relevant data from the ceiling pen requires knowledge of ceiling heights in rooms the subject is anticipated to occupy. Prior knowledge of subject's daily activities lead Mizuno et al.'s attempt at behavior recognition to achieve an 80% success rate. Such can be used for long-distance childcare services, employee productivity evaluation, or even security purposes where it may be necessary.

9.3.2 BIOMECHANICAL ANALYSIS AND GOLD STANDARDS

Biomechanics is an increasingly popular field of study for engineers. The same kinematic analyses methods used to calculate stress in beams and linkage systems can be used to find stresses seen in the human body, commonly in the musculoskeletal system. To perform such an analysis, kinematic data for a subject must be recorded. A common method for recording this, pointed out by Winter [7], includes the use of multiple high-speed cameras for gait analysis. The cameras record the positions of reflective markers on a subject to collect position information. Processing of this information can provide velocity and accelerometer value for the marked segments.

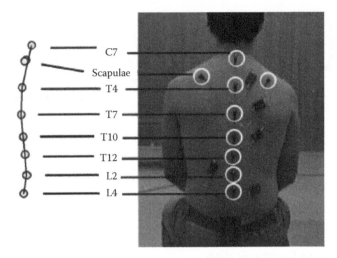

FIGURE 9.5   Marker system used with a camera for initial analysis by Dunne et al. [8].

This information, along with mass properties for different limb segments, provides enough information to find forces experienced in the body. Acquiring mass values for individual limb segments of a living subject is not entirely possible. Tables such as ones provided by Winter [7] allow for segment mass estimations to be made if the subject's overall weight is provided.

While cameras provide very accurate position data, their use is confined to where they can be set up without interference, and processing times can get lengthy. Studies such as one performed by Dunne et al. [8] (see Figure 9.5) explore the use of wearable sensors for biomechanical data collection. In this paper, a variety of sensors that could be worn to collect kinematic data are introduced—including strain sensors embedded in a stretchable material, piezoelectric or piezoresistive bend sensors, and the pairing of accelerometers with inertial sensors. Dunne et al. [8] provide an ample list of references for methods using each of these sensor types. In Ref. [8], a method for measuring spinal posture while working at a computer is explored. Because of their accuracy, a camera or marker system is used in an initial analysis to determine which contours on the back most effectively reflect spinal posture quality.

Once this is defined, Dunne et al. explore a variety of bend sensors to see which can most accurately recreate data collected by the camera or marker system. At the end of the study, Dunne et al. had recognized a local area along the spine (C7–L4) that accurately reflected good or bad spinal pressure from a majority of test subjects. In the end, Dunne et al. were able to create a wearable device capable of unobtrusively measuring spinal posture in the workplace. The study then compares their results to the accuracy of an "expert analysis," providing a case where data from a properly designed sensing system are more accurate than an expert analysis. Comparing a product in development to such *"Gold Standards"* (methods that have been proven and are used in the industry) such as the camera or marker system and expert analysts is a crucial step toward making a marketable product.

One of the largest concerns connected with collecting kinematic data from wearable accelerometer and inertial sensors is the previously mentioned artifacting. The common solution is to secure the sensors to tightly fitting clothing to prevent any independent movement. To avoid this, sensors need to either have a negligibly small mass or be secured tightly to the subject.

9.3.3 COGNITIVE STATE ANALYSIS

Studies regarding the anatomy and physiology of the *autonomic nervous system* (*ANS*) such as Ref. [9] have found links proving that emotional state directly affects the function of the ANS. The ANS innervates organs all through the body. Texts such as Ref. [9] explain the ANS by dividing it into two sections: the sympathetic nervous system, which triggers fight or flight responses, and the parasympathetic nervous system, which triggers rest and digest responses. The parasympathetic system triggers reactions in organs basically to divert energy toward digestion, while the sympathetic system diverts energy toward skeletal muscles for a fight or flight response. This is part of the body's overall effort to maintain homeostasis. Krassioukou and Weaver [9] list what organs are innervated by the ANS and what responses the two sections trigger.

Activation and inhibition of the two branches cannot be properly portrayed linearly. Before we continue, a clearer view of how the terms "activation" and "inhibition" are used regarding the ANS. Both branches are constantly sending signals at a frequency—that is to say they never stop sending signals. The term "activation" is used to describe the event where signal frequency increases, while "inhibition" is used to describe a decrease in signal frequency. While it is common for one branch to activate while another inhibits (called reciprocal activity), activation from one does not require inhibition from the other. There are instances where both activate (coactivation) and where both inhibit (coinhibition). This independence means that autonomic activity cannot be portrayed on a single axis. Each branch needs an axis, and complete independence means these axes are perpendicular to each other. The two-dimensional space created by these axes (see Figure 9.6) is an appropriate way to portray autonomic activity. A good definition of the ANS in a two-dimensional space with the sympathetic and parasympathetic systems as perpendicular axes is provided by Bernston et al. [10].

The purpose of the ANS is to help maintain overall body homeostasis—so how does it know when to trigger rest and digest or fight and flight? This is where the link between emotion and the ANS comes into play. It is common knowledge that feelings of fear leave people pale and their hearts racing. This is their body reacting to fear by preparing it for fight or flight. The increased heart rate allows the blood to deliver oxygen to the organs faster. The paleness is because the body constricts most blood vessels while dilating those only in skeletal muscle. This helps increase muscle performance to prepare us for strenuous activity. This proves that emotional states govern, if at the least play a role in the balance between sympathetic and parasympathetic control. Knowing this, it is easy to imagine that monitoring the physical state of organs innervated by the ANS would introduce the ability to measure human cognitive state. So different human emotional states affect the ANS differently,

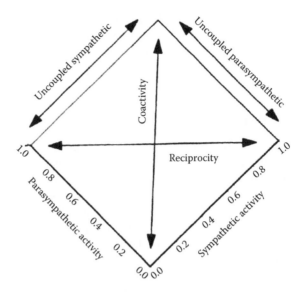

FIGURE 9.6 Two dimensional representation of autonomic space: sympathetic vs. parasympathetic.

which in turn affects the innervated organs differently. If studies are effectively performed to map a full reaction range of human emotions, the ANS would serve as a reliable and consciously passive expresser of the human's cognitive state (see Figure 9.7). Lisetti and Nasoz [11] provide a robust list that includes a variety of studies, elicitation methods they used, and biosignals they detected for cognitive state

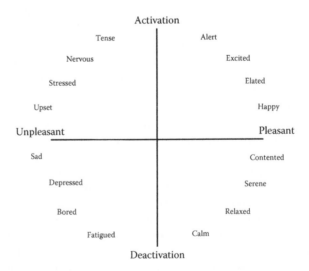

FIGURE 9.7 Two dimensional representation of emotional space—valence vs. arousal.

recognition. Almost all of these studies practice nonintrusive methods, which means only signals detected on the surface of the body can be collected. These include cardiovascular, respiratory activity, skin conductance, body temperature, and other signals. The data collected are commonly compared to subjective responses from users to map the reactions to the emotion felt. However, as previously mentioned, signal noise from these sensors can create false data. Comparing the nature of an experiment to that of different sensors can help identify sources and help filter out the false signals they create.

9.3.4 Biosensors for Human–Machine Interaction

Human–machine interaction is a regular part of our daily life: cars are driven to work, cellular devices are used to communicate and entertain, and computers are used for everything from writing reports to intensive design or calculation. Most technological interface developments leading up to the present focus on effective methods of sending information from the machine to the user. The machine is able to provide enough information to the user at once to allow the user to perform a desired task (write a paper, change lanes on the highway) while simultaneously considering the aforementioned parameters that reflect the machine's status. This allows the user to keep the machine properly maintained during use. The only way the user can communicate to the machine is through the designed interface (keyboard, wheel, mouse, touch screen). There is currently no equivalent type of human to machine communication. Picard et al. [12] compared the nature of human–human interaction to that of human–machine interaction. In a human–human interaction, if one feels annoyed or uncomfortable, the other will be able to notice and make decision based on this. Picard suggested that an ideal human–machine interaction would be modeled after such. Creating a machine awareness of user state creates the potential to increase the quantity and quality of information flow between the two. This increase introduces the potential to create a machine that can realize, recognize, and respond to the human state to create a less stressful, safer, and more productive working or interacting environment. Studies such as Ref. [13] use physiological factors to evaluate cognitive state during human–machine interaction (see Figure 9.8). In Brookhuis, heart rate is measured to estimate mental workload during a driving simulation to evaluate a driver support system. Factors such as mental workload, user drowsiness, and cognitive state can be powerful indicators of the effectiveness of interfaces in human–machine interaction.

Not only does human–machine interaction prove to be a field that could greatly benefit from biosensor application, it also provides an ideal platform for unobtrusive sensor implementation. Many human–machine interactions include almost constant human–machine contact. An ideally unobtrusive method for recording physiological signals can be achieved by placing physiological sensors on a machine surface that regularly experiences human contact. Lin [1,14] provides a good definition and examples for "natural contact" biosensing in human–machine interactions (see Figure 9.9).

FIGURE 9.8 Driving simulator used in Lin's lab at Northeastern University, Boston, Massachusetts.

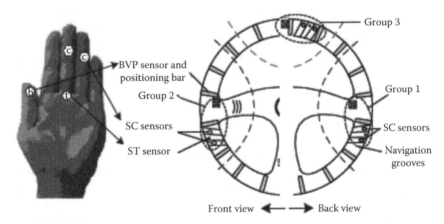

FIGURE 9.9 Design for physiological sensing through natural contact by Lin [1].

9.4 MEMS IN BIOSENSORS

Designing machines to house sensors for physiological measurement from human interaction does introduce some challenges. For accurate measurements, the sensors need to be located in areas where human contact is guaranteed. This contact area is often limited to the surface area of the human hand (or sometimes the fingertips), which can limit the number and variety of sensors that could be effectively used. micro-electrical mechanical system (MEMS) devices are created using micromachining methods commonly used for integrated circuit fabrication. This technology is implemented to build micro-scale switchers, motors, and sensors. Microsystem Design [15] is a full text providing methodology for and examples of MEMS devices.

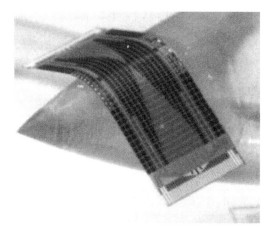

FIGURE 9.10   Flexible shear stress sensors by Jiang et al. wrapping around a conical object.

In Ref. [16], MEMS fabrication techniques were used to create a flexible "skin" housing an array of shear sensors (see Figure 9.10). The entire device has a thickness just shy of 100 μm. To put this in perspective, the average sheet of paper is two thousandths of an inch thick. This converts to about 51 μm, meaning the sensor and skin developed by Jiang et al. is just around two sheets of paper thick. Sensors built on a micro-scale embedded on skins such as this can line existing machine surfaces for unobtrusive sensing.

9.5   CONCLUSION

Implementation of biosensors in studies can provide feedback not previously available to help design with an advanced consideration for human factors. Developments in technology enable these sensors to be implemented outside of a laboratory setting and in our daily lives. The combination of this technology and psychophysiological responses under research can eventually lead to a world where machines are just as aware and considerate of humans as we are of them. In essence, advancement in the biosensor field will ultimately change human–machine interaction into human–machine cooperation.

ACKNOWLEDGMENT

The research work has been supported by the Sensors and Sensing Systems Program, National Science Foundation (NSF) through a CAREER Award (Grant # 0954579).

REFERENCES

1. Y. Lin, *Biosensor and Human Behavior Measurement*, Classpak, Copyright@2010 Yingzi Lin, Northeastern University, Boston, MA, Spring 2007–2010.
2. R. Shubert, U. Scheunert, G. Wanielik, Planning feasible vehicle manoeuvres on highways, *IET Intelligent Transport Systems*, 2(3): 211–218, 2008.

3. T. L. Vu et al., Pediatric body MR imaging: Our approach, *Applied Radiology*, 39(4): 8–19, 2010.
4. A. Guiseppi-Elie, S. Brahim, G. Slaughter, K. W. Ward, Design of a subcutaneous implantable biochip for monitoring of glucose and lactate, *IEEE Sensors Journal*, 5(3): 345–355, June 2005.
5. H. Mizuno, H. Nagai, K. Sasaki, H. Hosaka, C. Sugimoto, K. Khalil, S. Tatsuta, Wearable sensor system for human behavior prediction method, Solid-state sensors, Actuators and Microsystems Conference, 2007, *Transducers 2007 International*, 435–438, June 2007.
6. S. D. Kreibig, F. H. Wilhelm, W. T. Roth, J. J. Gross, Cardiovascular, electrodermal, and respiratory response patterns to fear- and sadness- inducing films, *Psychophysiology*, 44: 797–806, 2007.
7. D. A. Winter, Kinematics, in *Biomechanics and Motor Control of Human Movement*, 4th edn., Chapter 3, pp.50–52, John Wiley & Sons Inc., Hoboken, NJ, 2009.
8. L. E. Dunne, P. Walsh, S. Hermann, B. Smyth, B. Caulfield, Wearable monitoring of seated spinal posture, *IEEE Transactions on Biomedical Circuits and Systems*, 2(2): 97–105, June 2008.
9. A. V. Krassioukov, L. C. Weaver, Anatomy of the autonomic nervous system, *Physical Medicine/Rehabilitation-State of the Art Reviews*, 10(1): 1–14, Feb. 1996.
10. G. G. Bernston, J. T. Cacioppa, K. S. Quigley, Autonomic determinism: The modes of autonomic control, the doctrine of autonomic space, and the laws of autonomic constraint, *Psychological Review*, 8(4): 459–487, 1991.
11. C. L. Lisetti, F. Nasoz, Using noninvasive wearable computers to recognize human emotions from physiological signals, *EURASIP Journal on Applied Signal Processing*, 2004: 1672–1687, 2004.
12. R. W. Picard, E. Vyzas, J. Healy, Toward machine emotional intelligence: Analysis of affective physiological state, *IEEE Transactions on Pattern Analysis and Machine Intelligence*, 23(10): 1175–1191, Oct. 2001.
13. K. A. Brookhuis, C. J. G. van Driel, T. Hof, B. van Arem, M. Hoedemaeker, Driving with a congestion assistance mental workload and acceptance, *Applied Ergonomics*, 40: 1019–1025, 2008.
14. Y. Lin, Towards a natural contact sensor paradigm for non-intrusive and real-time sensing for bio-signals in human–machine interactions, *IEEE Sensors Journal.*, 11: 522–529, 2011.
15. S. D. Senturia, *Microsystem Design*, 1st edn., Kluwer Academic Publishers, Norwell, MA, 2001.
16. F. Jiang, Y.-C. Tai, K. Walsh, T. Tsao, G.-B. Lee, C.-M. Ho, A flexible MEMS technology and its first application to shear stress sensor skin, *Proceedings of the IEEE Tenth Annual Workshop on Micro Electro Mechanical Systems, 1997, MEMS'97*, Nagoya Japan, pp. 465–470, California Institute of Technology, Pasadena, CA, 1997.

10 Sweat Rate Wearable Sensors

Pietro Salvo

CONTENTS

10.1 INTRODUCTION

In recent years, public and private researchers have been looking at small and cost-effective wearable sensors for monitoring parameters of physiological interest. A wide and constantly growing sensors market is pushing toward the development of devices capable of gathering and processing information directly on human body. According to a study by ABI Research published in 2009 [1], wearable wireless sensors are expected to grow up to >400 million devices by 2014. Efforts and investments are attracted by the huge potential in the development of sensors to monitor users remotely and in real time while not altering their normal activities. Major demands come from >90% from healthcare, sports, and fitness applications as they move away from hospital and specialized structure-centered approaches. Therefore, users would be involved in their own treatment or training program and assisted in real time if any event requiring the intervention of a specialist occurs, for example, in the case of telemedicine monitoring. Strictly speaking, a wearable sensor implies that it has to be in direct contact with the garment and that the textile substrates, which are often referred to as smart textiles, become a support to connect the body and the sensor; moreover, it has to be user-friendly, unobtrusive, ergonomic, and, not less important, fashionable. The garment has to retain all its properties of flexibility and washability, and so does the sensor. There must be no discomfort so as to allow users to feel like they are wearing a normal garment and thus enable them to act normally. Like many wearable systems, sweat rate sensors are very complex devices since their reactive layer has to be exposed to the target sample in order for the reaction to occur. Hence, how to deliver the sample to the sensor is one of the most critical aspects involved as well as artifacts caused by movement. Since non-invasive procedures are desirable, tests that can determine the behavior of the system are absolutely essential. A physiological process potentially rich in information is perspiration, commonly known with its synonym sweating. In this chapter, after introducing sweat (i.e., its composition, mechanism of production, and role in human physiology), new and current techniques that are suitable for monitoring sweat rate are presented. Finally, the prototype of a novel sensor, which can be integrated in garments, is discussed in details.

10.2 SWEAT

The analysis of biological fluid underlies much of modern medicine. Blood is rich in information but there are a large number of pathologies that require frequent monitoring, sometimes for long periods or for a person's whole life. In these cases, blood sampling is too invasive for patients: it often requires long waiting times, it must be done in specialized laboratories or clinical structures and performed on an empty stomach, and it can cause infections. One possible way out may be to strengthen the analysis of other biological fluids whose sampling is less invasive and which are less chemically complex.

Sweat, in particular, seems very promising. It is composed of about 99% water and about 1% is made up mainly of sodium, chloride, potassium, bicarbonates, urea, lactic acid, glucose, and traces of other organic compounds [2]. pH is about 5 for low sweat rates but it can reach 6.5–7 for higher sweat rates [3].

The sodium concentration, which depends on the sweat rate, is about 20 mM in normal conditions, whereas can reach 100 mM for higher rates [3]. Potassium ranges from 5 to 6 mM [4] and chloride is about 35 mM [5] (see Table 10.1 for more details about concentrations). However, in presence of cystic fibrosis, which affects the

TABLE 10.1
Sweat Composition: Main Electrolytes and Compounds

Components	Range (mmol/L)
Sodium	20–100
Chloride	25–60
Potassium	1–15
Calcium	0.2–1.34
Magnesium	1.5–5
Urea	12–35
Lactate	10–15

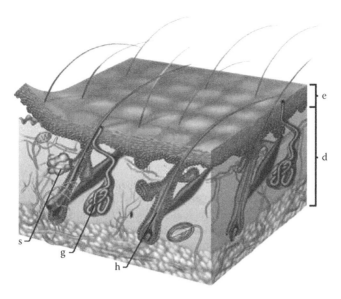

FIGURE 10.1 Sweat glands. The skin is composed of an epidermal layer (e) from which hair follicles (h), sweat glands (g), and sebaceous glands (s) descend into the underlying dermis (d). Sebaceous glands secrete sebum to moisten the skin. (Adapted from Kumar, V., et al., *Robbins and Cotran Pathologic Basis of Disease*, Chapter 25, Figure 25-1A, 8th edn., Copyright © 2010 by Saunders, an imprint of Elsevier Inc., 2010.)

lungs, pancreas, liver, and intestines sometimes leading to infections, poor growth, or infertility, sodium concentration can increase up to 200 mM and chloride can be two to five times greater than usual. It is worth noting that an increase in chloride is just an indication and not certain proof of the presence of cystic fibrosis. In fact, other diseases may also be responsible, such as Addison's syndrome or a renal dysfunction.

Sweat is excreted by two types of glands: eccrine and apocrine [7] (Figure 10.1). Eccrine glands are smaller than apocrine and are embedded in the dermis, that is, the layer of connective tissue lying under the epidermis. Their excretory part has a coil shape and is located in the reticular dermis or in the subcutaneous layer. A duct penetrates the epidermis and allows the sweat to reach the skin surface. The eccrine glands are under the control of the cholinergic sympathetic nervous system and respond to psychic, gustatory, and heat stimuli. Their dimensions can be very different—some people have eccrine glands five times bigger than other people do. Apocrine glands, which have the same structure of hair follicles and sebaceous glands, are active only at puberty and they do not have any role in thermoregulation. They are regulated by a neurotransmitter, adrenaline, and mostly concentrated in the axillary area. Apocrine glands are more numerous in female although they produce a more abundant excretion in male.

Eccrine glands share with the skin's capillary blood vessels the important function of thermoregulating the human body. Body cooling depends on the evaporation of water, that is, the most important component of sweat, and is typically 0.6 cal/mL of evaporated sweat [6]. The body sweats when skin temperature rises over 32°C–34°C. The flow of perspiration from the sweat glands to the ducts on the surface of the skin increases with the temperature and its maximum rate may reach 2 L/h in normal conditions or 4 L/h in extreme conditions (sustainable only for a short time). There are two main forms of perspiration: perspiratio insensibilis and sensibilis. Perspiratio insensibilis is a continuous process depending on the diffusion of water between the derma and epidermis. Its role is related more to the hydration of skin than to body thermoregulation. At 31°C, perspiratio insensibilis has a water flow of 6–10 g/(m²·h) from the skin of arms, legs, and trunk; up to 100 g/m²·h from the palms of hands and feet and from the skin of the face.

Perspiratio sensibilis is caused by the eccrine glands, which are present in the human body from birth. The average person has 2.6 million eccrine glands (varying from 1.6 to 4 million) in the skin, which are distributed, not uniformly, over the entire body except for the lips, nipples, and external genital organs [8,9]. Their density depends on the anatomic area: typical values are 64 glands/cm² on the back, 108 glands/cm² on the forearm, 181 glands/cm² on the forehead, and 600–700 glands/cm² on the palms of hands and feet [9,10].

Sweat rate is defined as the flow of water vapor emitted by the skin, that is, the amount of water emitted per unit area during a defined period of time. For adults, the literature reports an average sweat loss of about 500–700 mL/day in mild climate conditions ($T = 25°C$, relative humidity [RH] = 50%), but the excretion of about 1 L of sweat in 15 min is also possible in extreme conditions (e.g., in saunas). The maximum sweat rate ranges from 2 to 20 nL/min/gland [11].

10.2.1 APPLICATIONS

One of the main reasons for the fast development of wearable sensors is due to their high demand in sectors such as sports, fitness, and clinical medicine. Several devices are often used to provide information about the training session, for example, heart rate or electrocardiogram (ECG). For an athlete, it is important to monitor physiological parameters in real time in order to improve the training activity and in terms of health. This is true not only for professional sportsmen but also for all those who like to train seriously. For example, if a threshold is overcome, a sound or visual alarm like a beep or a light-emitting diode (LED) may be activated. Of course, it would be possible to use several indicators for different purposes, for example, normal conditions or when the range is close to a dangerous level. Data may also be stored or sent in real time to a personal computer for further processing by a specialist. An exhaustive evaluation of physiological functions may help in planning the best training program, for example, deciding on the correct duration and kind of exercises. An opportunity to test an athlete's performance in his or her typical training environment, wherever that may be, is clearly an added value.

During physical training, sweat evaporation is the body's most effective resource to dissipate excess heat that, otherwise, might damage the tissues and cause dysfunctions in the heart or lungs. However, an athlete must avoid dehydration since the human body needs fluids to correctly maintain normal physiological conditions. In case of long-duration exercise, if fluid is not replaced there is a continued reduction in plasma volume throughout exercise, which leads to an altered distribution of the cardiac output in the body.

While there are athletes who can tolerate fluid losses of 4%–5% of body mass [12], in some subjects even a 2% loss of fluid can cause a dysfunction of thermoregulation leading to a significant alteration in physical performance. The critical point is reached when the fluid deficit approaches 7% of the total amount [13], while a 10% loss can lead to heatstroke. There are not only volume losses, but also losses of electrolytes, primarily sodium, chloride, and potassium. Low electrolyte levels can cause gastrointestinal discomfort, headache, cramps, nausea, dizziness, and tachycardia. Reduced mental functions are observed in some cases. When the sodium concentration is low in blood, hyponatremia can occur. This abnormal condition, which is potentially life threatening, can be classified into three categories. In hypovolemic hyponatremia, the total body water (TBW) and sodium levels are both low due to different causes such as vomiting, diarrhea, renal losses, or insufficient rehydration during exercises. Hypervolemic hyponatremia is related to an excess of TBW, whose effect is a dilution of the sodium concentration. Possible reasons are heart failure, kidney dysfunction, or an excessive rehydration. In presence of drugs, adrenal insufficiency, emotional stress, or syndrome of inappropriate antidiuretic hormone (ADH) secretion, the euvolemic hyponatremia appears with an increment in TBW, whereas the sodium content is near normal.

Monitoring sweat rate can also be employed to assess the body thermoregulation of burn patients. It has been documented that the burned skin does not affect the efficiency to dissipate heat because it stimulates an overproduction of sweat in healthy skin areas to maintain the body thermoregulation [14,15]. Therefore, sweat rate sensors placed on burned and unburned skin could help to comprehend the activity of the sweat glands in the body and could provide some useful information on the healing process of damaged skin.

Obesity can be another possible application and, since it is strictly related to normal physical activity, it has many points in common with sport. It is normally associated with an excess of body fat but it should be more correctly associated with a body mass index (BMI) equal or higher than $30\,kg/m^2$ [16]. BMI is a statistical measurement that compares a person's weight and height, and is defined as the individual's body weight divided by the square of their height.

Obesity is often linked to hypertension, diabetes mellitus, high blood pressure, cardiovascular and neurological dysfunctions, and also to some types of cancer. It is caused not only by a sedentary lifestyle and a poor diet but it is also related to genetic mechanisms, the metabolism, or medicines abuse. The number of obese people is dramatically increasing in the western world and national governments have started several information campaigns to reduce its high socioeconomic impact.

Common therapies are based on diets and pharmacotherapy that are specific for each patient, but which have side effects and may only have a short period of applicability. Another option is bariatric surgery mainly based on reducing stomach size. In general, weight reduction can be accomplished, but there are operative risks including mortality. The best solution is prevention by constantly monitoring the evolution of obesity over time. This is extremely important when children are obese since it is very probable they will be obese as adults. As with sports applications, a wearable sensor can help not only the patient but also the doctor to keep under control the patient's health status by means of an indirect analysis from sweat production.

In the medical field, diabetic patients might benefit from sweat rate analysis. Most of the food we eat is broken down into glucose, that is, the form of sugar in the blood, which is the main source of fuel for the body. After digestion, glucose passes into the bloodstream, where it is used by cells for growth and energy. For glucose to get into cells, insulin must be present. Insulin is a hormone produced by the pancreas, a large gland behind the stomach. When we eat, the pancreas automatically produces the right amount of insulin to move the glucose from the blood into our cells. In people with diabetes, however, the pancreas either produces little or no insulin, or the cells do not respond appropriately to the insulin that is produced. The relation between sweat rate and diabetes has been widely studied since many diabetic patients have a reduced number of excitable sweat glands and a low volume of sweat per square centimeter of skin [17,18]. On the other hand, clinical observations show that an indication of the onset of a diabetic shock can be the excessive sweating. If correctly sensed, it can warn the patient to react promptly and take the necessary step to avoid severe consequences.

It is also worth mentioning that wearable sweat rate sensors could have a key role to establish the effectiveness of antiperspirant products designed to control undesirable wetness, in particular in the axillary area.

10.3 WEARABLE SWEAT RATE SENSORS: PRINCIPLES AND METHODS

Knowing the applications and the targets, it is possible to highlight some of the main characteristics of a wearable system. The electronics can be divided into two specific parts:

1. *Patch electronics*: It includes preprocessing electronics for the different sensors and physiological sensors. The aim is to locate the circuitry as close as possible to the sensors.
2. *Signal processing electronics*: This is the part responsible for data acquisition, multiplexing of the acquired signals, their conversion to digital, signal processing, local data storage, and data transmission to a PDA or a PC. These electronics may be located outside the garment for easy access, for example, both for battery replacement and better transmission performances, since it is the core of the system management and it does not need to be washable or flexible.

This splitting allows the system to be more user-friendly since it is easier to wash the garment, change batteries, and replace sensors if requested or needed. Low circuitry cost is another important aspect. The price of the electronic patch should be as low as possible and production costs should be comparable with the cost of the interfaced sensors. The cost of the detachable electronics is less limiting, since these devices should preferably be reusable. In order to have a system that can be used by patients, athletes, and specialists such as doctors and trainers, the user interface needs to be water resistant and as simple as possible, for example, operated using a single press button. Power consumption has to be preferably low since the system could be requested to work for several days. Close contact with the skin can be obtained through soft and flexible patches, which need to be disposable and removable from the garment.

Current techniques to monitor sweat rate mostly rely on an estimation of the difference in weight of the subject or hydrophilic patches applied on the body before and after stimulation either by physical activity or pilocarpine delivery via iontophoresis. These methods are impractical for frequent monitoring. Possible alternatives, suitable for wearable devices, regard to the measurement of changes in the conductivity of the skin or to the effect on resistive, optical, or capacitive sensors when water vapor is present.

10.3.1 SKIN CONDUCTIVITY SENSOR

Human skin is a good conductor of electricity and its electrical properties can be sensed and correlated to sweat production. It is made up of three principal layers:

1. Epidermis
2. Dermis
3. Subcutaneous layer

The epidermis is the outermost layer and it has a protective function. It is about 100 μm thick and its renewal is a continuous process. Dermis is thicker, about 2 mm, and, as already mentioned, is the stratum where sweat glands are located. The subcutaneous layer serves as a protective barrier for the organs and consists of connective tissue and elastin. Epidermis has low electrical conductance than the other two layers, but according to the model proposed by Edelberg [19], when sweat ducts fill, the epidermis conductance increases. The changes in conductance, measured in microsiemens (μS), are referred to as electrodermal activity (EDA) or galvanic skin response (GSR). EDA recordings that use an external current are defined as exosomatic. It is possible to distinguish between two types of skin conductance: tonic and phasic. The former, also known as skin conductance level (SCL), is the baseline level that describes the overall conductivity of the skin over time intervals ranging from tens of seconds to tens of minutes [20]. It generally depends on the degree of moisture in the upper epidermal layer. The latter can be seen as peaks occurring in SCL when a stimulus occurs. It depends to a great extent on the number of sweat gland ducts that are momentarily filled with sweat. Each peak represents an individual skin conductance response (SCR). SCR is uniquely quantified by four parameters: latency, rise time, half recovery time, and amplitude (Figure 10.2).

Historically, both SCL and SCR have been measured by the straightforward application of Ohm's law. When a direct current (DC) voltage source is applied between two electrodes, the skin conductance equals the current flowing through the skin divided by the voltage on the skin. A DC measurement that keeps the current constant returns the skin resistance.

Figure 10.3 shows the circuit proposed by Lowry [21] for measuring EDA. The resistor R_{skin} takes into account the skin resistance. The input voltage, V_{in}, is set to 0.5 V as recommended by Edelberg [22]. This voltage has to be accurate and

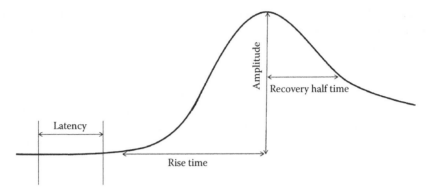

FIGURE 10.2   Raw SCR showing the components used to characterize SCRs.

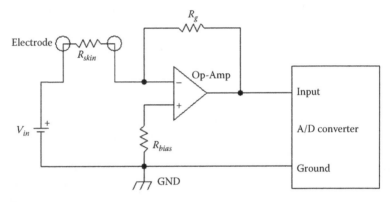

FIGURE 10.3   Basic circuit for EDA measurement. (Modified from Lowry, R., *Psychophysiology*, 14, 329, 1977.)

constant to ensure correct readings. A voltage regulator can be used for this purpose. A simpler solution consists of a low voltage precision diode coupled with a voltage divider made up by two resistors [23]. The output voltage, V_{out}, which is proportional to the skin conductance $V_{out} = -V_{in}(R_g / R_{skin})$, is pre-amplified by an operational amplifier (Op-Amp) and sent to an analog-to-digital (A/D) converter. Boucsein [24] suggests a coupler between the skin electrodes and the amplifier to avoid a difficult interpretation of the results. To increase the signal-to-noise ratio (SNR) and signal integrity, a bioamplifier with high input impedance, usually based on an instrumentation amplifier, should be used instead of an Op-Amp.

Although DC recording circuits are simple, they have several drawbacks. They cause electrode polarization, electrolyze the skin, and results can be affected by electromotive forces (EMFs) present in the circuit [25]. Figure 10.4 shows the solution proposed by Grimnes et al. where, by replacing DC voltage with alternating current (AC) voltage, the complex impedance (Z) is measured. A signal source (2.5 V rms at 22 Hz) is supplied by a lock-in amplifier and coupled by the resistor R_1 (100 MΩ) with the measuring electrode. The measured signal is sent to a noninverting Op-Amp and

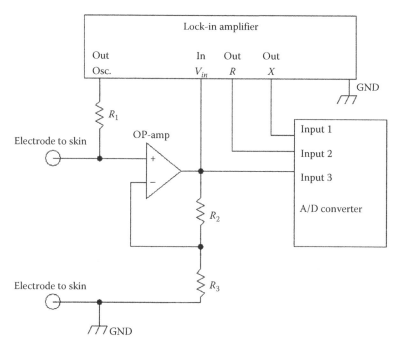

FIGURE 10.4 EDA measurement by AC conductance. (Modified from Grimnes, S. et al., *Skin Res. Technol.*, 17, 26, 2011.)

amplified by 100. The Op-Amp output is coupled to an A/D converter and the lock-in amplifier. AC conductance, G, is calculated from the impedance parameters; thus,

$$G = \frac{R}{R^2 + X^2} \tag{10.1}$$

where
R is the AC resistance (Ω)
X is the reactance (Ω)

EDA seems to be an efficient technique to have an indirect monitoring of sweat rate. However, skin conductance can change in response to parameters that are not necessarily related to an increase of sweat rate. Factors such as strong emotions, startling events, demanding tasks like a mental workload, environmental humidity, cold or skin relaxation can alter the output of the sensor. Therefore, only a highly controlled test can ensure that variations in EDA only depend on the activity of sweat glands.

Another aspect that has not to be underestimated regards to the skin electrodes. The electrode-to-skin impedance can be modeled as in Figure 10.5 [26].

To lower the mismatch between the skin and the electrode impedances, and minimize movement artifacts, a gel is frequently applied on the skin in many applications, for example, ECG and electroencephalography, and electrodes are referred to as *wet*.

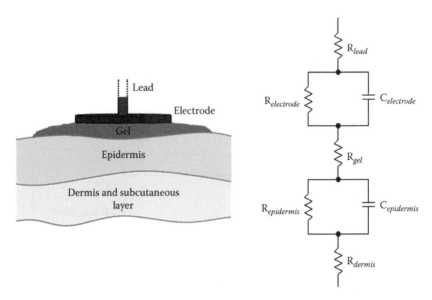

FIGURE 10.5 Equivalent circuit of the skin–electrode interface. At 10 Hz, the imped-ance of epidermis is typically in the range 100 KΩ–1 MΩ, whereas R_{dermis} is about 100 Ω. (Modified from Berson, S. and Pipberger, H.V., *Am. Heart J.*, 71, 779, 1966.)

For sweat rate monitoring, this practice could not be acceptable because it would modify the skin moisture. A solution is adopting *dry* electrodes. Most common wet electrodes are the silver/silver chloride type, which can be reusable or disposable. Dry electrodes are of different types, commonly gold, stainless steel, anodized aluminum, and silicon oxide or silicon nitride. Among dry electrodes, there is the category of insulating electrodes. They consist of a metal or semiconductor with a thin dielectric surface layer, for example, TiO_2 and Ti_2O_5, so that the coupling with the bioelectric signal is purely capacitive. The major drawback of dry electrodes is that, in order to compensate for the absence of the intermediate layer provided by the gel, they need to incorporate a high input impedance active buffer. This requirement leads to greater circuital complexity, power consumption, and overall larger dimensions of the sensor. Furthermore, the contact between electrodes and the skin cause a gradual buildup of a sweat layer that it is not stable over time and disturbs the measurement.

10.3.2 RESISTIVE WRIST WATCH

A practical application of sweat rate measurement is given in [27] where an on–off sweat rate sensor is described. The shape is similar to a wrist watch and is intended for use by diabetics. A peculiarity of this device is that the electrodes are not in direct contact with the skin, but they sense the increment in the thickness of the per-spiration layer on the skin. When the amount of sweat rate on the skin of the wearer exceeds a predetermined threshold level, the perspiration layer comes into contact with the electrodes and a speaker provides an audible warning. The electronic circuit is shown in Figure 10.6.

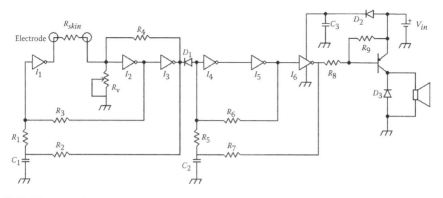

FIGURE 10.6 Schematic of the wrist watch for diabetics. (Adapted from Johnson, W.C., Perspiration indicating alarm for diabetics, U.S. Patent 4,365,637, December 28, 1982.)

When the resistance R_{skin} between the electrodes is very high, the output of the inverter I_3 is low and is fed back through R_4 to the input of the inverter I_2. Therefore, the output of I_2 and the input of I_3 are high. The inverter I_1 has a high output. When sweat rate increases, low impedance is seen between skin electrodes; thus, output of I_2 is low, whereas the output of I_3 is high. Capacitor C_1 starts charging at a rate determined by the RC time constant of R_2 and C_1. When the input of I_1 reaches a threshold level, its output is low. Afterward, the output of I_2 is high, and the output of I_3 is low. This sequence is then repeated. The oscillator circuit involving I_1, I_2, and I_3, works approximately at 5 Hz. The 5 Hz oscillator made up of inverters I_4, I_5, and I_6 switches on and off at 2.5 KHz. This unstable multivibrator drives the base of a PNP transistor whose collector is connected to an audio transducer. The power to the circuit is supplied by three or four 1.5 V batteries. Supplied voltage is filtered by D_2 and C_3 to prevent battery voltage fluctuations from affecting the circuit.

10.3.3 POLYMER FIBERS

A further step in the development of more precise and comfortable systems can be reached by developing a sensor that can be directly integrated into fabrics. As showed in [26], a possible solution to this demand is a polymer optical fiber (POF) made by polymethylmethacrylate -(PMMA) to sense sweat rate. The authors describe a sensor based on a 200 cm loop of 1 mm diameter POF fiber where a section of the cladding is substituted with a porous polymer and a fluorescent dye to create a moisture-sensitive coating (Figure 10.7).

Two different polymers have been tested: four layers of polyhydroxyethyl-methacrylate (P(HEMA)) coupled with a sodium fluorescein (NaFl) dye and a copolymer of HEMA and methylmethacrylate (MMA) couple with fluorescein isothiocyanate (FITC). When the polymer absorbs water, it swells and the refractive index is reduced by 1.5 compared with PMMA core, thus improving the waveguide efficiency. The POF sensor has maximum sensitivity between 98% and 100% humidity, 24 s response time for P(HEMA) coating and 9 s for copolymer HEMA and MMA.

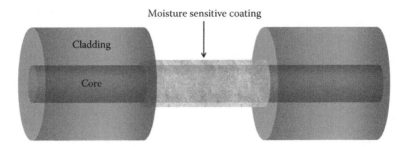

FIGURE 10.7   Optical fiber with moisture-sensitive coating.

Another technique is suggested in [29], where the amount of sweat rate is put in relation to variations in sweat pH values due to the loss of sodium during sweating. To monitor sodium, Hendrick et al. have used an electrospun fabric containing pH-sensitive fluorescent nanoparticles, C dots. The advantage of C dots is their brightness, which is claimed to be 20–30 times more than fluorescent dyes, and their resistance to quenching and photobleaching. The C dots are incorporated into nonwoven cellulose acetate fibers spun by electrospinning. FITC is chosen as a pH sensor because the pK_a of 6.4 perfectly fits in the biologically relevant range from pH 5 to 8.5. C dots use tetramethylrhodamine isothiocyanate (TRITC) dye core as an internal reference. Results have shown that the ratio of FITC/TRITC intensity increases with pH.

A different approach is adopted in [30] where doped polypyrrole (PPY) and polyaniline composite conductive hollow fibers (supplied by Santa Fe Science and Technology Inc.), which vary their resistivity when the polymer absorbs water vapor, are tested. Two different samples of PPY fiber doped with anthraquinone sulfonate (PPY/AS) are evaluated as humidity sensors, named fibers 1 and 2, respectively. The calibration curves for the two tested fibers are reported in Figure 10.8, where the data obtained in the ascending (A) and descending series (D) are plotted separately.

Both fibers show a linear characteristic in the region 0%–70% RH and a good reproducibility, thus allowing them to be used as humidity sensors in this range. A minimal hysteresis can be seen in fiber 1, while the best performance is obtained with fiber 2. The response time is greater than POF fibers (1 min), and there is an increasing trend with increasing humidity. If the mechanical properties of these fibers will be greatly improved, they might be weaved into garments. In fact, they undergo a limited bending before breaking and cutting is extremely dangerous and difficult since a short circuit soon occurs. The only solution to cut them is to freeze them in liquid nitrogen first. Presently, their intrinsic fragility makes them difficult to be used as wearable sensors.

10.3.4   CONDUCTIVE YARNS

Conductive yarns coated with hydrogel can be used to fabricate capacitive sweat rate sensor [30]. Hydrogel is a jelly-like polymer, which is water-insoluble and so absorbent that it can contain over 99% water. The considerable water content gives a high

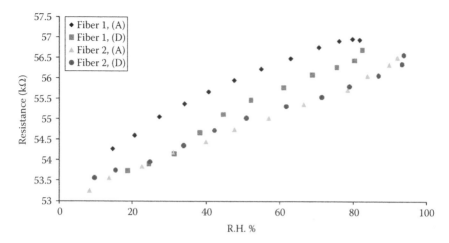

FIGURE 10.8 Calibration curves for the PPY/AS fibers, ascending (A) and descending (D) series. (Adapted from Salvo, P., Development of Wearable Sensors to Measure Sweat Rate and Conductivity, Ph.D. thesis, Interdepartmental Research Center "E. Piaggio," Faculty of Engineering, University of Pisa, Pisa, Italy, 2009.)

degree of flexibility to the sensor making it suitable for integration into garments. The idea is to use two conductive yarns as the plates of a capacitor whose dielectric is the hydrogel. It is known that capacitance is given by the general formula

$$C = \varepsilon_0 \varepsilon_r \frac{A}{d} \tag{10.2}$$

where

    C is the capacitance
    $\varepsilon_0 = 8.854 \times 10^{-12}$ F/m is the permittivity of free space
    ε_r is the relative static permittivity of the material between the plates, also called dielectric constant
    A is the area of each plate
    d is the distance between the plates

In its normal state, hydrogel has a low dielectric constant but when it absorbs water vapor, this value increases enormously. This increase is generally much greater than the increase in the distance of the plates; thus, the capacitance of the sensor increases as a result. A capacitive sweat rate sensor can be eventually created by weaving two coated conductive yarns. This technique is at a very early stage of development and its main drawback is the extremely low signal amplitude found in preliminary tests.

10.3.5 Humidity Gradient Sensor

Since the evaporation of water from any surface establishes a water vapor concentration gradient, Nilsson had the idea of using an open cylindrical chamber to measure RH (i.e., water concentration) at two points of a probe placed on the skin [31].

A quantity proportional to the difference in water vapor concentration could then be derived by applying Fick's first law of diffusion. As flow goes from regions of high concentration to regions of low concentration, it is proportional to the concentration gradient and depends on spatial position. In one dimension, Fick's first law is

$$\Phi = -D\frac{\partial C}{\partial x} \tag{10.3}$$

where
   Φ is the diffusion flow (mol/m²·s)
   D is the diffusion coefficient (m²/s)
   C is concentration (mol/m³)
   x is the position in one-dimensional space (m)

For water vapor, D is 2.49×10^{-5} m²/s at 25°C and increases by about 0.7% per degree. Although the chamber is not suitable for a wearable device because of its large dimensions, the measurement principle is adopted to create a sweat rate sensor in [32]. This device consists of a pocket created on two fabric nets where two RH sensors are inserted at different heights. The first humidity sensor is placed at a distance of 0.2 cm from the skin, whereas the second is 1 cm distant. The diffusion of the water vapor flow is favored by the negligible resistance offered by the large mesh of the net. The sensor is kept at the correct distance from the skin by an 8×4 mm gasket glued to the intermediate layer (Figure 10.9).

On the market, available RH sensors can be grouped into three categories [33]:

1. Resistive humidity sensors
2. Thermal conductivity humidity sensors
3. Capacitive humidity sensors

Resistive humidity sensors measure the change in electrical impedance of a hygroscopic medium such as a conductive polymer, salt, or treated substrate. The impedance change is typically an inverse exponential relationship (nonlinear) to humidity. Their sensing element is usually a salt or a conductive polymer that, when absorbs

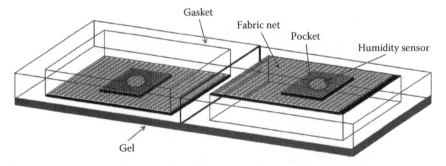

FIGURE 10.9 Sweat rate sensor based on the gradient between two humidity sensors placed at different distances from the skin. (From Salvo, P. et al., *IEEE Sensors J.*, 10(10), 1557, 2010.)

the water vapor, increases its electrical conductivity. The response time for most resistive sensors ranges from 10 to 30 s for a 63% step change. Thermal conductivity humidity sensors measure absolute humidity by quantifying the difference between the thermal conductivity of dry air and that of air containing water vapor. They consist of two matched negative temperature coefficient (NTC) thermistor elements in a bridge circuit. One is hermetically encapsulated in dry nitrogen and the other is exposed to the environment. The typical accuracy of an absolute humidity sensor is $3 \, g/m^3$, which converts to about ±5% RH at 40°C and ±0.5% RH at 100°C. Capacitive RH sensors consist of a substrate on which a thin film of polymer or metal oxide is deposited between two conductive electrodes. The sensing surface is coated with a porous metal electrode to protect it from contamination and exposure to condensation. The substrate is typically glass, ceramic, or silicon. The incremental change in the dielectric constant of a capacitive humidity sensor is nearly directly proportional to the RH of the surrounding environment. The change in capacitance is typically 0.2–0.5 pF for a 1% RH change, while the bulk capacitance is between 100 and 500 pF at 50% RH at 25°C. The response time ranges from 30 to 60 s for a 63% RH step change and the typical uncertainty is ±2% RH from 5% to 95% RH with a two-point calibration. Capacitive RH sensors are limited by the distance at which the sensing element can be located from the signal conditioning circuitry. This is due to the capacitive effect of the connecting cable with respect to the relatively small capacitance changes of the sensor.

Since the final product has to be integrated into a fabric, has to be comfortable for the user, and has to require a minimal circuitry for signal processing, it has been chosen a capacitive RH sensor, Philips H1 (radius ≈ 6 mm, thickness ≈ 20 mm, hysteresis ≈ 3%). In fact, thermal conductivity humidity sensors are not wearable and resistive sensors typically need a more complex circuitry (resistances >1 KΩ even for low RH values require expensive and complex electronics) than capacitive sensors.

The testing and calibration of sensor first entail defining a sufficient large area of the body where the sensor can be placed. Sweat samples, stimulated either by pilocarpine delivery via iontophoresis or by physical activity, have been collected using filter paper from healthy volunteers, a male and a female, on different body regions, that is, calf, forearm and lower back. Pilocarpine delivery is known to be the most practical method to induce sweating, but a sort of temporary adaptation has been observed resulting in a decreased amount of collected sweat. Tests show that sweat production can vary between different people, and even in the same person there may be different behaviors on different days. Differences amongst individuals who are at rest are relatively small, and there are no significant differences in sweat rates in various regions of the body. However, there are larger variations between these regions during intense sweating. Results indicate that there are areas of the body where sweat production is more intensive, such as the forehead and arm. Nevertheless, sweat production in the lower back is not particularly different from the majority of the other body areas; it has a wide surface and allows an easy placement of the sensor. On the basis of these considerations, the lower back is selected as the location for the sensor patch.

Artifacts caused by body movements are reduced fixing the sensor to the skin by attaching a thin sensing gel designed for ECG application (AG600 series, AmGel

Technologies) to the lower border of the gasket. To ease the reading of the humidity gradient, a textile semipermeable membrane could be interposed between the H1 sensors. In fact, with the membrane separating the two humidity sensors, a higher humidity gradient is expected. Skintech N25 and N400 membranes have been tested. The flow of water vapor through the membranes has been calculated by means of a mass balance after measuring the weight loss of a saturated salt solution, used as an emitting surface, in a defined duration of time. The bottom part of a flow-through chamber was filled with one such solution. In the top part, different humidity gradients were achieved by pumping air at a controlled degree of humidity. The comparison shows as the use of the membrane, because of its low permeability, saturates the air where the humidity sensor closer to the solution is placed. As a consequence, the life time of the sensor is shortened. Therefore, a structure based on a fabric net with a higher breathing capacity is preferable.

To calibrate each humidity sensor, a dedicated system has been prepared. It allows the flow of air at different degrees of humidity obtained by mixing dry air and saturated water vapor in different proportions. Air of chromatographic quality is supplied, at a pressure of 3 bars by a zero air generator (Domnick Hunter UHP-35ZA), to a distributor with several exiting gas lines. For each line, a valve and a mass flow controller (Brooks 5850S) provides a flow of perfectly dry air in the range 0–500 mL/min. In the saturated water vapor line, the dry air is bubbled into a glass vessel containing milliQ water at 50°C and then cooled in a pipe coil at 25°C. By mixing dry air and saturated water vapor in various proportions, all the possible RH values (RH 0%–100%) can be obtained. To avoid artifacts due to the inertia of the system (the time needed to reach a stable setup humidity value), a four-way valve is used to convey the same flow of both dry air and test vapor into the flow-through chamber housing the sensor. The RH of the outcoming air is checked by a thermohygrometer (Delta Ohm Digital DO9406), while sensor impedance was monitored by a precision LCR meter (Agilent E4980A), working at 1 kHz.

The calibration is one of the main problems in the development of the sweat rate sensor. A correct calibration is achieved only if a controlled flow of water vapor at a stable rate diffuses toward the sensor. An ad hoc procedure has to be defined to prepare a surface capable to simulate the complex structure of the skin. In an open chamber configuration, the humidity value on the top side of the sensor is fixed to the lab humidity value, and the only way to obtain different gradients is to change the solution inside the chamber. A sufficiently stable emitting surface may be obtained by soaking a sponge with deionized water or a saturated salt solution. In this case, flow values can vary by preheating the sponge at different temperatures. Unfortunately, this method proves lengthy and unsuitable for obtaining reproducible results. Excellent results are obtained with a slightly different approach.

A constant water flow is obtained by placing a sponge into a 1.2 cm high Petri dish, soaking it with deionized water, and controlling the temperature. The sweat rate sensor is at a height of 2 mm from the sponge. The flow rate was calculated from the weight loss over time in the Petri dish as measured using a laboratory scale (OHAUS Adventurer Pro) connected to a PC via a RS-232 interface and acquired by a program written in Labview. Different flow rates can be obtained by heating the chamber at different temperatures. Sensor capacitances are acquired by designing a scanning system that can return the two values simultaneously. An LCR meter (Agilent E4980A) is coupled to a switch (Agilent 3499A) by a multiplexer module (Agilent N2266A). To link the instruments, a GPIB connection is used, whereas to connect the devices under test (DUT), that is, the humidity capacitive sensors, an external board has been made. The BNC cable configuration used was the four-terminal pair (4TP), which guarantees the highest accuracy over wide impedance and frequency ranges. However, a discrepancy in measurement values among channels may occur. The measurement error increases due to the residual impedance and the stray admittance in the scanning system. To compensate for these errors, the channels used for the capacitive measurements can be calibrated by a reference capacitor. The scanning system is fully controlled in Labview. By software, it is able to monitor in real time the temperature and RH provided by the thermohygrometer D0-9406. The variation of the sweat rate sensor at different temperatures is reported in Figure 10.10. The resulting calibration equation is $\Phi = 115.76 \cdot \Delta C - 214.24$.

A good linear approximation of the dependence of flow from the capacitance (i.e., humidity) gradient was obtained. Tests have been carried out on several volunteers who have cycled for 25–30 min in an unventilated room kept at a temperature of 25°C and 50% RH. Data have been acquired by a wearable Bluetooth® interface developed by CSEM (Centre Suisse d'Electronique et de Microtechnique). As a gold standard, it is chosen the Vapometer (Delfin Technologies Ltd., Kuopio, Finland), a commercial instrument for the measurement of evaporation rate. Vapometer is presently used in many fields such as dermatology, academic skin research and testing laboratories, in the R&D of pharmaceutical, cosmetics, and personal care, and chemical industries as well as in veterinary sciences and zoology. Vapometer is not wearable and it does not perform measurement continuously. In fact, it has a delay that increases at high evaporation rates because of its long recover time quantifiable in about 2 min.

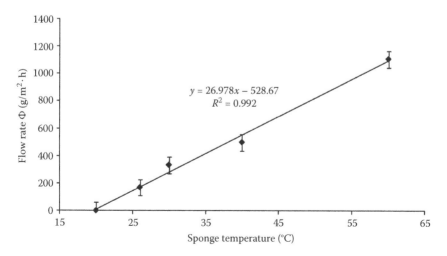

FIGURE 10.10 Water vapor flow from a sponge soaked in water versus temperature. (From Salvo, P. et al., *IEEE Sensors J.*, 10(10), 1557, 2010.)

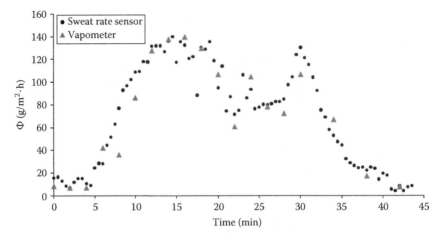

FIGURE 10.11 Sweat rate sensor and Vapometer: a comparison. (From Salvo, P. et al., *IEEE Sensors J.*, 10(10), 1557, 2010.)

A typical trial result is shown in Figure 10.11. The sensor allows a real-time monitoring and provides data coherent with the Vapometer, both during the exercise (the volunteer stopped cycling after ~25 min) and recovery. The average difference between the values of the two sensors is 4 g/m²·h with a maximum of 40 g/m²·h.

Even if at an early stage of research, this sensor seems very promising but tests to measure its fragility and behavior in different conditions are needed as well as to place it on body parts other than the lower back.

10.4 CONCLUSION AND FUTURE WORK

Developments in sweat rate sensors will have to address several key issues, such as system integration, remote connections, and the ability to operate in different areas of the body. Each of these scenarios will require specific technologies. A fully functional and reliable device is not yet available on the market, but the research is ongoing. In a close future, electronics miniaturization will allow integrating other sensors in the same patch, thus more precise measurements will be performed. For example, a temperature sensor can help correcting for wrong recordings of EDA, taking into account changes in the skin impedances that are not related to sweating. However, EDA-based devices do not seem to be able to compete with the other types of sweat rate sensors. More in general, sensors that can be integrated or woven into garments are the new frontier toward which the research efforts are heading. In fact, the ideal wearable sensor is the one that the user forgets to have worn. In this case, maximum comfort is achieved and results are not conditioned by the artificial nature of the user's behavior due to the presence of the sensing system. The capacitive sensor made by two conductive yarns coated with hydrogel described in Section 10.3.4 is one of the most attractive since it is suitable for integration in an industrial process by a textile company and would lead to fabricate a patch that is close to the ideal sensor. Probably, the main disadvantage in common among optical fibers and textile sweat rate sensors is the intrinsic fragility seen in early prototypes. To validate their use, they have to be evaluated in conditions more critical than tests conducted in labs with controlled operating conditions. Nevertheless, initial results show a considerable potential for practical applications.

Research work should also be centered on the design and characterization of new materials, preferably flexible and even stretchable. These materials should provide good mechanical characteristics to overcome the limitations of actual wearable sweat rate sensors and allow an accurate estimation of flow rates from low to high values. Polymers such as hydrogels or conducting compounds as PPY are possible starting points since they offer substantial benefits of manufacturability, lightness, and flexibility.

Another challenge is the development of a wireless sensor network that can report synchronous events from different areas of the body. Such an extensive body mapping would greatly improve the supervision of the health parameters of interest. This network should be remotely accessible and autonomous enough to collect real-time (or near real time) information and handling it with specific software without requiring an external intervention.

Hence, the future belongs to highly intelligent wearable systems capable to perform biological and physiological studies in a parallel mode in order to provide a macroscopic response on user's health status.

REFERENCES

1. Wearable wireless sensors, Report by ABI Research, http://www.abiresearch.com/research/1004149, 2009.
2. S. Robinson and A. H. Robinson, Chemical composition of sweat, *Physiol. Rev.*, 34:202–220, 1954.
3. K. Sato, W. H. Kang, K. Saga, and K. T. Sato, Biology of sweat glands and their disorders. I. Normal sweat gland function, *J. Am. Acad. Dermatol.*, 20:537–566, 1989.
4. K. Sato, Sweat induction from an isolated eccrine sweat gland, *Am. J. Physiol.*, 225:1147–1151, 1973.
5. C. A. Burtis, E. R. Ashwood, and D. E. Bruns, *Tietz, Fundamentals of Clinical Chemistry*, 6th edn., Saunders/Elsevier, St. Louis, MO, 2008.
6. G. Rindi and E. Manni, *Fisiologia umana*, Vol. 2, UTET, Torino, Italy, 2001.
7. V. Kumar, A. K. Abbas, N. Fausto, and J. Aster, *Robbins and Cotran Pathologic Basis of Disease,* 8th edn. Elsevier Saunders, Philadelphia, Pennsylvania, 2010.
8. W. Montagna and P. F. Parakkal, *The Structure and Function of the Skin*, 3rd edn., Academic Press, New York, pp. 376–396, 1974.
9. Y. Kuno, *Human Perspiration*, Blackwell Scientific Publications, Oxford, U.K., 1956.
10. K. Sato, The physiology, pharmacology and biochemistry of the eccrine sweat gland, *Rev. Physiol. Biochem. Pharmacol.*, 79:51–131, 1977.
11. K. Sato and F. Sato, Individual variations in structure and function of human eccrine sweat gland, *Am. J. Physiol.*, 245(2):203–208, 1983.
12. J. R. Brotherhood, Nutrition and sports performance, *Sports Med.*, 1:350–389, 1984.
13. Y. Epstein and L. E. Armstrong, Fluid–electrolyte balance during labor and exercise: Concepts and misconceptions, *Int. J. Sports Nutr.*, 9:1–12, 1999.
14. C. Ben-Simchon, H. Tsur, G. Keren, Y. Epstein, and Y. Shapiro, Heat tolerance in patients with extensive healed burns, *Plast. Reconstr. Surg.*, 67:499–504, 1981.

15. K. G. Austin, J. F. Hansbrough, C. Dore, J. Noordenbos, and M. J. Buono, Thermoregulation in burn patients during exercise, J. *Burn Care Rehabil.*, 24:9–14, 2003.
16. World Health Organization, Obesity: Preventing and managing the global epidemic, Technical report series 894, 2000.
17. W. R. Kennedy, M. Sakuta, D. Sutherland, and F. C. Goetz, Quantitation of the sweating deficiency in diabetes mellitus, *Ann. Neurol.*, 15(5):482–488, 1984.
18. W. R. Kennedy, M. Sakuta, D. Sutherland, and F. C. Goetz, The sweating deficiency in diabetes mellitus: Methods of quantitation and clinical correlation, *Neurology*, 34(6):758–763, 1984.
19. R. Edelberg, Electrodermal recovery rate, goal-orientation and aversion, *Psychophysiology*, 9:512–520, 1972.
20. B. Figner and R. O. Murphy, Using skin conductance in judgment and decision making research, in M. Schulte-Mecklenbeck, A. Kuehberger, and R. Ranyard (Eds.), *A Handbook of Process Tracing Methods for Decision Research*, Psychology Press, New York, pp. 163–184, in press.
21. R. Lowry, Active circuits for direct linear measurement of skin resistance and conductance, *Psychophysiology*, 14(3):329–331, 1977.
22. R. Edelberg, Electrical properties of the skin, in C. C. Brown (Ed.), *Methods in Psychophysiology*, Williams & Wilkins, Baltimore, MD, pp. 1–53, 1967.
23. D. C. Fowles, M. J. Christie, R. Edelberg, W. W. Grings, D. T. Lykken, and P. H. Venables, Committee report. Publication recommendations for electrodermal measurements, *Psychophysiology*, 18(3):232–239, 1981.
24. W. Boucsein, *Electrodermal Activity*, Plenum Press, New York, 1992.
25. S. Grimnes, A. Jabbari, Ø. G. Martinsen, and C. Tronstad, Electrodermal activity by DC potential and AC conductance measured simultaneously at the same skin site, *Skin Res. Technol.*, 17:26–34, 2011.
26. S. Berson and H. V. Pipberger, The low frequency response of electrocardiographs: A frequent source of recording errors, *Am. Heart J.*, 71(6):779–789, 1966.
27. W. C. Johnson, Perspiration indicating alarm for diabetics, US Patent 4365637, p. 1, December 28, 1982.
28. J. Vaughan, C. Woodyatt, P. Scully, and K. Persaud, Polymer optical fibre sensor to monitor skin moisture and perspiration, *Proceedings of the16th International Conference on Plastic Optical Fibers*, Turin, Italy, September 10–12, 2007.
29. E. Hendrick, M. Frey, E. Herz, and U. Wiesner, Cellulose acetate fibers with fluorescing nanoparticles for anti-counterfeiting and pH-sensing applications, *J. Eng. Fibers Fabrics*, 5(1):21–30, 2010.
30. P. Salvo, Development of wearable sensors to measure sweat rate and conductivity, PhD thesis, Interdepartmental Research Center "E. Piaggio," Faculty of Engineering, University of Pisa, Pisa, Italy, 2009.
31. G. E. Nilsson, Measurement of water exchange through skin, *Med. Biol. Eng. Comput.*, 15:209–218, 1977.
32. P. Salvo, F. Di Francesco, D. Costanzo, C. Ferrari, M. G. Trivella, and D. De Rossi, A wearable sensor for measuring sweat rate, *IEEE Sensors J.*, 10(10):1557–1558, 2010.
33. D. K. Roveti, Choosing a humidity sensor: A review of three technologies, www.sensorsmag.com, July 1, 2001.

11 Future of Medical Imaging

Mark Nadeski and Gene Frantz

CONTENTS

11.1 INTRODUCTION

There are those who fear technology is nearly at the physical limitations of our understanding of nature, so where can we possibly go from here? But technology is not where our limits lie. Integrated circuits (ICs) have always exceeded our ability to fully utilize the capacity they make available to us, and the future will be no exception, for technology does not drive innovation. In truth, it is innovation and human imagination that drive technology.

11.2   WHERE ARE WE GOING?

The broad field of medical imaging has seen some truly spectacular advances in the past half-century, many that most of us take for granted. Once a marvel only in the laboratory, advances such as real-time and Doppler ultrasonography, functional nuclear medicine, computed tomography, magnetic resonance imaging, and interventional angiography have all become available in clinical settings.

It is easy to sit back in wonder at how far the field of medical imaging has come. However, in this chapter, we will glimpse into the future. Some of this future is quickly taking shape today, though some of it will not arrive for years, if not decades.

Specifically, we will look at how advances in medical imaging are based on existing technology and how these technologies will provide more capacity and capabilities than we can conceivably exploit and finally lead to the conclusion that the future of medicine is not limited by what we know but rather by what we can imagine.

Let us begin by looking at the edge of what is real, that wonderful place where ideas are transformed into reality.

11.2.1   EyeCam*

For centuries, humanity has dreamed of being able to make the blind see. And, for as long, restoring a person's eyesight has been considered a feat commonly categorized as "a miracle."

About 10 years ago, Texas Instruments (TI) began collaborating with a medical team at Johns Hopkins well-known for its ability to make miracles happen. The team's goal was to develop a method to take the signal from a camera and turn it into an electrical impulse that could then be used to excite the retina as shown in Figure 11.1. If successful, they could return some level of vision to individuals who had lost their eyesight due to retinitis pigmatosa, a disease that affects more than 100,000 people in the United States alone.

Now at the University of Southern California (USC), this team continues to make significant progress. The project has evolved considerably over the years. Its initial conception consisted of mounting a camera on a pair of glasses that would require patients to rotate their heads in order to look around. Today the team is working to actually implant a camera module *within* the eye since it is much more natural to let the eye do the moving to point the camera in the right direction. However, it is one thing to say implanting a camera in a person's eye is more practical than mounting it on glasses and quite another to achieve it. A number of challenges come to mind:

Size: The complete camera module has to be significantly smaller than an eyeball in order to fit.

Power: The camera must have exceptionally low power consumption. At the very least, the energy needs to be scavenged from body heat, the surrounding environment, or a yet-to-be-invented wireless power circuit.

---

* *Source*: http://www.wrongdiagnosis.com/r/retinitis_pigmentosa/stats-country.htm

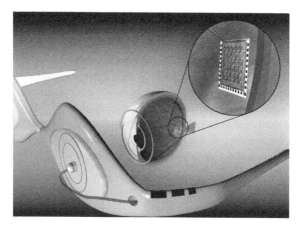

FIGURE 11.1 Example of the eyecam created and tested at the University of Southern California.

Heat: Initial cameras may rely upon a connected power source. Even so, it is critical that the camera does not produce much heat. To be practical, the camera must be able to dissipate enough power so as not to heat the eye to the point of discomfort.

Durability: The camera must be packaged in such a way as to be protected from the fluids in the eye.

Currently working with Georgia Tech University and experts at TI, the team at USC is busy making all this happen. Is such an ambitious project even possible? Although success has yet to be seen, the team envisions a successful completion of the project. And they have good reason to be confident for they are only just pressing at the edges of possibility.

Much of what lies ahead of us in the world of medicine is the identification of technologies and devices from other parts of our world that we can apply to medical electronics. For example, Professor Armand R. Tanguay, Jr., principal investigator on the "eyecam" project, acknowledges that they have many ideas about where else in the body they could implant a camera.*

Here a camera, there a camera,
In the eye a little camera.
Old Doc Donald had a patient.
E, I, E, I, O.

Certainly there is more than one verse to this song. The question we might ask ourselves is what do we imagine we need next?

---

* Unfortunately, we cannot print any of these exciting ideas without a nondisclosure agreement in place.

11.3 MAKING HEALTHCARE MORE PERSONAL

A device that can help the blind to see is a life-changing application of medical technology. Not all medical devices will have such a dramatic effect on the way we live. Most of the changes in medical care will have a much lower profile, for they will be incorporated into our daily lives. However, while their application may be more subtle, the end result will certainly still be quite profound.

The future of medicine is based upon a firm foundation of existing technologies. What is new, in many cases, is not the technology itself but rather how the technology is applied in new ways. Consider these key technologies:

- Digital imaging
- Telecommunications
- Automated monitoring

Each of these technologies is already firmly established in a number of disparate industries. Specifically applying them to medicine will still require creativity and hard work, but doing so will enable entirely new applications. Perhaps most importantly, for health care providers and their patients, the resulting advances will help shift health care into becoming a more routine part of daily life, creating a future where medical devices help us

1. Manage our chronic conditions
2. Predict our catastrophic diseases
3. Enable us to live out our final months/years in the comfort of our homes

11.3.1 ADVANCES IN DIGITAL AND MEDICAL IMAGING

Improving healthcare is the ultimate goal behind advances in medicine. As medical imaging advances, it will allow patients to have more personalized and targeted healthcare. Imaging, diagnosis, and treatment plans will continue to become more specialized and customized to a patient's particular needs and anatomy. We may even see therapies that are tailored to a person's specific genetics. Look at how far we have come already:

The migration to digital files: Photographic plates were once used to "catch" x-ray images. These plates gave way to film, which in turn are now giving way to digital radiography. Through the use of advanced digital signal processing, x-ray signals can now be converted to digital images at the point of acquisition while imposing no loss on image clarity. Digital files have a variety of benefits, including eliminating the time and cost of processing a film, as well as being a more reliable storage medium which can be transferred almost instantaneously across the world.

Real-time processing: The ability to render digital images in real time expands our ability to monitor the body. Using digital x-ray machines during surgical procedures, doctors can view a precise image at the exact time of surgery. Real-time processing also increases what can be done noninvasively. For example, the Israeli company CNOGA* uses video cameras to noninvasively measure vital signs such as blood pressure, pulse rate, blood oxygen level, and carbon dioxide level simply by focusing on the person's skin. Future applications of this technology may lead to identifying biomarkers for diseases such as cancer and chronic obstructive pulmonary disease (COPD).

The evolution from slow and fuzzy to fast and highly detailed: Today's magnetic resonance imagers (MRIs) can provide higher quality images in a fraction of the time it took state-of-the-art machines just a few years ago. These digital MRIs are also highly flexible, with the ability to image, for example, the spine while it is in a natural, weight-bearing, standing position. With diffusion MRIs, researchers can use a procedure known as tractography to create brain maps that aid in studying the relationships between disparate brain regions. Functional MRIs, for their part, can rapidly scan the brain to measure signal changes due to changing neural activity. These highly detailed images provide researchers with deeper insights into how the brain works—insights that will be used to improve treatment and guide future imaging equipment.

Moving from diagnostic to therapeutic: High-intensity focused ultrasound (HIFU) is part of a trend in health care toward reducing the impact of procedures in terms of incision size, recovery time, hospital stays, and infection risk. But unlike many other parts of this trend, such as robot-assisted surgery, HIFU goes a step further to enable procedures currently done invasively to be done noninvasively. Transrectal ultrasound,† for example, destroys prostate cancer cells without damaging healthy, surrounding tissue. HIFU can also be used to cauterize bleeding, making HIFU immensely valuable at disaster sites, at accident scenes, and on the battlefield. Focused ultrasound even has a potential role in a wide variety of cosmetic procedures, from melting fat to promoting formation of secondary collagen to eradicate pimples.

The portability of ultrasound: Ultrasound equipment continues to become more compact. Cart-based systems increasingly are complemented and/or replaced by portable and even handheld ultrasound machines. Such portability illustrates how, for a wide variety of health care applications, medical technology can bring care to patients instead of forcing them to travel. Portable and handheld ultrasound systems have also been instrumental in bringing health care to rural and remote areas, disaster sites, patient rooms in hospitals, assisted-living facilities, and even ambulances.

Wireless connectivity: Portability can be further extended by cutting cables. Putting a transducer, integrated beamformer, and wideband wireless link into an ultrasound probe will not only enable great cost savings by removing expensive cabling from the device, but also allow greater flexibility and

---

* www.cnoga.com
† www.prostate-cancer.org/education/novelthr/Chinn_TransrectalHIFU.html

portability. Further reducing cost and increasing portability enables more widespread use of digital imaging technology, enabling treatment in new areas and applications. A cable-free design also complements 3D probes, which have significantly more transducer elements and thus require more cabling, something that may become prohibitively expensive using today's technology.

The fusion of multiple imaging modalities: The fusion of multiple imaging modalities—MRI, ultrasound, digital x-ray, positron emission tomography (PET), and computerized tomography (CT)—into a single device provides physicians with more real-time information to guide treatment while reducing the time that doctors must spend with patients. For example, PET and CT are increasingly combined into a single device. While the PET scan identifies growing cancer cells, the CT scan provides a picture of the location, size, and shape of the cancerous growth.

Many of the real-time imaging modalities have greatly benefitted by advances in digital signal processors (DSP), devices that specialize in efficient, real-time processing. Specifically, the ability to exponentially increase the processing capabilities in imaging machines has enabled these advances to be useful in a hospital setting. However, to drive many of these applications into more widespread usage, another order of magnitude increase in processing capability will be necessary. This is a tall challenge.

Fortunately, silicon technology companies are now turning their attention to the world of medical electronics to meet these challenges. For instance, TI formed its Medical Business Unit in 2007 to address the needs of the medical industry. This type of partnership between technology companies and the medical industry will help ensure that the exciting possibilities of the future that we envision will be realized.

11.3.2 How Telecommunications Complements Medical Imaging

Advances in medical imaging are frequently complemented by advances in communications networks. Together, they have significantly improved patient care while also reducing costs for health care providers and insurance companies. The question is rapidly moving importance away from where we receive treatment to how we receive treatment.

Telemedicine is the concept where patient's medical data are transported digitally over the network to a medical professional. For example, 24/7 radiology has begun to emerge as a commonly available service. Instead of maintaining a full radiological staff overnight, a hospital emergency room can now send an x-ray via a broadband Internet link to NightHawk Radiology Services* in Sydney, Australia, or Zurich, Switzerland. NightHawk's staff then reads the x-ray and returns a diagnosis to the ER doctors.

---

* www.nighthawkrad.net

For some patients, the combination of imaging and communications enables diagnosis that they otherwise would not receive for reasons such as finance or distance. A prime example is the work of *Dr. Devi Prasad Shetty*, a cardiologist, who delivers health care via broadband satellite to residents of India's remote, rural villages who otherwise would not receive it simply because of the place they live in.* Today, one of Dr. Shetty's clinics can handle more than 3000 x-rays every 24 h. Shetty's telemedicine program has had a major impact in India, where an average of four people have a heart attack every minute.†

In contrast to telemedicine, telepresence‡ is where a medical professional virtually visits a patient through videoconferencing. Telepresence is increasingly used in both developed and developing countries to widen the distribution of health care. Videoconferencing is often paired with medical imaging systems, such as ultrasound, to enable both telepresence and telemedicine. Such applications frequently enjoy government subsidies because they bring health care to areas where it is expensive, scarce, or both.

One example of telemedicine is the Missouri Telehealth Network,§ whose services include teledermatology. Using this service, a patient at a rural health clinic can put his or her scalp under a video camera for viewing and diagnosis by a dermatologist hundreds of miles away. Videoconferencing equipment simulates a face-to-face meeting, allowing the doctor to discuss any conditions with the patient. In the case of someone with stage 1 melanoma, early detection via telemedicine may save his or her life.

Whether patients delay a doctor's visit because of distance, cost, available resources, being too busy, or even fear, telemedicine can mean the difference between suffering with a disease or receiving treatment. Virtual house calls, where physicians use videoconferencing and home-based diagnostic equipment to bring healthcare to a person without necessitating a visit to the doctor, can address most of these concerns.

Virtual house calls may be particularly attractive to patients in rural areas¶ or those in major cities with chronic traffic jams. Virtual house calls are also a way to bring health care to patients who otherwise would not be able to able see a physician, perhaps because they are bedridden, suffer from Iatrophobia (fear of doctors), or have limited means of transportation. Whatever the hurdle they are helping overcome, virtual house calls are yet another example of how advances in medical technology increasingly are bringing care to the patient instead of the other way around.

For example, let us take a look at how medical technology may have been able to help the nineteenth century poet Emily Dickinson who was reclusive to the point that she would only allow a doctor to examine her from a distance of several feet as she walked past an open door. If she were alive today, she would greatly benefit from advances in medical imaging that could accommodate her standoffishness while still diagnosing the Bright's disease that ended her life at the age of 55.

---

* www.financialexpress.com/news/Everyone-must-have-access-to-healthcare-facilities-Devi-Prasad-Shetty/42099/
† www.abc.net.au/foreign/stories/s785987.htm
‡ www.telepresenceworld.com/ind-medical.php
§ www.proavmagazine.com/industry-news.asp?sectionID=0&articleID=596571
¶ www.columbiamissourian.com/stories/2007/05/12/improving-care-rural-diabetics

Future medical technology will reach even further into our lives. Imagine a bathroom mirror equipped with a retinal scanner behind the glass that looks for retinopathy and collects vital signs. In the case of Dickinson, that mirror could have noticed a gradual increase in the puffiness of her face, a symptom of Bright's disease, and alerted her physician through an integrated wireless internet connection.

One of the underlying technologies behind medical imaging—DSPs—has a lot in common with that of telemedicine. DSPs play a key role in telemedicine. For example, DSPs provide processing power and flexibility necessary to support the variety of codecs used in videoconferencing and telepresence systems. Some of these codecs compress video to the point that a TV-quality image can be transported across low-bandwidth wired or wireless networks, an ability that can extend telemedicine to remote places where the telecom infrastructure has limited bandwidth. In the future, compression will also help extend telemedicine directly to patients' homes over cable and DSL connections.

DSPs also provide processing power necessary to support the lossless codecs required for medical imaging since compression could impact image quality and affect diagnosis. Another advantage DSPs offer is their programmability which allows them to be upgraded in the field to support new codecs as they become available, thereby providing a degree of future-proofing for hospitals and physicians.

11.3.3 AUTOMATED MONITORING

Consider this short list of medical devices that can operate noninvasively within our homes:

- Bathroom fixtures with embedded devices could monitor for potential problems, such as a toilet that automatically analyzes urine to identify kidney infections or the progression of chronic conditions such as diabetes and hypertension.
- A bathroom scale could track sudden changes in weight or body fat and then automatically upload this data to a patient's physician. The scale could even trigger scheduling of an appointment based on a physician's predetermined criteria.
- Diagnostic devices such as retinal scanners could be coupled with a patient's existing consumer electronics products, such as a digital camera, to provide additional diagnostic and treatment options. If the device can connect to the network, the medical data collected could automatically be made available to medical personnel.
- Sensors in the home could measure how a person is walking to determine if he or she is at risk for a medical episode such as a seizure.
- Equipment could be connected to a caregiver's network for remote monitoring. One example of such a product under development is a gyroscope-based device worn by elderly patients to detect whether they have fallen.* Near-falls could trigger alerts to caregivers while being documented and

---

* http://ieeexplore.ieee.org/xpl/freeabs_all.jsp?tp=&arnumber=1019448&isnumber=21925

reported to a patient's physician. This device could also track extended sedentary periods, which could be a sign of a developing physical or psychological problem.

- The term "personal area network" (PAN) may come to refer to a variety of devices that work together to regularly and noninvasively monitor and record a person's vital signs. Collected data could be automatically correlated to identify more complex medical conditions.

All of these examples will change how we approach practical medicine. Passive care of this nature becomes a round-the-clock service rather than something that occurs infrequently and disrupts our busy schedules. Constant monitoring also enables earlier identification of health problems before conditions can become irreversible, as well as eliminates the problem of patients not being conscientious about recording information about themselves. As a result, a patient may receive care that is more thorough than if he or she found time for an office visit every week or month.

This technology could be a viable way to improve care for patients who are too busy to schedule routine doctor's visits or, as is the case with iatrophobics, whose fear of doctor has them putting off regular checkups. For the rest of us, who are either too busy, too unconcerned, or too lazy to schedule regular checkups, this technology can help ensure that we do not go too long before any changes get needed care. And, as health care becomes a continuous service through automated processes, the cost of delivering care will be substantially reduced as well.

11.4   FUTURE OF TECHNOLOGY

This is quite an impressive list of what is just around the corner. And while many of these applications may sound like inventions from a future that we can only hope for and dream about, it is likely that the reality will be even more exciting. To understand why this is the case, let us now shift our attention to the underlying factors that enable and drive innovation.

In the decades since the invention of the transistor, the IC, the microprocessor, and the DSP, technology has significantly impacted every part of our lives. And during this time, we have seen computers shrink from being in large air-conditioned rooms to being fit into our pockets. Now we are seeing the next stage of this progression as computers move from being dedicated devices in our pockets to becoming small sub-systems that are integrated into other systems or even embedded in our clothing or our bodies.

Because it is technology that has brought us where we are, it is tempting to believe that technology is the driving factor behind medical imaging and innovation. We do not believe this is the case. For example, the reason that computers became smaller was not because technology allowed them to become smaller. The reason computers became smaller was because people found compelling reasons to make them that way. It is the need or desire for portable computing, not technology, that put computers like the Blackberry and IPhone into our pockets. This point is critical:

Technology does not drive innovation. Innovation drives technology.

11.4.1   REMEMBERING OUR FOCUS

It will be the wants and needs of the marketplace that determine the next technology that will go in our pockets as well as the things we can expect to be embedded in our clothing and bodies. In terms of advances in medical technology, it will be the needs of the patient that dictate what comes next.

Technology is exciting, and it is easy to forget what all of these amazing advances are really all about. For whether it is a retinal scanner in a bathroom mirror or a home ultrasound machine, it is the patient who is the greatest beneficiary. These advances enable health care to become more personal by bringing patients to doctors in their offices and doctors to patients in their homes. They also increase the effectiveness of health care by providing ways to identify diseases and other conditions before they become untreatable.

At the same time, these advances also allow health care to fade into the background and become a part of daily life. Imagine being scanned each morning while brushing your teeth instead of only during an annual checkup. That would be particularly valuable for patients with chronic or end-stage diseases because it may allow them to live their lives without having to move into a hospice.

Health care revolves around the patient, as it should. And technology in turn will help us address their key needs, as stated earlier in this chapter, to manage their chronic conditions, predict their catastrophic diseases, and allow them to live out the last days of their lives in the comfort of their own homes.

11.4.2   WHAT WE CAN EXPECT FROM TECHNOLOGY

Knowing that need drives innovation allows us to approach technology from a different perspective, one where it is more relevant to discuss what these advances will be rather than to discuss what will make them possible. For example, we could speculate on the new process technologies that will overcome current IC manufacturing limitations. However, this will tell us little about what will be built with these future ICs. Rather, it is by exploring what we can reasonably expect from technology that we can gain insight into the possible future.

Let us begin with a well-known tenant from the world of ICs, Moore's law. Moore's law [1] forecasts that the number of transistors that can be integrated on one device will double every 2–3 years. Made in 1965 by Intel co-founder Gordon E. Moor, this prediction has not only been adopted by the industry as a "law" but, as shown in Figure 11.2, has also accurately described the progress of ICs for the last 40 years.

In practical terms, Moore's law has allowed us to reduce the price of ICs. Advances in performance and power dissipation have also affected cost, but over the last decade it has been advances in IC technology that have been primarily responsible for driving cost down.

As we look into the future, we can expect Moore's law to continue to hold true; that is, we should be able to integrate about twice the number of transistors on a piece of silicon every 2–3 years for about the same cost. What does this mean in terms of medical imaging and innovation? By the year 2020, the price of an IC could be as

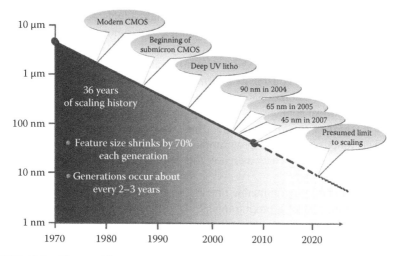

FIGURE 11.2   The trend for process technology over the last 40 years.

low as $1 for a billion transistors (a buck a billion). At that time, a high-end processor might cost $50.

Just imagine what could be done with 50 billion transistors.

Of course, this many transistors do not come without their issues:

1. The cost of developing a new IC may become prohibitively expensive.
2. Raw performance is no longer driven by Moore's law.
3. Power dissipation must be actively managed.
4. Digital transistors are not particularly friendly to the analog world.

11.4.3   DEVELOPMENT COST

Back in the mid-1990s, TI introduced a new technology node for ICs. With this particular new technology node, we were able to integrate 100 million transistors on an IC. To put this in perspective, most personal computers in the 1990s had less than 10 million transistors, not counting memory. (As a short aside, advances in new process nodes are generally 0.7 times the size of previous nodes, which yields twice the number of transistors for the same die area.)

Being able to build an IC with this density does not answer the question of what will be done with them. The more difficult question, however, was what it would take to design a product with 100 million transistors. Consider the math. If we estimate that a design effort could be so efficient that the average time to design each transistor were about 1 h, the project would take 50,000 staff years to complete. Clearly, starting a design of 100 million transistors from scratch would be virtually impossible. TI and other IC manufacturers solved and continue to solve this problem through intellectual property (IP) reuse. We will come back to this concept shortly.

On top of the excessive cost of an IC design, the cost of manufacturing tooling is on the order of one million dollars (we will overlook the billions of dollars spent on the IC wafer fab itself for the moment). For high-volume applications, this cost spreads out to a manageable number. For example, a design with a total build of one million units would reduce the per-unit tooling cost to $1. For low-volume applications, however, the cost can become prohibitive: an IC design for a product anticipated to sell 10,000 units during its lifetime has a per-unit tooling cost of $100.

The obvious conclusion is that only those IC designs with extremely high volumes can be justified. Most applications, however, and not just those in the medical industry, have a significantly smaller scope. Thus, these applications will not be based on ICs but rather on standard programmable processors with application-specific software. And as tooling costs increase, this will be true for virtually all products developed in the future.

Because standard programmable processors allow IP to be implemented in software, the tooling cost of a programmable device can be shared among many different products, even across industries (i.e., medical imaging, high-end consumer cameras, industrial imaging, and so on). Software customizes the processor, so to speak, and the more applications a processor can serve, the lower its cost.

The trade-off of implementing IP in software is, from an engineering perspective, a fairly "sloppy" way of designing a product, meaning that the final design will require far more transistors than if an application-specific IC is used. Given the advanced state of IC process technology, however, this does not matter. Back in the 1990s, we had more processors than we knew what to do with. Today we can build processors with more capacity than we can use. And in the year 2020, we will still have more transistors than our imaginations can exploit.

11.4.4 PERFORMANCE

For years, the performance of ICs has seemed to be driven by Moore's law, just as cost was. However, if we measure raw performance—that is, the number of cycles a processor could execute—it actually drifted from Moore's law in the early 1990s. Despite this, processors have still doubled in effective performance in accordance with Moore's law through sophisticated changes to processor architectures such as deeper pipelines and multiple levels of cache memory. These changes came at their own cost—lots of transistors. Fortunately, as stated before, we have plenty of those.

Improving performance through sophisticated changes to a processor's architecture is, in some ways, just a fancy way of saying that an architecture has been made more efficient. Deeper pipelines, for example, eliminate the inefficiencies of processing a single instruction by simultaneously processing multiple instructions. Eliminating inefficiencies only goes so far, though. Caches improved memory performance significantly, and caches for caches squeezed out a bit more performance, but having caches for caches actually begins to slow things down.

There are still many opportunities for increasing processor performance through architectural sophistication, but to achieve a major increase in performance, the industry is moving toward multiprocessing. Also referred to as multi-core, the central idea is that for many applications, two processors can do the job (almost) twice as fast as one. Multiprocessing, while yielding a whole new level of performance, also introduces a whole new level of complexity to processor architectures. And as we continue to add complexity, we then must create more complex development environments to hide the complexity of the architecture from developers.

11.4.5 MULTIPROCESSOR COMPLEXITY

To understand the effect of the complexity that multiprocessing imposes on design, we need to take a look at Amdahl's law [2]. Simply stated, Amdahl's law says that sometimes using more processors to solve a problem can actually slow things down. Consider a task such as driving yourself from point A to B. There is really no way to use multiple cars to get yourself there any faster. In fact, using multiple cars along the way will likely slow you down as you stop, switch cars, and then get up to speed again, not to mention the traffic jam you would have created.

The same problem applies to an algorithm that cannot be parallelized—that is, easily distributed across multiple processors. Splitting such an algorithm across equal performance processors will slow down overall execution because of the added overhead of breaking up the task across the multiple cores. Engineers describe these types of tasks as being "Amdahl unfriendly."

Amdahl-friendly tasks are those that can be easily broken into multiple smaller tasks that can be solved in parallel. These are the types of tasks for which DSPs are well-suited. Consider how a video signal can be broken into small pieces. Since each piece is relatively independent of the others, they can be processed in parallel or simultaneously.

The size of each piece depends upon the processors used. For example, TI developed the serial video processor (SVP) [3] for the TV market. The SVP contained 1000 one-bit DSPs each simultaneously processing one pixel in a horizontal line of video. Because Amdahl-friendly tasks like these can be parallelized, it is easier to determine how to architect the multiprocessing system as well as how to create the development environment with which to design applications that exploit it.

Amdahl-unfriendly tasks by far are much harder to solve as they prove difficult to divide into parts. Much of the research related to multiprocessing going on now in universities is focused on addressing how to approach such seemingly difficult problems (Figure 11.3).

11.4.5.1 Multiprocessing Elements

Even given the limitations of Amdahl's law, the best approach for increasing performance appears to be through multiprocessing. Before we can begin to consider how best to take advantage of multiple processors in the same system, however, we first need to ask "what is a processor?" There are, after all, many different types of processing elements:

- General-purpose processors: Examples include ARM cores, MIPS cores, and Pentium-class processors.
- Application-specific processors: Examples include DSPs and graphic processing units (GPUs).
- Programmable accelerators: Examples include floating point units (FPUs) and video processors.

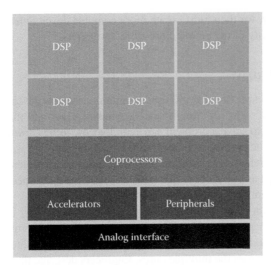

The trend continues
> More programmable DSP cores when generality is important.
> Add optimized programmable coprocessors.
> Use accelerators where the function is more fixed.

Look for even more programmable DSP cores in the future:
> $6 \rightarrow 32 \rightarrow ...$
> Stacking of chips for increasing integration

FIGURE 11.3 Higher performance through parallelism—more DSPs + flexible coprocessors.

- Configurable accelerators: These are similar to programmable accelerators in that they can perform a range of specific tasks such as filtering or transforms.
- Fixed-function accelerators: These are also similar to programmable accelerators with the exception that they perform only a single task such as serving as an anti-aliasing filter for an audio signal.
- Programmable hardware blocks: Examples include field programmable gate arrays (FPGAs), programmable logic devices (PLDs), etc.

In general, the term "multiprocessing" refers to heterogeneous (different elements) multiprocessing while "multi-core" refers to homogeneous (same elements) multiprocessing. The importance of this distinction is greater when talking about DSP algorithms than for more general-purpose applications. In a DSP application, there is more opportunity to align the various processing elements to the tasks that need to be accomplished. For example, an accelerator designed for audio, video/imaging, or communications can be assigned appropriate tasks to achieve the greatest efficiency. In contrast, very large, generic algorithms may be best implemented using an array of identical processing elements.

Despite all of its challenges, multiprocessing appears to be one of the next major advances that will shape IC, electronics, and medical equipment design. Today, multiprocessing is still not well understood and progress will likely be slow as advances percolate out of university research laboratories into the real world over the next decade or so. In the meantime, we have to get to used to the hit-and-miss nature of multiprocessing architectures and do our best to use them as efficiently as we can.

11.4.6 Power Dissipation

In relation to price and performance, power dissipation is the new kid on the block and where much research is starting to be focused. TI's first introduction to this important aspect of value goes back to the mid-1950s when its engineers developed the Regency radio [4] to demonstrate the value of the silicon transistor. This was the first transistor radio, and its obvious need for battery operation made it important to demonstrate low power dissipation.

Power dissipation again became important with the arrival of the calculator in the 1970s. Although most uses for these early calculators allowed for them to be plugged into a wall, sockets were not always nearby or convenient to use. The subsequent movement to LCD calculators with solar cells made low power an even more important requirement.

Lower power dissipation became a primary design constraint in the early 1990s with the arrival of the digital cellular phone. Early customers in this new market made it clear to TI that if power dissipation was not taken seriously, they would find another vendor for their components.

With this warning, TI began its now 20 year drive to reduce power dissipation in its processors. One of the results of this reduction in power is the creation of processors that have helped revolutionize the world of ultrasound by turning the once bulky, cart-based ultrasound systems into portable and even handheld systems.

Figure 11.4 shows how power dissipation has improved over the history of DSP development. Measured in units of DSP performance—the MMAC (millions of multiplies and accumulates per second)—it shows how power dissipation has been reduced by half every 18 months. As this chart was created by Gene Frantz, principal fellow at TI, this trend of power efficiency over time has come to be known as Gene's law.

It should be noted that the downward efficiency trend flattened a few years ago. This occurred because of issues with IC technology where leakage power was at parity with active power. As with any problem, once understood it was able to be resolved. Now we are back on the downward trend of power dissipation per unit of performance.

11.4.6.1 Lower Power into the Future

As we look to the future, the question is whether IC technology will continue to follow Gene's law. There are several reasons to believe it will. The two that seem to have the most promise are lower operating voltages and the availability of additional transistors.

Much of the downward trend for power dissipation has been, in fact, due to lowering the operating voltage of ICs. Over the last 20 years, device voltage has gone from 5 to 3 to 1 V. This is not the end of the line. Ongoing research predicts processors operating at 0.5 V and lower [5].

Just as more transistors can be used to increase the performance of a device, they can also be used to lower the power dissipation for a given function. The simplest method for lowering power consumption is using what is known as *the father's solution*. Who does not remember that loud voice resonating through the house, "When you leave a room, turn off the light!"

This wisdom is easily applied to circuit design as well—simply turn off sections of the device, especially the main processor, when they do not need to be in operation. A good example of this type of management is implemented by the MSP430 family of products [6]. And while it does require more transistors to turn sections on

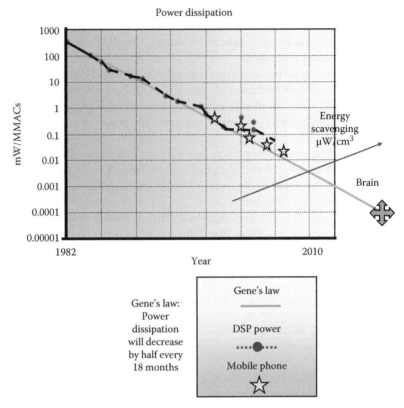

FIGURE 11.4 Gene's law. This graph shows the trend of power efficiency over time with efficiency measured as mW/MMAC. The MMAC is the base unit of DSP performance. The arrow pointing to the upper-right represents our ability to scavenge energy from the environment.

and off, fortunately, it is only a few, making this an extremely efficient approach to power management.

Power can also be managed through multiprocessing, although this approach requires many more transistors. Consider that a task performed on a single processor running at 100 MHz can be performed as quickly on two of the same processors running at 50 MHz. Given the nature of power, one processor running at 100 MHz will consume the same power as two processors running at 50 MHz so long as they are operating at the same voltage.

Due to the characteristics of IC technology, however, the 50 MHz processors can operate at a lower voltage than the 100 MHz processor. Since the power dissipation of a circuit is reduced by the square of the voltage reduction, two 50 MHz processors running at a lower voltage will actually dissipate less power than their 100 MHz equivalent.

The trade-off for managing power through multiprocessing is that two slower processors require twice the number of transistors as a single, faster processor. And again, given that we can rely upon having more transistors as we move into the future, multiprocessing is a feasible method for reducing power dissipation.

11.4.6.2 Perpetual Devices

Confident that IC power dissipation will continue to go down, we can begin to think about how we might take advantage of that. One interesting corresponding area of research which is receiving a lot of attention is energy scavenging. Energy scavenging is based on the concept that there is a plenty of environmental energy available to be converted into electrical energy [7]. Energy can be captured from light, walls vibrating, and variations in temperature, just to name a few examples of sources of small amounts of electrical energy.

Combining energy scavenging with ultralow-powered devices gives us the concept of "perpetual devices." Back in Figure 11.4, there is an increasingly wider arrow moving up and towards the right. This arrow represents our ability to scavenge energy from the environment. At the point this arrow crosses the power reduction curve, we will have the ability to create devices that can scavenge enough energy from the environment to operate without a traditional plug or battery power source.

Imagine the medical applications for perpetual devices. Implants once not feasible because of the need to replace batteries will be possible. Pacemakers will be able to support a wireless link to upload data, eyecams will be permanent once installed, and we may even see roving sensors that travel through our bodies monitoring our heart while cleaning our arteries.

11.4.7 INTEGRATION THROUGH SoC AND SiP

So far we have addressed the three "P"s of value: price, performance, and power. The final aspect of IC technology that will serve as a foundation of medical imaging into the future is integration. Integration refers to our ability to implement more functionality onto a single IC as the number of transistors increases. Many in the electronics industry believe that the ultimate result of integration will be what is referred to as an SoC or "system on a chip." There is no nice way to say this: They are wrong.

To understand this position, let us look back at history. When TI began producing calculators, the initial goal was to create a "single-chip calculator." This may be surprising, but such a device has never been produced by TI or anyone else. The reason is that we have never figured out how to integrate a display, keyboard, and batteries onto an IC. What we did develop was a "sub-calculator on a chip." It is important to catch the subtlety here. We did not develop the whole system, just a part of it. Perhaps it would be more accurate to use the term subsystems on a chip (SSOC).

The same is true when we look at technology today. No one creates complete systems on a chip, for many good reasons. The best, perhaps, is that in practical terms, by the time we develop an SoC, we find it has become a subsystem of a larger system. Put another way, once a technology makes sense to implement as an IC (i.e., it has passed the high-volume threshold required to reduce tooling to a reasonable per unit cost), it has likely been found to be useful in a great variety of applications.

And this leads us to the real focal point of system integration in the future—the system in package (SiP). Figure 11.5 shows that the roadmap of component integration can be simplified to three nodes. The first node—the design is built on a printed circuit board (PCB)—is well understood by system designers. At the second node,

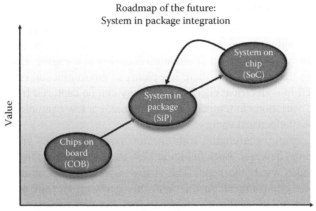

FIGURE 11.5   A simple roadmap of component integration.

SiP, all of the components are integrated "upward" by stacking multiple ICs into one package. It is at the third and final node that all of the components are integrated onto a single IC using SoC technology.

Again, once a system warrants its own SoC, invariably it is being designed back into larger systems. For this reason, technology is not stable at the SoC node, and it settles back either to being placed directly onto a PCB (node 1) or into a package (node 2). As we continue to increase the number of transistors available and lower their cost, these subsystems and the IP they represent will increasingly settle into a package (SiP) and less onto a PCB.

While compelling in itself, this is not the only reason for moving away from the SoC as the ultimate way to design systems. Consider that each advance in digital IC process technology delivers more transistors and perhaps operates at a lower voltage. But the real world is not digital, nor does it follow this trend, and analog circuits, as well as RF circuits, seem to favor higher voltages.

To create a whole IC system on a single piece of silicon requires using a single process technology. However, implementing digital circuits in an analog or RF process significantly increases the relative cost of the digital circuit. Likewise, implementing analog and RF in a digital process substantially reduces signal integrity. The only way we will be able to efficiently integrate all aspects of the system into one "device" is to implement each circuit in its appropriate process technology and combine the various ICs in a single package using SiP technology.

The fact that SiP is more efficient than SoC actually turns out to be really good news for we will be able to take "off-the-shelf" devices and stack them in one package that provides virtually all of the advantages an SoC would. The primary difference—and it is significant—is that by using SiP technology we will be able to manufacture devices within months rather than the years required to produce a new SoC design.

And, in much the same way that programmable processors can bring the cost economies of high volume to specialized applications, developers will be able to create highly optimized, application-specific SiPs as if they were standard ICs. This will give rise to a new product concept, the "boutique IC," where system designers can select from a variety of "off-the-shelf" ICs and "integrate" them into a single package for about the same cost as for the individual ICs. The resulting SiP will have the advantage of faster time-to-market, a smaller footprint, and a "living" specification where new SoCs can be integrated regularly to continually cost-reduce designs. This will certainly give new meaning to the concept of "one-stop shopping" for components.

11.5   DEFINING THE FUTURE

We have covered a lot of ground in our discussion about the future. When you first read about many of the medical applications we have suggested in this chapter, perhaps you thought they sounded more like science fiction than fact. However, most of these devices build upon existing, proven technologies such as digital imaging, telecommunications, and automated monitoring that will change how we approach medicine.

For the common person, and even latrophobics such as Emily Dickinson, new, noninvasive techniques that are increasingly available in the comfort of their home will make the difference in diagnosing and treating diseases before they become debilitating or life-threatening. For people who live in remote locations, telemedicine will bring doctors and patients together in new and powerful ways.

The underlying IC technology required to bring many of these devices to reality is already, or will soon be, available. Moore's law will continue to give us more transistors than we can conceivably use. More general-purpose devices based on software programming models will enable the volumes that result in these transistors being available at a reasonable cost. Using some of those extra transistors to create multiprocessing circuits will provide higher performance and lower power dissipation. Finally, SiP technology will make possibly smaller, more integrated designs as well as potentially enable an entirely new way to design ICs.

As a result, technology is not the limiting factor defining the future of medical imaging. Quite the opposite, technology is more the sandbox in which we can design the creations inspired by our imagination. Innovation is driven, not by the fact that we can shrink a computer to fit in our pocket or our body, but rather by the fact that we need or want to shrink that computer.

For the medical industry, it is all about the patient. And what applications will arise and how fast they will manifest depends upon what patients need. Technology, for its part, will comply by providing us everything we need to make what we envision become real. So what will the future of medical imaging bring? The future will be whatever we want to make it.

The miracles are just beginning.

REFERENCES

1. G.E. Moore, Cramming more components onto Integrated circuits, *Electronics*, 38(8): 114–117, April 19, 1965.
2. T. Lewis and H. El-Rewini, *Introduction to Parallel Computing*, Prentice-Hall, Inc. 1992.

3. SVP For Digital Video Signal Processing, Product overview, Texas Instruments, 1994, SCJ1912.

4. M. Riordan and L. Hoddeson, *Crystal Fire: The Invention of the Transistor and the Birth of the Information Age*, W.W. Norton and Company, 1998.

5. A.P. Chandrakasan et al., Low-power CMOS digital design, *IEEE J. Solid-State Circuits*, 27(4): 473–484, April 1992.

6. C. Nagy, Embedded Systems Design Using the TI MSP430 Series (Embedded Technology), Newnes, 2003.

7. S. Roundy, P.K. Wright, and J.M. Rabaey, *Energy Scavenging for Wireless Sensor Networks*, Kluwer Academic Press, Norwell, MA, 2003.

12 Spatial and Spectral Resolution of Semiconductor Detectors in Medical Imaging

Björn Heismann

CONTENTS

12.1 INTRODUCTION

Medical imaging devices commonly use gamma and x-ray radiation to generate internal images of the human body. Single photon emission tomography (SPECT) and positron emission tomography (PET) systems detect gamma emissions of radionuclide tracers. Computed tomography (CT), radiography, and mammography systems measure the x-ray attenuation of the human body. Figure 12.1 outlines the modes of operation of SPECT, CT, and radiography devices.

The image quality and dose usage of these systems are strongly influenced by the employed radiation detectors. From the early stages onward, scintillator detectors

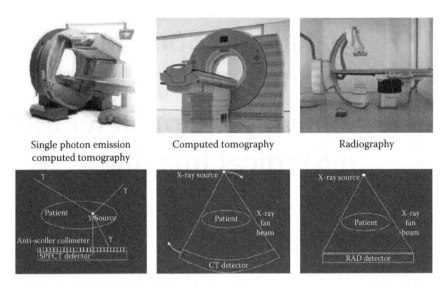

FIGURE 12.1   SPECT, CT, and radiography medical imaging devices.

based on materials such as NaI, BGO, LSO, GOS, and CsI performed the first step of radiation detection [1]. Over the last years, an increasing number of scientific and commercial activities have used conversion semiconductor detectors for medical imaging. For lower x-ray energies, amorphous selenium detectors are routinely employed in mammography detectors. For higher x-ray and gamma-ray energies, CdZnTe and CdTe have come into focus. SPECT prototypes in cardiology, scintimammography, and small animal imaging have been presented (see, e.g., [2–4]). The authors report the improved spectral resolution and underline the potential to perform dual-isotope imaging. For CT, direct conversion counting electronics and prototype systems have been built and evaluated [5–8]. The high x-ray flux of $>10^8$ quanta per second and square millimeter is found to be a major challenge. A main reason for this is attributed to the dynamic material properties of CZT. It has been shown that defects such as Te inclusions and subsequent inferior hole mobility lead to polarization in CZT detectors under medical imaging x-ray fluxes [9–12]. The main mechanism is seen in the creation of a dynamic space charge in the semiconductor bulk, degrading the charge transport properties. The potential benefits of semiconductor detectors in medical imaging rely mainly on their spatial and spectral resolution. In this chapter, we analyze the signal transport in both a scintillator and a semiconductor detector. As an application example, we focus on a CT detector. The pixel geometry, scintillator material, and thickness as well as the electronic readout are chosen accordingly. As figures of merit, we use the modulation transfer function to quantify the spatial resolution and the detector response function (DRF) $D(E,E_0)$ to analyze the spectral behavior. It should be noted that the results indicate an upper performance limit since degradations by, for example, material defects are not included.

12.2 DETECTOR PHYSICS

12.2.1 INDIRECT AND DIRECT CONVERSION DETECTORS

The indirect conversion scintillation detector in Figure 12.2 is based on a GOS scintillator bulk material. Each pixel is enclosed by an epoxy compound filled with back-scattering TiO_2 particles. Typical pixel dimensions of around 1 mm and below are obtained. A registered photosensor detects the secondary light photons at the bottom surface of each pixel. The primary interaction in a detector pixel is given by absorption of an incoming x-ray quantum by a gadolinium atom. The x-ray energy is converted into light photons. The energy conversion rate is around 12% [13]. Secondary light photon transport takes place. Photons that reach the photosensor contribute to the output energy signal E_0. Radiography and mammography detectors follow similar designs. CsI is usually employed as a scintillator. Due to its vertical needle structure, CsI has the advantage of providing intrinsic light-guiding properties; thus, no back-scattering septa are required. This allows for an improved detector resolution at the expense of a reduced stopping power and signal speed.

Two main physical effects influence the spatial and spectral resolution in pixelized scintillator detectors: first, the primary energy deposition is not perfectly localized. For the high Z atom gadolinium, absorption is governed by the photoelectric effect. This generates fluorescence escape photons with mean free path lengths on the order of several $100\,\mu m$. They might be reabsorbed in the pixel volume, become registered in a neighboring pixel, or leave the detector volume completely. Second, light transport is affected by optical cross talk. Septa walls are

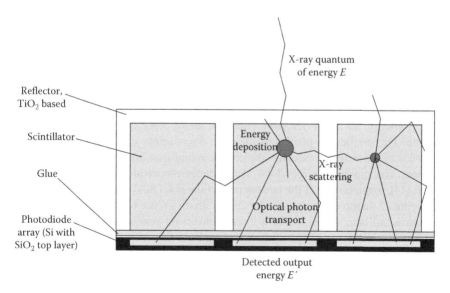

FIGURE 12.2 Schematics of a CT scintillation detector as an example for an indirect conversion detector.

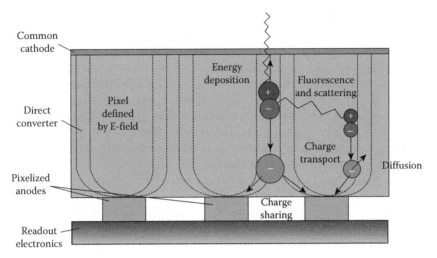

FIGURE 12.3   Schematics of a common cathode CZT direct conversion detector.

designed with a limited thickness to optimize overall dose usage and light yield. As a consequence, a significant portion of the light is transferred to adjacent "false" pixels (see Figure 12.2).

The direct conversion CZT scheme is shown in Figure 12.3. A common cathode design with pixelized anodes on the bottom surface of the semiconductor bulk is typically used. Pixels are established by funnel-shaped electrical fields of several 100 V/mm. The physics of the primary energy deposition are comparable with the indirect conversion detector. However, the deposited energy is converted to charges instead of optical photons. The holes and electrons are separated and accelerated by the electrical field. Electrical pulse signals are induced on the electrodes. The main signal pulse is generated when the electrons follow the stronger curved electrical field in the bulk region close to the anodes.

The main signal degradation mechanisms are comparable with indirect conversion scintillator detectors: first, fluorescence scattering takes place. Due to the lower K-edge energy, the mean free path lengths of fluorescence quanta in CZT are about 100 μm. The smaller the pixel size, the more fluorescence cross talk will affect the behavior of the detector. Second, the charge signal transport is affected by charge sharing. The moving charge cloud also induces electrical pulses on neighboring pixels [14], again mostly at the bottom part of the pixel field configuration.

Figure 12.4a through e summarizes the main difference between an indirect conversion scintillator detector and a direct conversion semiconductor detector. The scintillation detector is an optical device using light photons as intermittent information carriers. A direct conversion detector omits the conversion to light and directly generates charge carriers. It is an electrical device that employs electrons and holes to transfer the event information to the electrodes.

Both detector types can be operated in an integrating or a counting mode. In integrating mode, the charge information is sampled over an integration time and converted to a digital signal. In counting mode, the total number of events

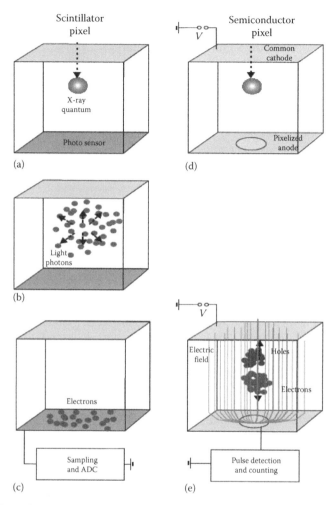

FIGURE 12.4 Signal conversion steps in an indirect conversion detector (a–c) and a direct conversion detector (d and e).

is measured by counting the charge pulses. In addition to this, the energy of each absorbed quantum can be obtained by measuring the total charge or pulse amplitude of each quantum. Counting detectors thus offer spectral resolution of the input quantum field.

The detector parameters for our comparison are summarized in Table 12.1. For the scintillator detector, a 1.4 mm thick GOS with a pixel size of 1.2 mm has been chosen. The direct conversion detector has a 2 mm thick CZT at 700 V bias with a quadratic pixel size of $(450\,\mu m)^2$. For this pixel size, fluorescence cross talk contributes significantly. The choice reflects mostly the high-resolution case. The spatial resolution is not directly comparable with the scintillator detector. The setting is chosen to investigate if a direct conversion detector can provide improved spatial resolution at a reasonable spectral resolution.

TABLE 12.1

Detector Simulation Parameters

	Indirect Conversion	Direct Conversion
Material	Gd$_2$O$_2$S:Pr	CdZnTe
Thickness (mm)	1.4	2.0
Pixel length (mm)	1.2	0.450
Acceleration voltage (V)	n.a.	700
Threshold noise (keV RMS)	n.a.	3

We neglect electronic noise in the indirect conversion detector. It does not influence the spatial and spectral resolution comparison significantly. A threshold noise level of 3 keV (root-mean-square) is assumed for the direct conversion approach.

12.2.2 SIGNAL TRANSPORT PROCESSES

Figure 12.5a and b outlines the cascaded system theory (CST) model of an indirect conversion integrating and a direct conversion counting detector. CST models have been applied to a number of detector evaluations, especially for flat panel radiography and mammography detectors [15].

The indirect conversion detector in Figure 12.4a has the following signal conversion steps: first, the x-ray or gamma quantum is absorbed and its energy is converted to light photons. Second, the light photons travel through the scintillator setup. Third, light photons are detected as an electrical current by the photosensor. Finally, the current signal is digitized by a sampling ADC. The first step of the direct conversion detector in Figure 12.4b also consists of the x-ray energy deposition. A cloud of electrons and holes is generated. Second, the generated charges travel to the electrodes, forming current pulses. Finally, the current pulses are detected by a readout electronic.

The signal transport can be simulated by a cascade of independent conversion steps. For the indirect conversion detector, these steps can be done as follows [13]:

1. *Primary energy deposition.* The primary energy deposition is modeled by a Monte Carlo simulation tool based on the GEANT4 particle interaction simulation framework [16–18]. For each incoming x-ray quantum, the spatial energy deposit is simulated on a $10 \times 10 \times 10$ sub-voxel grid. Rayleigh and Compton scattering and escape processes generally lead to multiple deposition sites per event. Each portion of the primary energy deposit is converted into a number of optical photons, taking into account the main scintillator photon emission energies. The energy to optical photons conversion gain for Gd$_2$O$_2$S:Pr is taken from measurements as $E_c = 0.12$ with a standard deviation of $\sigma(E_c) = 0.04$. As a result of the first step, we obtain a lookup table of individual energy deposition events. By using a high number of events (10^6 and more), systematic errors are avoided.

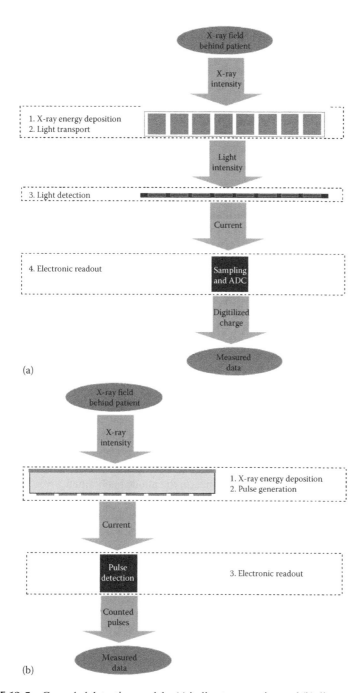

FIGURE 12.5 Cascaded detection models: (a) indirect conversion and (b) direct conversion.

2. *Light transport.* The second step describes the light transport to the photosensor pixels. A photon-tracking Monte Carlo simulation [13] traces the photon paths in the entire detector system until they are detected at a pixel of the photosensor array or lost by bulk absorption or scintillator escape. Photon interaction processes such as optical scattering, photon reabsorption and diffuse, and specular reflection at pixel septa borders are included. The corresponding optical parameters are taken from experimental results. For a specific photon starting position, the average detection probability of an optical photon in a photosensor pixel is obtained.
3. *Light detection.* The light photons that have reached a photosensor pixel are converted to electrical charges. The wavelength-dependent quantum efficiency $\beta(\lambda)$ of the photosensor is taken into account. As a result of the third step, we obtain a photocurrent for each pixel.
4. *Electronic readout.* In the final step, the photocurrent is sampled to charge and digitized. For medical x-ray applications, sigma–delta ADCs are common ADC designs. Direct current measurements by a charge-coupled oscillator are also employed. The electronic readout usually has limited linearity and additional offset noise. For the results in this chapter, nonlinearity and electronic noise do not play a role and are neglected.

For a given detector geometry, x-ray quantum input spectrum, and field distribution, this scheme yields the average signal of the scintillator detector.

The signal chain of the direct conversion detector in Figure 12.4b is modeled as follows [19]:

1. *Primary energy deposition.* The primary energy deposition step is equivalent to the scintillator model. Instead of a GOS material, a CZT absorber is used.
2. *Pulse generation.* A detailed charge transport model can be based on the work of Eskin et al. [20]. A local weighting potential allows to calculate the signal pulse shape for arbitrary charge starting positions in the detector [21]. A time-resolved pulse signal on the anode is obtained.
3. *Electronic readout.* Depending on the priority of spectral or spatial resolution, two main electronic design schemes for direct conversion detectors can be selected. Spectrally resolving detectors in SPECT and PET require a precise measurement of the energy of each quantum.

   Due to this, the anode signals are usually filtered with comparably long shaping times. The signal is integrated and digitized. High-resolution detectors on the other hand address applications in mammography, radiography, and CT. The corresponding electronics employ shorter shaping times close to the primary pulse duration. The filtered pulse signals are usually detected by amplitude threshold triggering [5,6].

In the following, we assume the second case of a high-flux x-ray detector. The threshold noise due to electronic noise contributions in the electronic readout is included in the model.

12.3 SPATIAL RESOLUTION

The spatial resolution of x-ray detectors is mainly given by pixel pitch and aperture. The pixel pitch defines the Nyquist frequency. The smaller the pixel aperture, the larger the spatial resolution will become. However, in practical imaging systems, defining the spatial resolution of a detector is a trade-off with dose usage and detector cost. In particular, scintillator detectors are often limited by the required septa walls and the cost of the required number of electronic digitization channels.

12.3.1 Definition of the Modulation Transfer Function

The modulation transfer function $MTF(f)$ is commonly used to describe the spatial resolution of pixelized detectors [15]. It is given by the normalized absolute value of the Fourier transform of the detector pixel point spread function. The MTF evaluation scheme is commonly applied to pixelated scintillator and CZT detectors (see, e.g., [22]).

12.3.2 Simulation and Measurement of the MTF

For both detector types, the MTF is determined by the "slanted slit" method. Figure 12.6 shows a slanted slit image for the indirect conversion detector. A tungsten plate with a slit of 0.1 mm width is placed on top of the scintillator array with a slit angle of ~3° with respect to the fundamental directions of the pixel lattice. The slit is illuminated by an x-ray flat field. Summing the image along the line direction yields the line spread function. It is an oversampled representation of the point spread function. The procedure is repeated with various angles of the slit toward the axes to obtain a two-dimensional MTF of the detector. It has been shown that

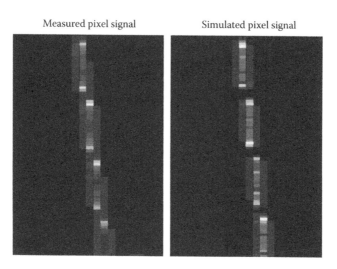

Measured pixel signal   Simulated pixel signal

FIGURE 12.6 Measured and simulated slanted slit images for the indirect conversion GOS detector.

measured MTFs have an excellent agreement to simulated MTFs for both indirect conversion detectors and direct conversion detectors [22]. In the following, we use the simulation framework described in the previous section to obtain the slit images required for the MTF calculation.

12.3.3 PROPERTIES OF THE MTF

Figure 12.7 shows the MTF comparison between an indirect and direct conversion detector. The straight lines in red and blue are simulated curves, whereas the corresponding dashed curves show the respective ideal sinc functions. The indirect conversion detector shows a mid-frequency drop in comparison with the ideal sinc function. This is mainly due to optical cross talk that leads to a low-pass signal filtering in the detector. In principle, the mid-frequency drop can be recovered by appropriate inverse filtering at the expense of amplified electronic noise in the signal. For high and medium flux medical applications, this has no major impact. The signal-to-noise ratio is mainly affected in low-flux screening applications.

In comparison to this, the direct conversion detector is close to the ideal sinc behavior. The remaining deviations are mainly due to fluorescence escapes between adjacent pixels. Despite the fact that the pixel aperture has been more than halved,

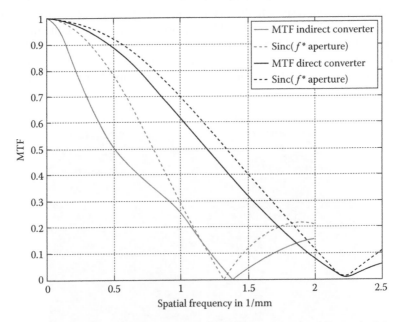

FIGURE 12.7 Modulation transfer functions for indirect conversion 1.2 mm pixel detector and direct conversion 450 μm detector. Dashed lines reflect the ideal sinc functions. (The data have been reproduced from Wirth, S. et al., Simulations and measurements of the modulation transfer function of scintillator arrays, M06-257, *IEEE Nuclear Science Symposium Conference Record*, 2008.)

charge sharing plays only a minor role compared with the effects of optical cross talk. Note that in both detector systems a small deviation in the zero frequency position is visible. This is due to the fact that fluorescence cross talk leads to smaller signal contributions close to the pixel borders, effectively shrinking the pixel aperture.

12.4 SPECTRAL RESOLUTION

In nuclear physics and medical imaging applications such as PET and SPECT, the spectral resolution of the detector is commonly described by the pulse height spectrum (PHS). A typical PHS of a CZT pixelized detector and an NaI Anger camera is shown in Figure 12.8. For x-ray applications, the detector has to register a whole range of input energies. Figure 12.9 shows a 80 and 140 kV tungsten tube spectrum. The generalization of the PHS to a range of input energies E leads to the DRF (see [13]).

12.4.1 Definition of the DRF

The DRF $D^{(i,k)}(E,E')$ yields a probability density to measure the output energy E' for an incoming quantum of energy E. The incoming quantum flux is directed at a central reference pixel. Its lateral position is equally distributed across the pixel area. The output energy is detected at a photosensor pixel with the position (i,k). The pair $(0,0)$ marks the center position, $(1,0)$ the horizontal neighbors, $(0,1)$ the vertical neighbors, and so on (see Figure 12.10).

The DRF allows us to express the statistics of the microscopic signal transport processes as a macroscopic probability function. We can simplify its variable dependencies for medical imaging applications. Here, the pixel-to-pixel variation of the projected anatomical input signal is usually below 1%. This is close to a flat-field

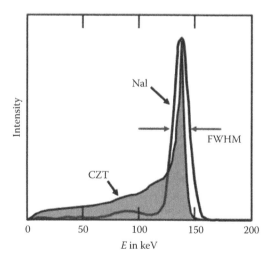

FIGURE 12.8 Pulse height spectrum (PHS) of CZT pixelized detector and an NaI Anger camera. (Picture taken from *Emission Tomography*, Wernick, M.N. and Aarsvold, J.N., Copyright (2004), from Elsevier.)

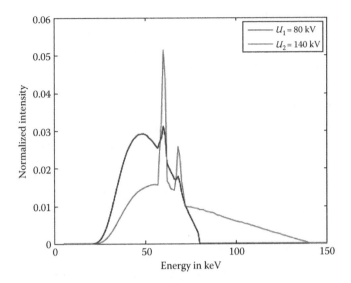

FIGURE 12.9 80 and 140 kV tungsten tube spectra.

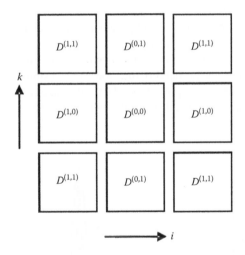

FIGURE 12.10 Spatial indices (i,k) of the detector response function. The symmetry arises for pixels unaffected by border effects.

irradiation of the detector. In this case, the mean signal cross talk between pixels is symmetrical. We realize the flat-field approximation by irradiating the detector surface homogeneously. The simplified $D(E,E')$ function is used to describe the results.

12.4.2 COMPARISON OF THE DRFS

The DRFs of the indirect and direct conversion detector setups are shown in Figure 12.11a and b. In both cases, the probabilities are normalized to 1 for each

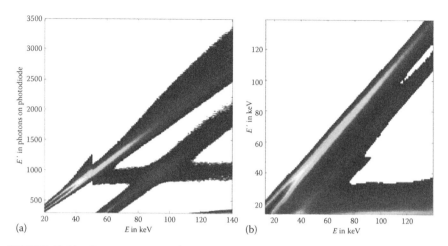

FIGURE 12.11 Detector response function for (a) indirect conversion GOS detector and (b) direct conversion CZT detector.

input energy E. This leads to the respective color codings. Below 15 keV output energy E', the electronic noise in the counting direct conversion detector dominates the output behavior. The respective range is omitted for better clarity.

The indirect conversion $D(E,E')$ in Figure 12.11a consists of the following structures: up to the gadolinium K-edge energy $E_K \sim 50.2$ keV, a linear branch $E \sim E'$ is visible. Its broadening is explained by the energy conversion gain variance. The output energy peak has a tail toward higher output energies E' for increasing input energy E. This light tailing effect is due to the fact that the light transport yield increases with the interaction depth, which in turn increases with the input energy E. Above the K-edge energy, a secondary branch occurs. The events are formed by absorption of the primary energy with a fluorescence energy loss to the surroundings. The corresponding reabsorbed fluorescence events are found in the third, approximately vertical branch starting at around 50 keV output energy. Its slight inclination is again due to the increase of the interaction depth with input energy. The overall absorption probability of the quanta is reduced with increasing input energy E. The low-energy output events including Compton and Rayleigh scatter depositions are not shown; see [13] for a more detailed discussion of these effects.

The direct conversion detector $D(E,E')$ in Figure 12.11b has a more pronounced linear branch. Its stronger relative signal content is explained by the about two times higher intrinsic conversion gain of CZT and the reduced depth-dependency due to the small pixel effect. The fluorescence branches appear at the lower Cd, Zn, and Te fluorescence energies of 23–28 keV. The differential branches are consequently closer to the main linear branch. Charge-sharing events create a low-energy tail increasing toward lower-output energies and overlapping with the fluorescence branches.

The spectral behavior described by $D(E,E_0)$ has consequences for both detector schemes. In the following, we consider the cases of an integrating indirect conversion detector and a counting direct conversion detector as prominent examples.

12.4.3 Integrating Indirect Conversion Detectors

For integrating indirect conversion detectors, it has been shown that the output signal variance increase leads to a Poisson excess noise [13]. Following the work by Rabbani et al. [23], a formula for the noise amplification is established as

$$f(E) = \frac{1}{\sqrt{\alpha(E)}} \frac{SNR_{out}}{SNR_{in}} = \frac{\langle E' \rangle}{\sqrt{\langle E' \rangle^2 + \sigma^2(E')}} \tag{12.1}$$

where
   $\alpha(E)$ is the quantum detection efficiency
   $<E'>$ is the average output energy
   $\sigma(E')$ is the output energy variance
   $f(E)$ is a generalized energy-dependent Swank factor

The Poisson excess noise shown in Figure 12.12 is most pronounced around the K-edge. A noise increase of about 15% is visible. This is due to the fact that the output signal variance increases strongly beyond the K-edge. For continuous input spectra, a typical excess noise of 5%–10% can be estimated, depending on the input spectra and the patient attenuation.

12.4.4 Counting Direct Conversion Detectors

For a counting direct conversion detector, we can distinguish between the full energy resolution required in SPECT or PET and the binned energy resolution required for dual-energy CT or radiography. In the case of full energy resolution, $D(E,E_0)$ contains directly the normalized PHS for specific input energies.

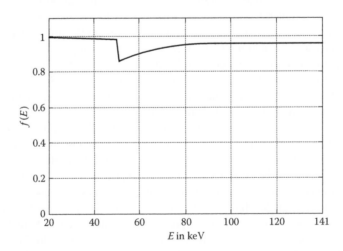

FIGURE 12.12 Generalized Swank factor $f(E)$. (Reproduced from Heismann, B.J. et al., *Nucl. Instrum. Methods Phys. Res. A*, 591, 28, 2008.)

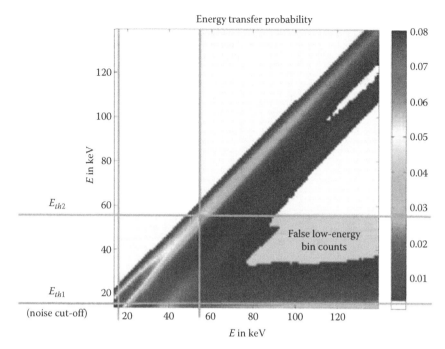

FIGURE 12.13 Schematics of energy binning for the detector response function of Figure 12.11a. Two energy threshold levels $E_{th1} = 15\,\text{keV}$ and $E_{th2} = 55\,\text{keV}$ are used.

In the following paragraphs, we focus on the case of a two-bin energy resolution. Similar to shown in Figure 12.13, this is commonly achieved by using two threshold levels in the electronic readout. The first threshold E_{th1} discards noise events. The second threshold E_{th2} separates the output energy range into two separate bins. The diagonal rectangular sections mark the quanta events that are correctly assigned. The lower right region contains high-energy bin primary events that are falsely assigned to the low-energy bin.

Figure 12.14 shows the consequence of the low-energy shift. We have assumed a 140 kV tungsten x-ray tube input spectrum (see the shaded gray curve). The two detected spectra in the respective energy bins are given by the two straight curves.

Two effects are visible: first, the low energy detected spectrum loses low-energy events. This is due to the fact that fluorescence events can carry away enough energy from a primary event to reduce the detected energy below the first energy threshold. Second, the two detected spectra overlap significantly. When we normalize both detected spectra to 1, we obtain the system weighting functions of the two energy bins. The overlap reaches around 60% and is thus comparably larger than the 40%–50% overlap of dual kVp or dual-source CT [24]. This indicates that the dual-energy measurement capabilities of direct conversion detectors are significantly affected by low-energy shift mechanisms due to fluorescence.

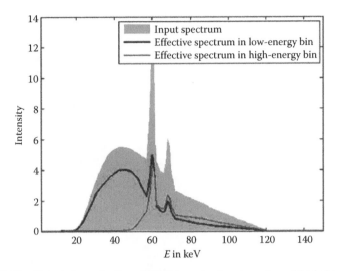

FIGURE 12.14 Detected spectra for a 140 kV tungsten input spectrum (shaded gray).

12.5 CONCLUSIONS

From the MTF results, we can deduce that the spatial resolution of semiconductor detectors is a clear potential benefit for medical imaging devices. The direct conversion of the primary x-ray field information into charge pulses omits the inter-pixel cross talk of scintillator detectors almost completely. The direct conversion into charges demands for a strict control of electrical semiconductor defects. Low-flux screening applications in CT and other medical x-ray devices probably benefit the most from the improved spatial resolution as it requires less image filtering for the same obtained image resolution.

The intrinsic energy resolution of a counting detector readout is a second potential benefit of a direct conversion semiconductor detector. For gamma-ray emission applications such as SPECT and PET, the registered charge is a direct measure for the primary quantum energy. X-ray applications usually require only two or three energy bins defined by threshold energies. The DRF results indicate that CZT semiconductor detectors are prone to a shift of quantum detection to lower-energy bins due to inter-pixel fluorescence cross talk.

The required spatial and spectral resolutions in semiconductor detectors are defined by the targeted medical device. For mammography and radiography detectors, spatial resolution is essential. CT relies on the detection of high x-ray fluxes at intermediate spatial and spectral resolution. SPECT and PET detectors mainly require a superior energy resolution. For each of these applications, detector parameters such as the pixel size and the electronic readout have to be balanced accordingly. For mammography and radiography detectors, spatial resolution is essential. CT relies on the detection of high x-ray fluxes at intermediate spatial and spectral resolution. SPECT and PET detectors mainly require a superior energy resolution. For each of these applications, detector parameters such as the pixel size and the electronic readout have to be balanced accordingly.

REFERENCES

1. G. F. Knoll, *Radiation Detection and Measurement*, 3rd edn. John Wiley & Sons, New York, 2000.
2. I. M. Blevis, M. K. O'Connor, Z. Keidar, A. Pansky, H. Altman, and J. W. Hugg, CZT gamma camera for scintimammography, *Phys. Med. Biol.*, 21 (Suppl. 1), 56–59, 2006.
3. K. B. Parnham, S. Chowdhury, J. Li, D. J. Wagenaar, and B. E. Patt, Second-generation, tri-modality pre-clinical imaging system, M06-29, *IEEE Nuclear Science Symposium Conference Record*, San Diego, CA, pp. 1802–1805, 2007.
4. D. J. Wagenaar, J. Zhang, T. Kazules, T. Vandehei, E. Bolle, S. Chowdhury, K. Parnham, and B. E. Patt, In vivo dual-isotope SPECT imaging with improved energy resolution, MR1-3, *IEEE Nuclear Science Symposium Conference Record*, San Diego, CA, pp. 3821–3826, 2007.
5. E. Kraft, P. Fischer, M. Karagounis, M. Koch, H. Krueger, I. Peric, N. Wermes, C. Herrmann, A. Nascetti, M. Overdick, and W. Ruetten, Counting and integrating readout for direct conversion X-ray imaging: Concept, realization and first prototype measurements, *IEEE Trans. Nucl. Sci.*, 54 (2), 383–390, 2007.
6. D. Moraes, J. Kaplon, and E. Nygard, CERN DxCTA counting mode chip, *Nucl. Instrum. Meth.* A591, 167–170, 2008.
7. Y. Onishi, T. Nakashima, A. Koike, H. Morii, Y. Neo, H. Mimura, and T. Aoki, Material discriminated x-ray CT by using conventional microfocus x-ray tube and CdTe imager, M27-2, *IEEE Nuclear Science Symposium Conference Record*, Honolulu, HI, pp. 1170–1174, 2007.
8. J. P. Schlomka, E. Roessl, R. Dorscheid, S. Dill, G. Martens, T. Istel, C. Baeumer, C. Herrmann, R. Steadman, G. Zeitler, A. Livne, and R. Proksa, Experimental feasibility of multi-energy photon counting K-edge imaging in pre-clinical computed tomography, *Phys. Med. Biol.*, 53, 4031–4047, 2008.
9. S. A. Soldner, D. S. Bale, and C. Szeles, Dynamic lateral polarization in CdZnTe under high flux X-ray irradiation, *IEEE Trans. Nucl. Sci.*, 54 (5), 1723–1727, 2007.
10. E. Bolotnikov, N. Abdul-Jabber, S. Babalola, G. S. Camarda, Y. Cui, A. Hossain, E. Jackson, H. Jackson, J. James, K. T. Kohman, A. Luryi, and R. B. James, Effects of Te inclusions on the performance of CdZnTe radiation detectors, R27-2, *IEEE Nuclear Science Symposium Conference Record*, Honolulu, HI, pp. 1788–1797, 2007.
11. G. S. Camarda, A. E. Bolotnikov, Y. Cui, A. Hossain, S. A. Awadalla, J. Mackenzie, H. Chen, and R. B. James, Polarization studies of CdZnTe detectors using synchrotron x-ray radiation, R27-3, *IEEE Nuclear Science Symposium Conference Record*, Honolulu, HI, pp. 1798–1804, 2007.
12. L. Abbene, S. D. Sordo, F. Fauci, G. Gerardi, A. L. Manna, G. Raso, A. Cola, E. Perillo, A. Raulo, V. Gostilo, and S. Stumbo, Study of the spectral response of CZT multiple-electrode detectors, N24-298, *IEEE Nuclear Science Symposium Conference Record*, Honolulu, HI, pp. 1525–1530, 2007.
13. B. J. Heismann, K. Pham-Gia, W. Metzger, D. Niederloehner, and S. Wirth, Signal transport in computed tomography detectors, *Nucl. Instrum. Methods Phys. Res. A*, 591, 28–33, 2008.
14. T. Michel, G. Anton, M. Boehnel, J. Durst, M. Firsching, A. Korn, B. Kreisler, A. Loehr, F. Nachtrab, D. Niederloehner, F. Sukowski, and P. T. Talla, A fundamental method to determine the signal-to-noise ratio (SNR) and detective quantum efficiency (DQE) for a photon counting pixel detector, *Nucl. Instrum. Methods Phys. Res. A*, 568, 799–802, 2006.
15. I. A. Cunningham, Applied linear system theory, in *Handbook of Medical Imaging*, Vol. 1, J. Beutel, H. L. Kundel, and R. L. van Metter, Eds. SPIE, Bellingham, WA, pp. 79–159, 2000.

16. J. Giersch and J. Durst, Monte Carlo simulations in X-ray imaging, *Nucl. Instrum. Methods Phys. Res. A*, 591, 300, 2008.
17. S. Agostinelli et al., G4—A simulation toolkit, *Nucl. Instrum. Methods Phys. Res. A*, 506 (3), 250–303, 2003.
18. J. Allison et al., Geant4 developments and applications, *IEEE Trans. Nucl. Sci.*, 53 (1), 270–278, 2006.
19. B. J. Heismann, D. Henseler, D. Niederloehner, P. Hackenschmied, M. Strassburg, S. Janssen, and S. Wirth, Spectral and spatial resolution of semiconductor detectors in medical x- and gamma ray imaging, R03-1, *IEEE Room Temperature Semiconductor Workshop*, Dresden, Germany, pp. 73–83, 2008.
20. J. D. Eskin, H. H. Barrett, and H. B. Barber, Signals induced in semiconductor gamma-ray imaging detectors, *J. Appl. Phys.*, 591, 647, 1999.
21. B. Kreisler, J. Durst, T. Michel, and G. Anton, Generalised adjoint simulation of induced signals in semiconductor X-ray pixel detectors, *J. Instrum.*, 3, 11, 2008.
22 T. Michel, Energy-dependent imaging properties of the Medipix2 X-ray-detector, *Proceedings of Science on the 16th International Workshop on Vertex Detectors*, Lake Placid, NY, pp. 044, 2007.
23. M. Rabbani, R. Shaw, and R. van Metter, Detective quantum efficiency of imaging systems with amplifying and scattering mechanisms, *J. Opt. Soc. Am. A*, 4, 895–901, 1987.
24. B. J. Heismann and S. Wirth, SNR performance comparison of dual-layer detector and dual-kVp spectral CT, *IEEE Medical Imaging Conference Record*, pp. 3820–3822, 2007.

13 Advances in CMOS SSPM Detectors

James F. Christian, Kanai S. Shah,
and Michael R. Squillante

CONTENTS

Progress in science and technology is often limited by the ability to detect and accurately characterize low levels of visible light. One of the most demanding and scientifically important technology areas limited by photodetector capabilities is the detection and characterization of ionizing radiation. The most common approach to detect ionizing radiation is to detect the scintillation emitted by certain materials. Crystalline materials, such as single crystal NaI, are efficient at converting ionizing radiation into visible light photons and produce an optical signal that is related to the energy of the incoming radiation. Detecting this signal has been a challenging

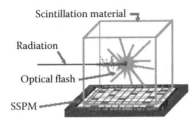

FIGURE 13.1 Illustration of an optical pulse generated in a scintillation material on an SSPM detector.

problem since the 1920s and continues to be a key driving force for improving photodetectors. Developed in the 1930s [1,2], photomultiplier tubes (PMTs) have remained the most sensitive, practical technology for detecting low levels of photons. In the 1940s, PMTs were applied to the detection of ionizing radiation [3–5]. The PMTs in use today are fundamentally the same as those developed in the 1930s and they have significant drawbacks in many applications. To address the limitations of PMTs, researchers have searched for a solid-state alternative for decades. Sensitive silicon semiconductor devices such as p-i-n photodiodes, avalanche photodiodes (APDs), and drift diodes have been developed, which have successfully replaced PMTs in certain specific applications but each has its own limitations. Over the past few years, a new silicon detector technology based on standard silicon complementary metal-oxide-semiconductor (CMOS) processing has been developed, which may finally challenge PMTs in a much wider range of applications.

CMOS solid-state photomultiplier (SSPM) detectors are ideal devices for converting small, fast optical pulses into electrical signals. Many developments in SSPM detectors have been made in the context of nuclear detection applications, which use a scintillation material to convert radiation, such as x-rays, gamma-rays, neutrons, and charged particles, into an optical signal, as Figure 13.1 illustrates. This conversion represents the starting point for many instruments for applications in nuclear detection, spectroscopy, dose measurements, and medical imaging [6–10].

This chapter presents an overview of selected characteristics of photodetector performance, particularly in the context of nuclear detector applications, and describes operating characteristics of emerging CMOS SSPM devices. This chapter also presents some recent advances in scintillation detectors, which have stimulated the development of CMOS SSPM detectors.

13.1 BACKGROUND

A variety of detectors convert the optical signals from scintillation detectors into electrical signals for subsequent processing by the readout components, whose design depends on the application. The earliest type of scintillation-based detectors consisted of a phosphor screen or a scintillation crystal (typically NaI) coupled to a PMT, which provides a large gain and bandwidth for fast light pulses [11]. The PMT is still consistently used, as it provides excellent signal to noise performance. Due to the advent of solid-state devices, other photodetectors have been edging toward the performance characteristics of the PMT. These consist of photodiodes, APDs, and silicon photomultipliers. Each of these solid-state photodetectors has different modes of operation, providing a viable replacement for the PMT within a specified context. The characteristics of these photodetectors, particularly those relevant to scintillation applications, are discussed; however, many references provide a more detailed discussion and derivation of efficiency and noise terms [12–14]. An important part of the nuclear detector consists of the scintillation material, where a brief discussion is given to describe basic characteristics of advanced scintillation materials.

Relevant signal and noise considerations will be discussed in terms of charge, or equivalent charge per event, not current. The emission time of the scintillation event defines the time interval of the measurement. We try to consistently use units of detected photons, photoelectrons, or equivalent electrons to describe signal and noise terms.

13.1.1 Photodetector Overview: PMTs and Silicon Detectors (pins, APDs, and SSPMs)

The basic characteristics of photodetectors include the efficiency for converting photons to photoelectrons, the quantum efficiency (QE) or (single photon) detection efficiency (DE), dark current for a given area, whose fluctuations introduce noise, gain, and excess, or gain, noise. The quantum efficiency, η, refers to the efficiency for converting optical photons into photoelectrons and includes contributions from the optical reflection of the device and the collection of the photoelectron charge within the device, as described in Equation 13.1 [12]:

$$\eta = (1-\Re) \cdot \varsigma \cdot (1-\exp[-\alpha \cdot d]), \tag{13.1}$$

where
$\Re$ is the reflectance
ς is the charge collection efficiency
α is the absorption coefficient, in cm^{-1}
d is the thickness of the detection material, in cm

The product of the quantum efficiency and incident light intensity quantifies the magnitude of the detected optical signal. A signal of interest in scintillation detection applications is the number of photons produced in the scintillation material.

Depending on the mode of operation and the readout electronics, the visible light photons can be measured as an integrated photocurrent or as individual pulses. For gamma-ray spectroscopy applications, pulse measurement, or "pulse counting," is required so that each event can be fully characterized. Photocurrent measurements are inherent measurements of rates of energy absorbed; the description in this chapter is given on a "per event" basis, which is applicable for scintillation detectors. The time dependence is inherently associated with the definition of the "event."

Scintillation pulses of light have time durations ranging from nanoseconds to microseconds depending on the scintillating material. Collecting this signal involves an integration time long enough to collect the emission signal. The signal is described as the number of optical photons, n_p, or the number of photoelectrons, n_e, for an event, often without explicitly considering the time dependence, although it is inherently present as the integration time, or time interval needed to collect the signal.

Maximizing the signal, or collection of optics photons, often represents the first and most important step in selecting a detector. The quantum efficiency generally depends on the wavelength of the optical photons; therefore, matching the response of the photodetector to the wavelength of the optical photons optimizes the signal amplitudes. A long integration time increases the noise due to dark current fluctuations and reduces the maximum event rate.

Once the magnitude of the signal is optimized, the noise terms associated with various photodetectors are considered. At room temperature, most photodetectors produce dark current (or dark events), which is a source of current produced from thermally induced signals in the photodetector. The rate of dark events (dark current) depends on temperature, and the random fluctuations associated with dark events introduce noise. The thermally generated events can be thought of as an "equivalent background" of events. The input-referenced dark event rate (current) is the event rate relative to no amplification. Considering the inherent integration, or accumulation time, associated with the signal, we will generally refer to the amount of dark charge accumulated for a given scintillation event, n_d. The statistical fluctuations associated with this input-referenced background charge affect the signal-to-noise performance of the photodetector.

Another important characteristic of many optical detectors is their ability to multiply the signal and provide gain. Internal gain in a photodetector is needed to surmount the input noise of nonideal amplification electronics. Internal gain also reduces the significance of Ohmic, or surface (non-bulk) leakage current. The noise in the readout electronics trades off with the complexity, speed (bandwidth), and cost. The gain of the photodetector, M, is given on a per event basis, where the average is $\langle M \rangle$. The multiplication process introduces noise associated with statistical fluctuations in M. The excess noise factor, F, defined in Equation 13.2, quantifies the fluctuation associated with the detector in relationship to the multiplied input signal, or number of detected photons:

$$F \equiv \frac{\langle M^2 \rangle}{\langle M \rangle^2} = \frac{\sigma_M^2}{\langle M \rangle^2} + 1, \qquad (13.2)$$

where
$\langle M^2 \rangle$ is the second moment
σ_M^2 is the variance in the gain

Although inhomogeneities in the multiplication materials and in the scintillation materials will introduce variations, these are generally not considered in the excess noise term. The excess noise factor is essentially a multiplicative factor for the variance of the gain distribution, or a ratio of the measured distribution in the pulse amplitudes relative to the idealized expected distribution from the input signals. It effectively represents a scaling for the signal-to-noise ratio (SNR) relative to the quantum efficiency, as described in Equation 13.3:

$$\text{SNR} \propto \frac{QE}{F}. \tag{13.3}$$

To calculate F, the following expression [11] relates measured pulse-height distributions for the measured signal, $\langle n_s \rangle$, to σ_M^2 in Equation 13.2 for a photodetector with gain, $\langle M \rangle$:

$$\langle n_s \rangle = \langle M \rangle \cdot \langle n_e \rangle, \tag{13.4}$$

and

$$\sigma_s^2 = \langle M \rangle^2 \cdot \sigma_{ne}^2 + \langle n_e \rangle \cdot \sigma_M^2 \tag{13.5}$$

where
  $\langle n_s \rangle$ is the average signal
  $\langle M \rangle$ is the gain of the photodetector
  $\langle n_e \rangle$ is the initial number of photoelectrons, or detected photons
  σ_s is the deviation due to fluctuation in the average signal
  σ_M is the deviation due to gain fluctuations
  σ_{ne} is the deviation due to fluctuations associated with the detection of photons

The deviation in the average signal is not simply multiplicative error propagation but is determined by the statistics associated with cascading the fluctuation in the incident signal with those of the gain, as seen by the absence of a square term for the average number of photoelectrons, $\langle n_e \rangle$. In an ideal amplifier, the noise of the signal arises from the Poisson statistics associated with the number of electrons in the signal, $\sigma_s^2 = \langle M \rangle^2 \cdot \langle n_s \rangle^2$ and $\sigma_{ne}^2 = \langle n_e \rangle$. With these substitutions, Equation 13.5 gives $\sigma_M^2 / \langle M \rangle^2 = 0$, which means that $F = 1$ for an ideal amplifier.

The detectors mentioned in the following are described in terms of their signal and noise characteristics for detecting events. The detector capacitance affects the noise performance, that is, larger detector capacitance increases the magnitude of the readout noise [10], but the relevance depends on the magnitude of the readout noise relative to the signal associated with the detection signal amplitude, and as a limiting case, the detection of single optical photons. For some applications, the speed of the photodetector is important. The readout electronics, especially in photodetectors with little to no gain, often limits the time response, or bandwidth, of the detection system.

13.1.1.1 Photomultiplier Tubes

PMTs are excellent low-light, high-bandwidth, large-area photodetectors. They are vacuum tubes that convert incident light into an electrical signal, and the Photomultiplier Handbook [11] describes the fundamental principles and operating considerations. In a PMT, visible light photons are converted into electrons by the photoelectric effect in the "photocathode" of the device. These photoelectrons are accelerated in an electric field such that they create additional electrons by secondary emission upon impact with "dynodes." Figure 13.2 shows a simplistic illustration of a PMT detecting and amplifying an incident light signal. In brief, the photocathode converts the incident light signal to an electrical signal, based on the photoelectric effect. The electrons, represented by straight arrows in the figure, are accelerated into the dynodes where they generate additional current. Finally, the total charge signal strikes the anode and is readout with subsequent electronics. In the figure, open circles at the dynodes represent the "equivalent" of holes, which is simply current from the power supply. The figure illustrates two separate paths for the charge, the electron path in the vacuum and the current path for the power supply.

The gain of the PMT depends on the number of dynodes and the bias. PMTs typically provide gains on the order of 10^6. Although the gain can be decreased by lowering the bias, there is a practical lower limit, which constrains their use as low or unity gain detectors. The photocathode and initial gain stage represents a critical step and a major factor in determining the signal and noise performance. Equation 13.6 shows the relative multiplication noise expected as a function of the gain for each stage in the PMT, where $\langle m_1 \rangle$ represents the average number of photoelectrons emitted by the photocathode. The excess noise factor for PMTs is typically ~1.2:

$$\frac{\sigma_M^2}{\langle M \rangle^2} = \frac{\sigma_1^2}{\langle m_1 \rangle^2} + \frac{\sigma_2^2}{\langle m_1 \rangle \cdot \langle m_2 \rangle^2} + \cdots + \frac{\sigma_k^2}{\langle m_1 \rangle \cdot \langle m_2 \rangle \cdots \langle m_{k-1} \rangle \cdot \langle m_k \rangle^2} \quad \sim \frac{1}{\langle m_1 \rangle}. \quad (13.6)$$

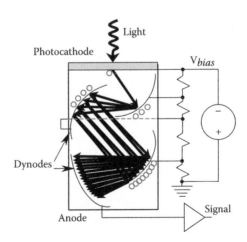

FIGURE 13.2 Simple illustration of a PMT detecting and amplifying an incident optical signal.

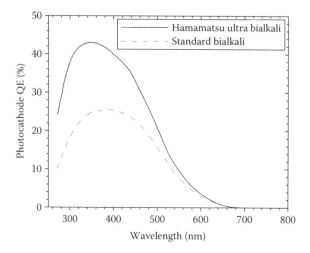

FIGURE 13.3 QE from the Hamamatsu Ultra Bialkali photocathode compared to the standard bialkali photocathode. (Data from the Hamamatsu catalog.)

The wavelength dependence of the QE depends on photocathode material. Bialkali photocathodes are very common and provide good detection for UV to green light. Recent advances in the QE performance of bialkali photocathode materials by Hamamatsu, referred to as "super" and "ultra" bialkali photocathodes, provide a substantial improvement, a factor of 2 in QE, as illustrated in Figure 13.3, data of which are from Hamamatsu catalog [15].

The dark current associated with PMTs and bialkali photocathodes (S-22) is relatively low, ~ 0.005 fA/cm^2 (equivalent to about 30 photoelectrons/s/cm^2). The dark current produces a "background" number of photoelectrons, which is the "input-referenced" dark current, that is, dark current in equivalent photoelectrons (i.e., divide out the gain), multiplied by the time interval of the measurement, or the integration time. The square-root of the number of input-referenced dark electrons yields contribution from the fluctuation in the dark charge to the noise, assuming Poisson statistics. The low dark current is important because its fluctuations represent a noise source and the dark current consumes power, which is a consideration for battery operation.

Generating high voltage for the bias at the dynodes also consumes power, ~ 300 mW, as described by Cunha et al. [16]; however, charge pumps, such as the Cockroft–Walton circuit, decrease this consumption, down to better than 60 mW as described by Brunner et al. [17]. The capacitance of the anode is ~ 5 pF for a 2′-diameter PMT, (Burle 8575). Due to the high gain of the PMT, the readout noise is generally negligible. Miniaturization of PMTs reduces the capacitive load, the spread of the electron signal, and the cost of PMTs.

Microchannel plates: Other types of devices that combine photoelectric effect in a photocathode with secondary emission are the microchannel plate (MCP) and the related channel electron multiplier array (CEMA) [18–20]. Instead of individual dynodes, the secondary emission takes place in a continuous channel where the electric field drives the electrons down the length of the channel, permitting numerous

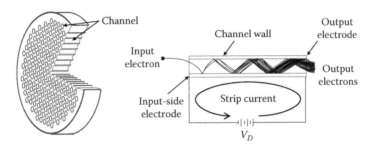

FIGURE 13.4 Operating principle of an MCP, illustrating the secondary multiplication of the input electron, or photoelectron, in the channel. The small size of the channel electron multiplier provides excellent time resolution and the array configuration provides imaging capabilities.

multiplications as shown in Figure 13.4. Individual MCPs have gains of 10^3–10^4; multiple layer "chevron" MCPs can have gains in excess of 10^6. Developed in the 1980s, these devices have high gains similar to PMTs. Using microchannels provides two benefits. First, they are very compact, permitting their use in small portable equipment; second, because the signal is constrained to the microchannel, high-resolution imaging is possible for applications such as night vision.

Hybrid devices: Photocathodes can be coupled to various electron multipliers, such as MCPs and APDs [21–23]. In these devices, the photocathode defines the QE characteristics and the electron multiplier defines the gain and noise characteristics.

13.1.1.2 Silicon Detector Technology

The physical basis for detecting visible light photons in semiconductors is the photovoltaic effect. Observations of the interaction of photons with semiconductors began in the 1800s first with the observation of a photovoltage in AgCl by Becquerel in 1839 [24] and later with the observation of photoconductivity in selenium by Smith in 1873 [25]. The photovoltaic effect is similar to the photoelectric effect except that rather than being emitted from the material, the electrons located in the semiconductor valence band are excited, by the photon, to the conduction band. The development of junction devices allows the use of the photovoltaic effect to measure the intensity of the light striking the detector. A variety of silicon detectors have been invented; the most sensitive of these are p-i-n photodiodes, APDs, Geiger photodiodes (GPDs), and silicon drift detectors.

Photodiodes: Photodiodes are semiconductor diodes designed such that absorbed light photons generate photoelectrons that enter the depletion region of the device. They usually comprise semiconducting materials with p-type and n-type layers, introduced by doping, and sometimes with a third intrinsic layer, a structure referred to as a p-i-n diode. The p-i-n structure enhances sensitivity due to larger sensitive volume and lowers noise due to the low doping, thus high resistivity, of the intrinsic layer. There are many references that provide a thorough, pedagogical description of the physics of semiconductor materials [12,26]. This section simplistically summarizes the basic concepts in the context of photodetection.

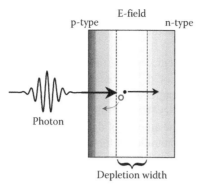

FIGURE 13.5 Illustration of a photodiode, showing the internal electric field at the p–n junction, characterized by the depletion width. The depletion width is the electric field region between the p-type and n-type doping.

In a photodiode, the n and p doping forms a junction which generates an internal electric field. The region near the p–n junction where the charge migrates is called the depletion region. The definition of a "depletion width" is a simplistic, phenomenological average because the internal electrical field is often not uniform and the tails often extend to the end of the device. The width of the internal electric field region, the depletion width, determines the junction capacitance, as illustrated in Figure 13.5, and is affected by an externally applied bias. Forward biasing the diode reduces the depletion width and collapses the field, whereas reverse biasing the diode increases the electric field and extends the depletion width. The internal electric field generates photocurrent by separating the electron–hole pair created by the absorption of a photon, preventing their recombination in the diode. A more complete description of p–n junctions and diodes is given by Sze [26], and a thorough description of photodiodes, and other optical detectors, is given by Saleh and Teich [12].

The bandgap, Eg, of the semiconductor material determines the energy cutoff where photons are absorbed and converted to an electron–hole pair. Analogous to photocathodes, the selection of the material specifies the spectral sensitivity through the wavelength dependence of the QE. The most common and mature semiconductor material is silicon, which has a 1.1 eV bandgap and which can achieve QEs of ~80% for visible photons, 400–700 nm. Silicon is an indirect bandgap material, which means that transitions must be mediated by a phonon. It also means that the flow of current in the material does not efficiently generate photons, unlike light emitting diodes (LEDs).

As indicated in Equation 13.1, the QE is determined by reflectivity, charge collection, and absorption. The reflectivity is a function of the material and the ambient environment. Reducing the difference in the refractive index change or using an antireflective coating will mitigate reflection losses. The absorption is another intrinsic term of the material, and the material thickness affects the probability for stopping an incident photon. The charge collection is related to the recombination lifetime and the internal electric fields near the surface. Although the efficient collection of charge near the surface improves the QE performance for photons absorbed near the surface (e.g., $\lambda < 400$ nm for silicon), it can also degrade the noise performance because of the increased generation of carriers at the surface, which increases the dark current.

The amount of dark current depends on the collection volume, the doping concentration, and the number of defect sites. Both CMOS and pin diodes generate about ~ 1 nA/cm^2 at room temperature. A Boltzman distribution, with an exponent $1/2Eg/(k \cdot T)$, given by Shockley–Reed–Hall [26], describes the temperature dependence of the dark current. The amount of dark current generally scales with the area of the detector; large-area photodiodes for scintillation detectors may require cooling or use high-bandgap materials to reduce the noise from the dark current fluctuations. Of course, high-quality wide-bandgap materials represent the best recourse for low-power, low-noise devices.

In general, photodiodes are operated at zero bias, or with a small reverse bias, where small is relative to the breakdown voltage. Under these conditions, the photodiodes provides no internal gain, so external readout electronics often defines gain, bandwidth, and noise performance. The overall capacitance, which contributes to the readout noise, depends on depletion width and device area. As described in the following, current–voltage (I–V) and capacitance–voltage (C–V) measurements are commonly used to characterize the QE, dark, and breakdown characteristics of diodes, with I–V measurement, and the junction capacitance with the C–V measurements.

APDs: A unique device that combines the advantages of solid-state photodetectors with those of high gain devices such as PMTs. APDs have internal gain that provides a high SNR and have high quantum efficiency. They are fast, compact, and rugged; and these properties make them suitable detectors for many applications, including nuclear detectors and imagers.

An APD is basically a diode operated at a very high reverse bias. The physical mechanism upon which avalanche gain depends, impact ionization, occurs when the electric field is strong enough to generate collisions that ionize bound valence electrons in collisions with the photocurrent electrons. This creates an additional electron–hole pair which accelerates to create more charge pairs through further collisions, resulting in current gain.

In its simplest form, an APD is a p–n junction formed in a semiconductor wafer structured in such a way that it may be operated near breakdown voltage under reverse bias. When either photons or charged particles are absorbed in the semiconductor, electron–hole pairs are generated and are accelerated by the high electric field. These electrons gain sufficient velocity to generate free carriers by impact ionization, resulting in internal gain [27,28]. Small area APDs have been in routine use in the telecommunications industry for some years [29–31] and in a variety of other applications such as optical decay measurement, time domain reflectometry, and laser ranging [32]. The standard large-area avalanche diode is a single element sensor made on a thin silicon wafer. Large-area silicon APDs are typically operated with a gain of 200–2,000 but can be operated with gain up to 10,000. This allows them to operate with a better SNR than standard silicon photodiodes [26].

Initial research on large-area silicon APDs was carried out in the late 1960s [33,34], but large-area APDs attracted little attention until the 1980s when improved fabrication techniques led to higher gains and better reproducibility [35]. These devices have been used in a wide variety of nuclear spectroscopy applications [36–39]. In comparison to PMTs, these sensors are smaller, more rugged, and more stable and use less power. They are also far more sensitive and have inherently better SNRs than the other semiconductor photosensors. In addition, they may be operated without cooling and are insensitive to magnetic fields and vibration. These detectors may also be fabricated to directly sense low-energy beta particles and consequently provide a very attractive sensor for numerous applications.

APDs are photodiodes operated with a reverse bias large enough to produce gain through impact ionization. The internal gain provided by APDs reduces the relative contribution of readout noise when detecting small optical pulses from scintillation materials or small pulses are generated by the direct interaction between x-ray and the photodiode. Figure 13.6 illustrates the process of avalanche multiplication in the high-field region of a reverse-biased photodiode. It is not only possible for the electrons to impact ionize and contribute to the gain but it is also possible for the holes to ionize, which is an important consideration in the noise performance. The trapezoidal shape illustrated in the figure represents a bevel at the edge of the device, which prevents the premature breakdown at the edges. This enables the application of bias voltages large enough to provide ample gain. A comprehensive discussion on the operating principles of APD devices is provided by McIntyre [27,28] and Saleh and Teich [12,40], as well as other references, and the previous description is a simplistic summary.

Analogous to photocathodes in PMTs, fabricating APDs in materials other than silicon [41,42] provides sensitivity over a range of wavelengths. The maturity and availability of silicon, and its wavelength response make this material a prime candidate for many scintillation detection applications.

The QE may have a slight dependence on the bias [43] but is often approximated by the QE at unity gain, that is, when operated as a photodiode. Equation 13.7 describes bias dependence of the gain in terms of incident signal, I_0, which could have units

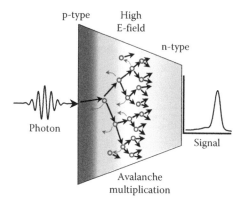

FIGURE 13.6 Illustration of avalanche multiplication in an APD. The bevel shape eliminates premature breakdown at the size of the device, enabling the use of reverse-bias voltages that provide gain through impact ionization.

of current or charge, the avalanche breakdown voltage, V_b, and a phenomenological parameter, alpha, related to the impact ionization probability:

$$M = \frac{I}{I_0} = \frac{1}{1 - (V / V_b)^{\alpha}}. \tag{13.7}$$

The excess noise associated with the ideal impact ionization of a single carrier, for example, electrons, can be evaluated using Equation 13.6 as a PMT with many stages that have a gain of 2 [44]. Substituting $\sigma_k^2 = \langle m_k \rangle$ leads to a series that converges to 1 for the relative gain variance, which gives 2 for the excess noise factor. The excess noise factor increases when the ionization of holes contributes to the signal. The magnitude of the hole-multiplication contribution depends on the doping profile and the strength of the internal electric field. Therefore, the hole-multiplication probability increases with bias until the self-propagating generation of current ultimately leads to avalanche breakdown. The limiting value for the APD excess noise of 2 effectively halves the QE when considering the SNR. In summary, impact ionization produces an excess noise factor of 2, $F = 2$, and hole ionization increases the excess noise factor, $F > 2$, in APDs.

Figure 13.7 shows an APD device in silicon, along with the QE, and typical gain versus bias voltage curve, measured at three temperatures. The dark current is equivalent to photodiode when referenced to the input; and the scaling with area and temperature is also similar to a photodiode. The breakdown voltage also exhibits a dependence on temperature: the breakdown voltage decreases with decreasing temperatures, as seen in the right plot of Figure 13.7. The regulation of the operating bias is important. The power consumption associated with the dark current increases due to internal gain and operating bias.

Geiger photodiodes: Equation 13.7 shows that the gain of an APD increases with increasing bias voltage and as the bias voltage approaches the breakdown voltage, the gain goes to infinity. This trend can be seen in Figure 13.7. At the breakdown voltage, both electrons and holes have sufficient velocity to cause knock on multiplication, which leads to a self-propagating avalanche, or Geiger avalanche. At this voltage, the device effectively becomes a short circuit. When an external resistor

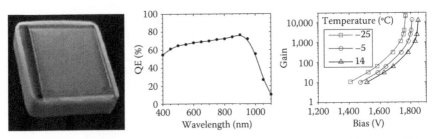

FIGURE 13.7 Picture of a 13 mm × 13 mm silicon APD fabricated at Radiation Monitoring Devices, Inc., Watertown, MA. The center plot shows the typical QE for the APD, and the plot at the right shows the gain as a function of bias voltage for three temperatures.

FIGURE 13.8 Photograph of three chips containing round GPD elements in sizes ranging from 5 μm in diameter (on the center chip) to as large as 180 μm in diameter on the chip at the left (top right). The chip at the right was fabricated with a custom M-R-S process, *circa* 1997, the chip in the center uses a commercial CMOS process and was fabricated around 2004, and the chip at the right was also fabricated in a commercial CMOS process around 2005. The photographs are roughly to scale: the size of the square contact pads at the perimeter of each chip is 100 μm × 100 μm.

limits the current, the GPD devices function as a Geiger–Mueller tube that discharges the junction capacitance, which determines the gain that can be 10^6–10^8. A single charge generated in the device causes breakdown, either from a photon or thermally generated. If the device is small, it is possible to raise the bias above breakdown for macroscopic periods of time between thermally generated charges (McIntyre) [45,46]. In this over-biased state, the device becomes sensitive to single visible light photons. Such devices are APDs operated in Geiger mode, called Geiger photodiodes by Cova et al. [47,48] and Vasile et al. [49,50]. Figure 13.8 shows a photograph of GPD devices designed by Vasile et al. [50] and Christian et al. [51] at Radiation Monitoring Devices, Inc., Watertown, MA.

Solid-state photomultiplier: An SSPM is an array of photodiode, or APD, elements operated in Geiger mode, that is, above the reverse-bias avalanche breakdown voltage, with integrated quenching components [52–54]. The elements contain a quenching component, such as a resistor, that serves the same purpose as the ballast resistor in a Geiger tube. Each SSPM element is a Geiger detector for optical photons, and proportional information, that is, the number of photons detected is proportional to the number of triggered GPD elements when the array is uniformly illuminated, as illustrated in Figure 13.1. Important characteristics of the SSPM detectors are that they provide single photoelectron resolution, that is, they exhibit low excess noise factors, and they can withstand illumination by room lights when biased without damage, that is, they are robust.

Single GPD elements were introduced by McIntyre [45,46] and advanced by Cova et al. [47,48] and Vasile et al. [49,50]. Buzhan et al. [52], who refers to SSPM as SiPMs, introduced a process where arrays of GPD elements with an integrated quenching resistor could be fabricated. Jackson and coworkers [55] also developed analogous sensors, which is the foundation for devices manufactured by SensL. RMD [51] and Rochas et al. [56] developed SSPMs using commercial CMOS foundry processes. We prefer to call these devices SSPMs because we are interested in generalizing the concept of an array of GPD elements to materials other than silicon. Hamamatsu has introduced the multi-pixel photon-counter (MPPC), which is an equivalent device to SSPM. Following is a list of companies that offer SSPM or equivalent devices: Avo Photonics (Horsham, PA), AdvanSiD (Italy), Hamamatsu (Japan), Philips

Suggested range of operating bias	27.5–32.5 V
DE_{max}	15% at 500 nm
Breakdown voltage	27.2 ± 0.2 V
Temperature sensitivity	50 mV/°C
Gain	~V_x × (10⁶)
Typical dark current (output referenced)	100 µA (V_x = 2 V, 1 quadrant)
Equivalent dark count rate (input referenced)	~350 MHz (~56 pA) (V_x = 2 V, 1 quadrant)

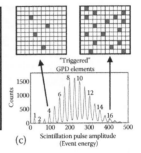

(a)   (b)   (c)

FIGURE 13.9 Photograph of 1 cm × 1 cm SSPM devices (a) and a table of properties (b). The illustration on the (c) shows the dependence of a pulse-height spectra on the number of triggered GPD elements.

(the Netherlands), RMD (Watertown, MA), SensL (Ireland), FBK (Italy), Photonique SA (Switzerland), Excelitas (New York), and Voxtel (Beaverton, OR).

Christian et al. [51,57,58] at Radiation Monitoring Devices, Inc. have designed CMOS SSPM devices that range in size from 1.5 mm × 1.5 mm to 1 cm × 1 cm. Figure 13.9, left, shows a photograph of the 1 cm × 1 cm SSPM devices, next to a table of characteristics for the 1 cm × 1 cm devices, center, along with an illustration of a pulse-height spectra, showing the proportionality to the number of triggered GPD elements.

The amount of charge, q, produced by the illumination, such as a scintillation event, is proportional to the number of triggered GPD elements, $\langle n_t \rangle$, as described in Equation 13.8:

$$q \propto \langle n_t \rangle \sim DE \cdot \langle n_p \rangle \cdot G_q, \quad \text{where } G_q \sim C_{jn} \cdot V_x. \quad (13.8)$$

Equation 13.8 simply embodies the dependence of the signal charge on the number of triggered elements and the number of incident photons. The gain associated with the Geiger discharge, G_q, is assumed to have negligible statistics, unlike the gain term in an APD, designated as M in Equation 13.7. Each triggered GPD element produces a charge pulse, ~10⁶ electrons, that is proportional to the product of the junction capacitance and the excess bias, which is the bias above the breakdown voltage. The variance associated with the charge generated in the Geiger discharge is generally very small, relative to the variance associated with the incident photon statistics, as illustrated by the resolution of the number of triggered elements in the amplitude of scintillation pulse amplitudes in Figure 13.9, right. In the following discussion, the relative variance of the Geiger gain, σ_G/G_q, is neglected, that is, assumed to be very small compared to other terms, which affects the "envelope" of the peaks illustrated in Figure 13.9.

The QE and P_g characteristics of the single GPD element determine the detection efficiency performance of the SSPM detector, as described by Equation 13.9:

$$DE_{SSPM} = QE_{GPD} \cdot ff \cdot P_g \quad (13.9)$$

where

DE_{SSPM} is the detection efficiency of the SSPM

ff is the geometric fill factor, which is the active area of the device

P_g is the Geiger probability

In general, the QE term contains the wavelength dependence and the P_g term contains the P_g dependence; however, the composition is not exclusive: there is a small bias dependence to the QE and a small wavelength dependence to the P_g. Figure 13.10, left, plots the wavelength dependence for the QE of a single GPD element, when operated as a photodiode.

The center plot of Figure 13.10 shows the dependence of the Geiger probability, P_g, on the excess bias for two different incident wavelengths. The small dependence is attributed to the different contributions of electron- and hole-initiated [60] Geiger discharges. The plot on the right shows the detection efficiency for the CMOS SSPM devices given in Figure 13.10 at two excess bias voltages. In scintillation detection applications, where scintillation material is wrapped in reflective material, the scintillation light reflected from the detector may be subsequently detected, which effectively increases the DE.

The large gain associated with the Geiger avalanche introduces additional processes, such as after pulsing and cross talk, which contribute to the signal, especially when operating the SSPM at high excess bias. After pulsing refers to the delayed release of charge from traps in the GPD element that produces subsequent Geiger pulses, and cross talk refers to the triggering of adjacent elements from photons generated by hot carrier emission [60,62]. The bias dependence of these additional processes often limits the maximum operating bias and thus the maximum DE. These processes contribute as an additional multiplication mechanism for the number of triggered GPD elements, as illustrated in Equation 13.10, where the approximation denotes the use of substitutions assuming a linear response for the SSPM to the incident number of photons, which is accurate for small signals, where the number of triggered GPD elements is <30% of the total elements in the array:

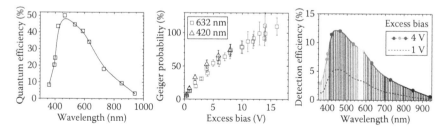

FIGURE 13.10 Plot of the quantum efficiency of a GPD element as a function of wavelength and the Geiger probability as a function of excess bias, $V_{bias} - V_b$, where V_b is the breakdown voltage. The quantum efficiency is a measure of the photoelectron current generated in the diode, while the Geiger probability, P_g, is the probability to generate a self-sustained avalanche above breakdown (excess bias). The plot on the right shows the DE of a CMOS SSPM device at 1 and 4 V excess bias.

$$\langle n_t \rangle = \frac{\langle q \rangle}{G_q} = M_a(\tau) \cdot M_x \cdot \langle n_t^0 \rangle \sim \langle M(\tau) \rangle \cdot \left(DE \cdot \langle n_p \rangle + \langle n_d \rangle \right). \tag{13.10}$$

In this expression, the product of the DE and $\langle n_p \rangle$ represents the average number of pixels triggered by the incident optical signal plus input-referenced dark events, $\langle n_d \rangle$, as the primary event, $\langle n_t^0 \rangle$, and the additional elements in the SSPM that are triggered through either after pulsing or cross talk represent a secondary event characterized by a multiplicative increase in the number of triggered pixels, n_t. The distinction between a primary number of triggered GPD elements and a secondary number of GPD elements is important in developing an expression for the excess noise, or variance, for the device due to contributions from the secondary multiplication terms, where n_t^0 corresponds to n_e in Equation 13.4 and n_t corresponds to n_s. It is often convenient to scale the signal charge, $\langle q \rangle$, by the elements gain, G_q, to present the histogram of the pulse amplitudes in terms of the number of triggered GPD elements.

The contribution of the after pulsing depends on the integration time, τ. The Geiger pulses associated with after pulsing often exhibit variations in the charge produced by the GPD discharge because they often occur immediately after an initiating Geiger event, which means the GPD element may not have been fully recharged to the operating bias.

The statistics associated with the number of triggered pixels is expected to be binomial [62]. The following expressions describe a binomial probability distribution, $P_{n,p}(v)$, and the expected moments of the pulse-height spectrum in terms of the incident light intensity and the detection efficiency of the SSPM:

$$P_{n,p}(v) = \frac{n!}{v! \cdot (n-v)!} \cdot (p)^v \cdot (1-p)^{n-v},$$

$$n \to n_{ttl}, p \to \frac{\langle n_t \rangle}{n_{ttl}} = M \cdot \left(1 - \mathrm{Exp}\left[-\frac{DE \cdot \langle n_p \rangle}{n_{ttl}} \right] \right), v \to n_t$$

$$P(n_t) = \frac{n_{ttl}!}{n_t! \cdot (n_{ttl} - n_t)!} \cdot \left(\frac{\langle n_t \rangle}{n_{ttl}} \right)^n \cdot \left(1 - \frac{\langle n_t \rangle}{n_{ttl}} \right)^{n_{ttl} - n_t}, \tag{13.11}$$

$$\left(\frac{\sigma_{nt}}{\langle n_t \rangle} \right)_{det}^2 \cong \frac{F_{SSPM} \left[\langle n_t \rangle \left(1 - \left(\langle n_t \rangle / n_{ttl} \right) \right) + \langle n_d \rangle \right]}{\langle n_t \rangle^2}.$$

In Equation 13.11, $P_{n,p}(v)$ is the probability of counting v successes, in n trials with a success probability of p. The equation also shows the mapping of this probability distribution to the number of triggered GPD elements in an SSPM, where M is the trigger multiplier, which is the product of M_a and M_x, and the expected variance in the number of triggered GPD elements, where F_{SSPM} is the excess noise term for the SSPM detector. The last expression assumes that binomial statistics describes the variance in the number of triggered elements, and the variance of the after pulsing and cross talk processes is embodied in the excess noise term.

Poisson distributions conveniently describe the detection statistics for small scintillation pulses; however, when the light pulse is bright enough to trigger >30% of the GPD elements, a binomial distribution is more accurate. Nominally, n_{ttl} represents the total number of GPD elements in the SSPM device. Slow scintillation pulses, slow relative to the response of the GPD element, which is ~50 ns in the CMOS devices shown in Figure 13.9, may retrigger the same GPD element, so the value used for n_{ttl} may be "adjusted" to include the affected number of GPD elements that could be triggered during the event, which introduces a τ dependence to the "effective" n_{ttl}.

For large signals, where the number of triggered GPD element is >30% of the total number of elements in the array, the signal from the SSPM exhibits a nonlinear response relative to the incident number of photons. Equation 13.12 describes the relationship between the number of GPD elements triggered by the photons, which excludes cross talk and after pulsing contributions, and a uniform, Poisson-distributed incident photon pulse [63]:

$$\left\langle n_t^0 \right\rangle = n_{ttl} \cdot \left(1 - e^{-DE \cdot \langle n_p \rangle / n_{ttl}} \right). \tag{13.12}$$

Although other functions describe the saturation of the SSPM by Poisson-distributed photons, this function conveniently fits experimental measurements, as illustrated in Figure 13.11, left. Other "saturation functions" that include the dependence on the integration time have been proposed [64].

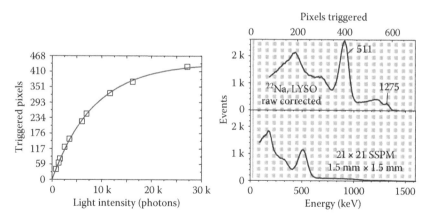

FIGURE 13.11 Plot of the number of triggered GPD elements, or pixels, as a function of the incident light intensity, per event. Notice the nonlinear "saturation" response associated with detecting a Poisson-distributed ensemble of photons with an SSPM containing a finite number of elements. The plot on the right shows the same 400-element SSPM with an LYSO ($Lu_{1.8}Y_{0.2}SiO_5$: Ce) scintillation detector irradiated with ^{22}Na. The plot at the top shows the scale in the number of triggered element, which extends above 441 because the integration time is 100 ns, and the signal above 441 is attributed to the retriggering of elements in the 50 ns event. The plot on the right shows corrected energy spectrum.

The saturation of the SSPM detector also distorts the pulse-height spectrum, as illustrated in Figure 13.11, right. The correction is relatively "straightforward" [65] but is important when quantifying the energy resolution performance for small SSPM detectors or large scintillation signals. Obviously, increasing the number of elements in the SSPM detector also reduces the nonlinear effects for a given scintillation pulse amplitude.

Dark current is uncorrelated, random triggering of GPD elements, analogous to a constant background. Input-referenced, $1\,mm \times 1\,mm$ CMOS SSPM exhibits $\sim 10\,pA$ of dark current at room temperature: with 10^6 gain, the measured currents are $\sim 10\,\mu A$. As with photodiodes and APDs, the input-referenced dark current scales with temperature and area, and the fluctuations associated with the room temperature dark current ultimately limit the size of the detector in applications where pulse amplitude, or energy, resolution is important.

The breakdown voltage of APDs, and the GPD elements in SSPM detectors, also depends on temperature. In an SSPM, the gain, G_q, and the Geiger Probability, P_g, depend on excess bias, $V_{bias} - V_b$, and, therefore, the breakdown voltage. The breakdown voltage typically changes by $50\,mV/°C$. Operating at large excess bias voltages can reduce the temperature dependence in the SSPM signal; however, there is a trade-off between excess bias and noise that must be optimized. Adjusting the bias to keep the excess bias constant provides an approach to stabilize the SSPM against temperature fluctuations.

As shown in Equation 13.8, the gain depends on the junction capacitance, and the capacitance of the SSPM device scales with the area, or the number of GPD elements. A $30\,\mu m \times 30\,\mu m$ CMOS GPD element typically has a junction capacitance, under bias, of $\sim 130\,fF$.

Additional terms that may contribute to the signal charge include an Ohmic term, q_0, and the current that flows through the quenching resistor before the Geiger avalanche is quenched, which is the integral term given in the following. These are generally negligible and assumed to be zero:

$$q_{SSPM} = M_A \cdot M_x \cdot G_q \cdot n_t + q_0 \quad \left(+ \int_0^{\tau_q} i_{R_q} dt \right). \tag{13.13}$$

In analog photodetectors, such as PMTs and APDs, the noise associated with the gain of the optical signal, which is G_q in the digital SSPM device, is relatively large and widens the distribution in the number of detected photons, which is the envelop of the single photoelectron events illustrated in Figure 13.9. In the SSPM detector, the noise associated with G_q is small and negligible compared to terms from processes that can trigger more GPD elements such as cross talk and after pulsing.

13.2 CHARACTERISTICS OF SSPMS

The parallel connection of the GPD elements in an SSPM sums the charge pulses from each element to produce quantized analog output with gains, which can be adjusted through the bias voltage, equivalent to that achieved in PMTs. In this section, we discuss basic electrical characteristics of SSPM devices. The basic model of the GPD element has been described in detail in various works [45,48,66], and the following description provides a summary of simple simulations of circuits with GPD elements.

13.2.1 Fundamental Electrical Characteristics and Model

The absorption of an optical photon initiates a self-propagating avalanche, referred to as a Geiger avalanche, in the GPD element that discharges the junction capacitance. The discharge proceeds until the bias across the diode drops, due to the presence of the quenching, or ballast resistor, to the breakdown voltage. The charge produced in the discharge, therefore, is proportional to the product of the junction capacitance and the excess bias. The charge pulse produced by the Geiger avalanche can be modeled as a current source.

Figure 13.12, left, shows the schematic representation of the SSPM detector, composed of GPD element in series with a quenching, or ballast resistor, connected in parallel. The figure in the center shows an "equivalent" current source representation of the SSPM element, comprising GPD element connected to quenching resistor, R_q. The figure on the right shows the pulse of the current source for simulating the charge from the Geiger avalanche. Much of the "first-order" modeling of the electrical characteristics is based on the work of Cova et al.

The electronic model generates a charge pulse with an exponential rise time, τ_r, and a fast exponential fall, which corresponds to the evacuation of charge from the depletion after the Geiger discharge is quenched. Due to the presence of the field when quenched, where the bias across the GPD drops to V_b, the time associated with sweeping the charge from the depletion region is assumed to be on the order of 10–100 ps. In modeling the charge pulse, the exact value is generally not critical because these time scales are fast compared to the 50 ns recharge time, which is proportional to the product of C_{jn} and R_q.

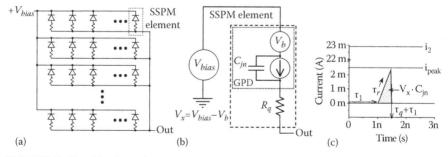

FIGURE 13.12 (a) schematic representation of SSPM detector, showing an array of GPD elements with a series quenching resistor connected in parallel. (b) model of the SSPM element consisting of a transient current source, a junction capacitor, a voltage source for the breakdown reference, a voltage source for the bias, and the quenching resistor. (c) illustration of the transient pulse generated by the current source modeling the charge produced by the Geiger avalanche.

The following list summarizes the "first-order" characteristics of the SSPM elements and input parameters for using a current source to model a simple charge pulse equivalent of the GPD signal:

- V_b: Breakdown voltage
- R_q: Quenching resistance
- C_{jn}: Junction capacitance
- R_{on}: Diode "on," or series, resistance
- V_{bias}: Bias voltage

These parameters can be determined from traditional I–V and C–V measurements of the GPD and SSPM element, as described in the following. The "on," or series, resistance represents the resistance of the diode when it is conducting, which should be related to the sheet resistance associated with the device, and can be recovered from the forward-bias I–V curve. These characteristics can be used to calculate the following parameters needed for using a current source to generate the GPD charge pulse:

- τ_r: Rise time
- τ_q: Quench time
- i_2: Equivalent current

Equation 13.14 relates the GPD characteristics to the parameters needed for modeling the GPD charge pulse with a transient current source:

$$\tau_r = C_{jn} \cdot R_{on}$$

$$\tau_q = -\tau_r \cdot LN\left(\frac{V_b}{V_{bias}} + \frac{V_b}{V_{bias}} \cdot \frac{R_{on}}{R_q} - \frac{R_{on}}{R_q} \right)$$

$$i_2 = \frac{C_{jn}\left(V_{bias} - V_b\right)}{\int_0^{\tau_q} f(t)dt} = \frac{C_{jn}\left(V_{bias} - V_b\right)}{\tau_q + \tau_r\left(\mathrm{Exp}\left(-\tau_q / \tau_r\right) - 1\right)},$$

(13.14)

where $f(t) = 1 - \mathrm{Exp}(t/\tau_r)$. In some versions of SPICE, the parameters listed in Equation 13.14 can be plugged into an exponential pulse function for the current source in circuit simulations; however, there is a complication that depends on the specific version of SPICE.

Modeling the electrical characteristics of the SSPM element as a current pulse with an exponential rise and fall seems straightforward, but alas, many versions of SPICE do not provide an adequate function for the current source. Certain "legacy" versions of SPICE implement the following pulse function that is not suitable for this application:

$$i(t) = I_1 + (I_2 - I_1) \cdot \left(1 - \mathrm{Exp}\left(\frac{-(t - \mathrm{Td}_1)}{\tau_1}\right)\right) + (I_1 - I_2) \cdot \left(1 - \mathrm{Exp}\left(\frac{-(t - \mathrm{Td}_2)}{\tau_2}\right)\right).$$

(13.15)

This function does not produce a simple pulse with an exponential rise and fall when τ_2 is fast compared to Td_2, which is the case for Geiger pulses, where τ_1 is the rise time, Td_1 is an arbitrary start time, Td_2 is the quenching time, and τ_2 is the time needed to sweep the charges from the depletion region upon quenching. Some versions of SPICE, such as Star-HSpice, provide the following function for the time-dependant current source needed to simulate the GPD charge pulse:

$$i(t) = I_1 + (I_2 - I_1) \cdot \left(1 - \mathrm{Exp}\left(\frac{-(Td_2 - Td_1)}{\tau_1}\right)\right) \cdot \mathrm{Exp}\left(\frac{-(t - Td_2)}{\tau_2}\right). \quad (13.16)$$

A current source based on a behavioral model, or a piecewise linear current source, can be implemented as a "work around" for the "legacy" versions of SPICE. The "first-order" description given earlier can be used in simple simulations of circuits with GPD and SSPM components, such as the incorporation of a resistive, charge division network for a position-sensitive (PS) readout of an SSPM in an imaging application.

13.2.2 EXPERIMENTAL CHARACTERIZATION OF SSPMs

The experimental characterization of SSPMs is divided into two sections. One section presents measurements needed to characterize the electrical properties of SSPM devices, and the other section presents measurements that recover the distribution in the number of triggered GPD elements, which provides information needed to quantify the multiplication associated with the average number of triggered GPD elements, and the associated excess noise, due to cross talk and after pulsing.

13.2.2.1 Electronic Characterization of SSPM Devices

Traditional $I-V$ and $C-V$ measurements characterize the basic electrical properties of SSPM devices. The forward bias of a single GPD element provides information on the total series resistance, which includes the quenching resistance and the series, or on, resistance of the GPD. Small variation in the forward-bias barrier complicates the analysis of the forward-bias $I-V$ curve for the SSPM device.

The recovery of the "on," or series, resistance of the GPD requires an independent measurement of the resistance from the quenching resistor, or a measurement of the series resistance for the GPD element. A nonlinear least-squares fit, using a standard menu selection in origin, is used to recover the parameters listed in the plot at the left of Figure 13.13. Due to the form of the fitting function, where an analytic expression for I as a function of V cannot be found when a term for the series resistance is included in the $I-V$ expression for the diode, the fit is performed on a $V-I$ plot, as opposed to an $I-V$ plot, because an analytic expression for V as a function of I can be found.

The reverse-bias $I-V$ curve of the SSPM device provides information on the dark current, the "apparent" detection efficiency, when using a calibrated illumination source ("apparent" refers to the inclusion of signal from after pulsing and cross talk), the breakdown voltage, and the maximum operating bias. The $C-V$ measurement of the SSPM yields the junction capacitance, when divided by the number of GPD elements in the array.

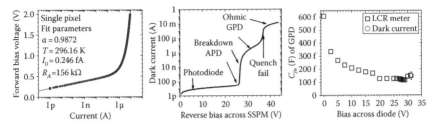

FIGURE 13.13 Plot of the forward-bias dark $I–V$ curve for the SSPM. The forward-bias $I–V$ curve is used to evaluate the silicon series resistance terms. Plot of the reverse-bias dark $I–V$ curve for the SSPM at room temperature. The reverse-bias $I–V$ curve shows diode breakdown voltage, range of Geiger operation, and transition where passive quenching fails; and it labels locations of the curve where the elements operate as photodiodes, APDs, GPDs, and a resistor, labeled Ohmic. Within the range of Geiger operation, the curve begins to turn up, indicating a nonlinear increase due to nonlinear behaviors of the gain, dark current, Geiger-mode multiplier. After passive quenching fails, the device turns on with a current limiting resistance. Plot of the capacitance versus voltage of the SSPM. The square points are measured using the LCR meter. As the entire SSPM is used, the recorded capacitance is divided by the number of pixels in the SSPM, which is 961. The point labeled with a circle and error bar is a measure of the junction capacitance by measuring the dark current and dividing by the number of triggered events per second and assuming the gain is the product of the junction capacitance and the excess bias.

The $I–V$ curve, in the center of Figure 13.13, of an SSPM measured with an HP4145A parameter analyzer shows the region where the elements behave as a photodiode, an APD, a GPD, and a resistor, labeled "Ohmic" in the plot. The plot also shows the increase in current due to avalanche breakdown, around 27 V, and a second increase in current when passive quenching fails, or quenching breakdown, around 38 V. The failure of passive quenching can be seen in Figure 13.14, which shows the $I–V$ curve for a single GPD element that is illuminated with an LED and the waveforms seen on an oscilloscope in the quenching breakdown region. The large quenching resistor protects the GPD from damage at high bias voltages, as well as under strong illumination.

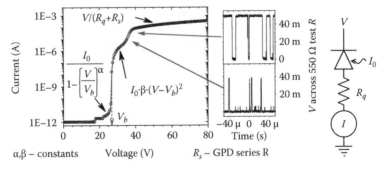

FIGURE 13.14 $I–V$ curve for a single GPD element in an illuminated SSPM, illustrating the avalanche breakdown, V_b, and the quenching breakdown, where passive quenching fails and the device becomes Ohmic.

Figure 13.14 also shows the functions used to recover the breakdown voltage using a nonlinear least-squares fit of the data, which is as a solid curve. Notice that the current near 0 V does not exhibit an improvement in collection efficiency with the application of bias, between 0 and 5 V, because the measured current is below the sensitivity of the picoammeter.

The junction capacitance determines the gain associated with the Geiger discharge, and the gain is useful in scaling histograms to the number of triggered GPD elements. The plot on the right of Figure 13.13 shows a C–V measurement of an SSPM using HP4274A LCR meter. The point labeled with a circle and error bar on this plot is from a measure of the dark current at bias. It measures the total number of triggered GPD elements over a few microseconds, giving a total dark count rate. Dividing the count rate and excess bias from the current measurement yields an estimate of the junction capacitance at the operating bias.

In SSPM devices where test elements are provided, these separate elements can be used to characterize the after pulsing multiplier [67] and a cross talk multiplier [69] using digital, or universal, counters and a pulsed LED. The cross talk multiplier can be estimated by comparing the dark count rate in the separate element when the rest of the array is unbiased and when it is biased. The use of separate elements is convenient because the digital signature of the signals from the SSPM device can be obscured by pileup of the dark events in devices with many elements. Readout electronics that enable the analysis of the waveform minimize the effect of pulse pileup and facilitate the characterization of SSPM devices.

In summary, the electronic characterization provides parameters relevant to the electronic signal generated by the SSPM, which can be used in the "first-order" model described earlier. This is useful when integrating readout electronics on the chip, especially when using standard, commercially available CMOS process. The photodetection characteristics, which include the DE and noise terms that depend on after pulsing and cross talk multiplication, are important in assessing the anticipated performance of the SSPM sensor.

13.2.2.2   Waveform and Transform Analysis

The DE is a key radiometric property of SSPM photodetector and has been described earlier. The excess noise associated with M_a and M_x represents another important characteristic. To quantify this noise term, the multiplication in the number of triggered elements by cross talk and after pulsing must be measured. The presented waveform analysis, many aspects of which are similar to the analysis of Du and Retiere [69], provides a means for extracting the distribution associated with the number of trigger GPD elements, or pixels. Digitizing and analyzing the waveform from the SSPM detector provides a convenient means for characterizing the SSPM device because it eliminates pileup effects. The relatively large gain provided by the SSPM can drive the input of many high-speed amplifiers. Figure 13.15 shows a setup used by Johnson et al. [71] to digitize and store the waveforms from SSPM detectors.

Figure 13.16 shows a plot of the waveform from a 1.5 mm × 1.5 mm SSPM at room temperature. The Geiger pulses are generated by dark events in the device, and the operating bias is ~8 V excess, that is, 8 V above the avalanche breakdown, V_b. The event labeled "Single Event" illustrates a cleanly resolved single Geiger avalanche

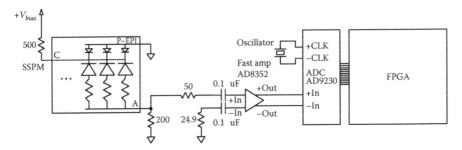

FIGURE 13.15 Block diagram of the setup used to digitize and store the waveform data from a CMOS SSPM detector for subsequent analysis. The setup is essentially a high-speed amplifier coupled to a matching ADC.

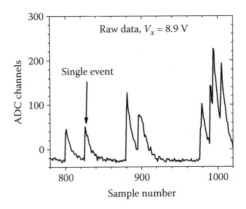

FIGURE 13.16 Plot of the waveform for a 1.5 mm × 1.5 mm SSPM recorded by the setup for the dark events at room temperature.

from a thermal event, which is the single-electron, input-referenced event from the dark current. In the plot, pulses that contain more than one event are present. The purpose of the waveform analysis is to decompose the measured waveform as a superposition of Geiger pulses, where each Geiger pulse is characterized by an amplitude and time. Using this approach, a histogram of pulse amplitudes can be generated for evaluating the distribution associated with the number of triggered GPD elements to quantify the cross talk and after pulsing multipliers and determine the excess noise of the SSPM.

To analyze the waveforms, the GPD response function is generated by averaging selected "good events" in the waveform. The selection process may introduce bias, but the use of the measured waveforms, versus model functions, easily accommodates distortions in the waveform due to impedance matching or the readout electronics. Singular valued decomposition is applied to invert the matrix composed of the representative GPD response waveform, where the starting point of the GPD response waveform is incremented across each column. The inverse is multiplied by the raw data vector to generate a vector that describes the amplitude of the representative

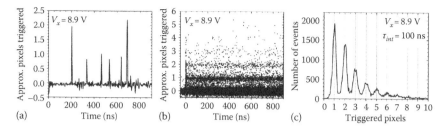

(a)   (b)   (c)

FIGURE 13.17   (a) Plot of amplitude associated with the representative GPD response pulse as a function of time, where the GPD response pulse is essentially deconvolved from the waveform. The (b) plot shows a histogram of pulse heights for 1 μs time intervals with a dark Geiger event defining the origin. The histogram of the number of events as a function of the triggered pixels, which is shown in the plot at the (c), is generated by integrating in time with a selected integration time of 100 ns over a 1 μs time interval. Additional processing has been applied to eliminate the zero amplitude peak and align the peaks to the number of triggered pixels.

GPD response waveform for each step in time. Figure 13.17, left, shows a plot of this transform, which is amplitude, in the approximate number of triggered GPD elements, or pixels, as a function of time. This transformation essentially deconvolves the representative GPD response function from the waveform data.

A histogram of amplitudes and times associated the Geiger events in the waveform, from dark current, is generated by locating the peaks in the transformed data. The events from 1 μs intervals are binned in the center plot of Figure 13.17 such that an initial Geiger pulse, or dark count, defines the time origin. The distribution in the number of triggered GPD elements arises from the statistics of the dark events and contains contributions from cross talk.

A histogram for the number of Geiger events as a function of the amplitude, in triggered GPD elements, or pixels, is generated by integrating the data using a selected time interval. The plot at the right of Figure 13.17 shows the resulting spectrum, using a 100 ns integration time. The deviation in this spectrum from the expected binomial distribution provides a means to quantify the multiplier associated with after pulsing and cross talk, as well as the associated excess noise.

The collection of pulse heights is used to calculate the expected output-referenced dark current by summing the charge associated with each event and dividing by the total collection time. Figure 13.18 compares the calculated dark current at four operating biases to the measured dark current. The agreement confirms that the number and the amplitude of Geiger pulses recovered in the analysis are consistent with the measured current.

The center plot in Figure 13.18 shows the multiplier associated with after pulsing and cross talk at four bias voltages as a function of the integration time. The increase with integration time to a limiting value is attributed to after pulsing, while extrapolating to zero integration time provides a convenient estimate of the cross talk multiplier. The plot at the right of Figure 13.18 shows the excess noise associated with the trigger multiplier, which is the product of M_a and M_x. The excess noise was calculated assuming that the trigger multiplier is a cascaded, or secondary, process

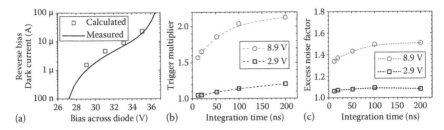

FIGURE 13.18 (a) Plot of dark current calculated from the histogrammed waveforms compared to the measured dark current; note that the reverse-bias avalanche breakdown voltage is ~27 V. The (b) plot shows the multiplier associated with cross talk and after pulsing as a function of integration time, where the cross talk contribution can be extrapolated at zero integration time, and variation with integration time is attributed to after pulsing. The plot at the (c) shows the excess noise factor as function of the integration time.

that is initiated by the primary process associated with the optical or thermal triggering of a GPD element.

SSPM are high-gain, low-noise photodetectors. The excess noise term in SSPMs is associated with fluctuations in the multiplication processes that increase the number of triggered GPD elements, such as after pulsing and cross talk. The analysis of the SSPM waveform provides a method for characterizing the multiplication processes, as well as the fluctuations, or excess noise term.

13.3 OVERVIEW OF EMERGING SCINTILLATION MATERIALS FOR NUCLEAR DETECTION

SSPMs are popular as a detector for scintillation materials because the noise from the dark current is minimized in event counting measurements. The dark events occur randomly in time, whereas the scintillation light illuminates the SSPM in a pulse, so the GPD elements triggered by the scintillation event are essentially in coincidence. The performance of recently developed scintillation materials, in terms of energy resolution, rivals that of room temperature semiconductor detectors, in addition to providing improved capabilities in detecting neutrons and recovering the linear-energy transfer from charged particle measurements. Table 13.1 compares the characteristics of NaI: Tl to that of selection of emerging scintillation materials.

The table shows a selection of materials in the listed classes, as well as a selection of classes of recently developed scintillation materials. Not only do these materials provide unprecedented luminosity but also exhibit a highly proportional response, which yields a high energy resolution, especially for low-energy, <100 keV, events. For coupling to an SSPM detector, the fast scintillation materials enable the use of larger-area detectors, for a specified energy resolution, because the number of dark events that are integrated with the signal can be reduced with short integration times.

13.3.1 GAMMA-RAY SPECTROSCOPY

In scintillation detection applications, the energy of the interaction with the scintillation material determines the amplitude of the light pulse. A histogram of pulse

TABLE 13.1

List Comparing NaI: Tl to Emerging Scintillation Materials, by Category, with Advanced Light-Yield, Proportionality, and Neutron Detection Capabilities

Scintillation Material	Luminosity (Photons/MeV)	Pulse Duration (ns)	Peak Emission (nm)	References
NaI: Tl	38,000	230	415	[6]
Alkaline-Earth Halide Related				
CaI_2: Eu	100,000	1,150	470	[71,72]
SrI_2: Eu	92,000	900	430	[71,72]
BaBrI: Eu	93,000	500	435	[72,73]
Rare-Earth Iodides				
YI_3: Ce	100,000	34	530	[72]
GdI_3: Ce	99,000	33	550	[72,74]
LuI_3: Ce	110,000	30	475	[75]
Elpasolites				
CLYC	20,000	37	370	[75]
CLLB	55,000	55	420	[75]
CLLI	55,000	50	450	[75]
Organics				
TPB	20,000	4–20	440	[76,77]
DPA	20,000	16	470	[76,77]
Li-salicylate	5,000	14	425	[76,77]
II–IV and Others				
ZnSe: Te	55,000	50,000	645	[78]
ZnTe: O	110,000	1,000	645	[79]
$LiGdCl_4$: Ce	97,000	50–9,900	435	[80]

amplitudes versus frequencies describes the detected radiation field. Figure 13.19 shows a block diagram of the components, and a pulse-height spectrum from a standard bialkali PMT, where the amplitude of the single photoelectron event is resolved. This amplitude enables the direct calculation of the number of detected photons, which gives the number of scintillation photons through the QE.

The number of detected photons is useful for characterizing the light yield from the scintillation detector; however, nuclear spectroscopy applications generally convert the pulse amplitude into the energy of the detected radiation event, in keV. In gamma-ray spectroscopy applications, the width of the peak in the histogram represents the energy resolution. Minimizing the energy resolution improves the extraction of information from the spectrum, in the presence of background, which is important in identifying the isotopes that generate signatures in the spectrum. For scintillation detector, the measured energy resolution can be written as the sum of three terms, as expressed Equation 13.17:

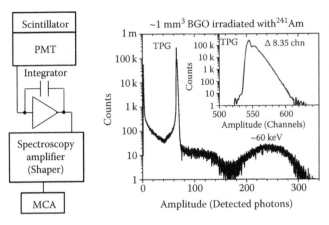

FIGURE 13.19 Block diagram of the components used to characterize the light yield from a scintillation detector and perform nuclear spectroscopy measurements. The measured pulse-height spectrum can be quantified in terms of the number of detected photons when single optical photons are resolved, as shown in the inset with a fit to a Poisson distribution. "TPG" labels a tail-pulse generator signal, which is a charge pulse injected into the input of the integrator for calibration. The data in the inset were taken without the ^{241}Am source but are to the "TPG" peak labeled in the spectrum.

$$\left(\frac{\sigma_E}{E}\right)^2 = F \cdot \left(\frac{\sigma_E}{E}\right)^2_{Detector+Dark} + \left(\frac{\sigma_E}{E}\right)^2_{Scintillation} + \left(\frac{\sigma_E}{E}\right)^2_{Readout} \qquad (13.17)$$

where

F is the excess noise factor

the term "Detector + Dark" represents the contribution from the detected photon statistics and the dark current

the term "Scintillation" includes the effect of the non-proportional response for the scintillation material

the term "Readout" contains the contribution from the readout electronics

The equation shows that the excess noise has a direct impact on the energy resolution. The average energy of the interaction, E, is proportional to the number of photons generated in the scintillation material. The average interaction energy, and the variance, is directly related to the number of detected photons, and a nonlinear response in the detector, for example, SSPMs, needs to be included when calculating this term [65]:

$$\left(\frac{\sigma_E}{E}\right)^2_{Detector+Dark} \sim \left(\frac{\sigma_s}{\langle n_s \rangle}\right)^2_{Detector+Dark} , \text{ with linear response.} \qquad (13.18)$$

In Equation 13.18, the signal contains a contribution from the detected scintillation light and a contribution from the dark current. The contribution of the dark current can be assessed from the width associated with the peak from a tail-pulse

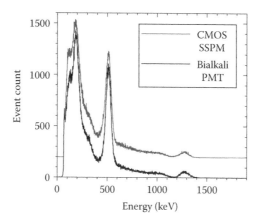

FIGURE 13.20   Gamma-ray spectrum from a (5 mm)³ LYSO crystal irradiated with a ²²Na source coupled to a (5 mm)² CMOS SSPM and a bialkali PMT, illustrating comparable energy resolution performance.

generator (TPG); however, the contribution from readout noise to the TPG width must be subtracted.

A nonlinear, or non-proportional, response of the scintillation material, where an amount of scintillation light is not linearly dependent on the interaction energy, or energy deposited in the scintillation detector, contributes to the energy resolution of the detector. This term may be dominant for scintillation materials such as LYSO and NaI. Figure 13.20 shows the gamma-ray spectrum from a 5 mm × 5 mm × 5 mm LYSO crystal irradiated with a ²²Na source when coupled to a 5 mm × 5 mm CMOS SSPM, compared to the spectrum measured when coupled to a bialkali PMT. The energy resolution of the 511 keV peak in the spectra from the two detectors is comparable: the width of the peak is 13% for the CMOS SSPM and 12% for the PMT.

The third term relevant to the energy resolution describes the contribution of the noise from the readout electronics. In gamma-ray spectroscopy, the noise from the integrating preamplifier generally determines the magnitude of the readout noise. The internal gain of the detector increases the signal amplitude relative to the readout noise. The magnitude of the readout, or electronic, noise is often measured with the TPG signal, when the detector is operated at low bias. Many references discuss the minimization of the readout noise in the context of gamma-ray spectroscopy [10,81,82].

13.3.2   OTHER APPLICATIONS: NEUTRON DETECTION, CHARGED PARTICLE DETECTION, AND MEDICAL IMAGING

Other applications for scintillation detectors include neutron detection, for thermal and energetic neutrons, charged particle detection, and medical imaging. The capture of a neutron by doping materials, such as ⁶Li, ¹⁰B, and ¹⁵⁷Gd, in certain scintillation material, which has a high probability at thermal energies, produces an exothermic reaction, which produces light in the scintillation material. For these applications, the ability to distinguish between light produced by neutron capture and light produced by the interaction of gamma-rays is important. Energetic neutrons are often detected by the scintillation light produced from recoil protons.

Charged particles also deposit energy when interacting with scintillation material to produce light. The relationship between the light yield and the mass and energy of the particle has been described by Birks [83,84]. At high particle energies, when the particle travels through the scintillation detector, the energy is deposited, and the amount of light decreases. Applications for charged particle detection include measuring dose due to space radiation [85,86].

Medical imaging applications [7,87] require the detection and location of radioisotopes, used for labeling. Imaging techniques such as SPECT and PET require imaging, or PS, detectors, for images when coupling to scintillation detectors. These nuclear imaging techniques generally label physiological targets. Other imaging modalities, such as MRI or CT, provide a morphological reference. Medical imaging instruments that combine physiological and morphological imaging are emerging. Since MRI provides excellent soft tissue contrast, nuclear imaging detectors that can be operated in the bore of the magnet are needed. PS photomultiplier detectors, described in the following, have been developed for this application.

13.4 CMOS SSPMS FOR SPECIALIZED SCINTILLATION APPLICATIONS

The design and fabrication of SSPM detectors using commercial CMOS technology provides a convenient platform for integrating components to accommodate specialized applications. The first application described in the following is the use of scintillation detection for a dosimeter chip [88,89], where small silicon SSPM detectors represent an ideal, low-cost, low-power, compact scintillation detection platform. The second application describes the integration of a comparator with the GPD elements to produce a pixel-level conditioned signal [90,91] or a digital SSPM. The third application integrates components to produce an imaging sensor, called a PS SSPM [92–94].

13.4.1 Real-Time and Spectroscopic Dosimeters

Many dosimeters integrate, or accumulate, dose from the interaction of radiation with the material. Often these detectors require a readout step, where they must be heated or illuminated. The use of scintillation materials for dose applications provides a sensitive, real-time detection platform. Incorporating CMOS SSPM detectors enables the event by event detection of radiation, similar to a Geiger tube, but unlike a Geiger tube, the amplitude of the event signal, for example, the scintillation pulse, describes the energy of the interaction.

Figure 13.21 shows a diagram of a dosimeter chip, fabricated in a CMOS process, at the left, above a photograph of the dosimeter chip. The chip contains an SSPM detector coupled to an analog-to-digital converter and random-access counters to record the number of detected events within an energy range. Information on the energy of the interactions is needed to provide a human-equivalent dose when using scintillation materials that are denser than human tissue.

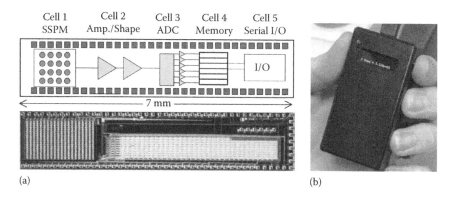

Cell 1   Cell 2   Cell 3   Cell 4   Cell 5
SSPM   Amp./Shape   ADC   Memory   Serial I/O

I/O

← 7 mm →

(a)   (b)

FIGURE 13.21   (a) Diagram and picture of a dosimeter chip using a CMOS SSPM detector. The photograph on the (b) shows a battery-operated dosimeter unit using the dosimeter chip.

The picture on the right of Figure 13.21 shows a photograph of a prototype dosimeter unit that houses the dosimeter chip. The unit is battery operated and displays the absorbed dose. The integration of the components on the chip in the application-specific integrated circuit minimizes the power consumption, resulting in an extended battery life. The dosimeter chip can accommodate a number of different scintillation materials.

13.4.2   Pixel-Level Conditioned (Digital) SSPM

The resolution of the number of triggered GPD elements in an SSPM degrades as the number of triggered elements increases due to the propagation of fluctuations in the GPD element gain, G_q. It should be noted, however, that this "blurring" of the resolution is substantially smaller than the envelope of the distribution associated with the number of triggered GPD elements. To preserve the resolution of the number of triggered GPD elements, Christian et al. at Radiation Monitoring Devices, Inc. fabricated an array that integrated a comparator with each of the GPD elements [90,91]. Figure 13.22 shows a photograph of four SSPM devices, where each GPD element in the device contains a comparator, and the digital signal from the comparator is connected in parallel for the output. The fill factor of the RMD devices is 29%, compared to 50% for the devices illustrated in Figure 13.9.

The signal conditioning not only preserves the resolution of the number of triggered GPD elements but also eliminates the temperature dependence in the gain. Philips has also fabricated a similar device, which they refer to as digital SSPM [95].

13.4.3   Position-Sensitive SSPM

When imaging with a scintillation material, the location of the event as well as the energy of the event must be resolved. In an Anger camera, the scintillation light is distributed over a number of PMT detectors. In some incarnations, the signal from the PMT detectors is summed in a charge dividing resistive network. The ratio of the

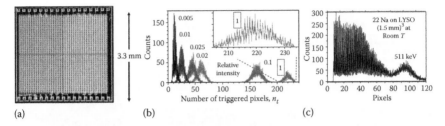

FIGURE 13.22 (a) Picture of CMOS SSPM design by Christian et al. at Radiation Monitoring Devices, Inc., Watertown, MA, with pixel-level conditioning, referred to as a digital SSPM by Phillips, next to a histogram, (b), from a pulsed LED with different incident intensities. Notice the resolution of the multiplicity in the number of triggered elements for all intensities. The plot at the (c) shows the spectrum from an LYSO scintillation detector irradiated with ^{22}Na, clearly resolving the number of detected photons in the 511 keV peak.

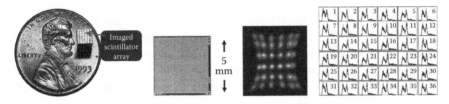

FIGURE 13.23 Photograph of a segmented LYSO scintillation array (from whom) next to picture of PS SSPM and image generated for nuclear medicine.

signal at the four contacts of the resistive network yields the location of the scintillation event, while the sum yields the energy.

An analogous charge sharing resistive network can be integrated on an SSPM detector [92–94]. The ratio of the signal at four readout contacts determines the location of the event [96] and the sum of the signals determines the energy of the event. Figure 13.23 shows a photograph of segmented LYSO scintillation array, next to a photograph of a 5 mm × 5 mm PS SSPM. When coupled to the PS SSPM, the segmented scintillation detector produced the image shown in the figure when irradiated with a flood field of gamma-rays from a ^{22}Na source.

As seen in the image, the location of each of the scintillation segment is cleanly resolved. The plot to the right of the image shows the energy spectrum obtained for each of the scintillation segments resolved in the image.

The high-gain, small number of electrical connections and its insensitivity to magnetic fields make the PS SSPM detector an attractive technology for PET MRI. Of course, for PET imaging, preservation of the timing resolution is important, and readout circuits have been developed for this purpose [97].

13.5 SUMMARY

In summary, photodetectors commonly used with scintillation materials are essentially characterized by their efficiency for detecting optical signals, the noise associated with internal gain, as given by the excess noise factor, and fluctuations in the dark current. In scintillation detection applications, the signal is a pulse of charge produced by the scintillation event on the charge produced by thermal events, or dark current. PMTs represent an excellent low-noise high-gain detector and have seen recent improvements in the photocathode QE. PMTs exhibit very small dark currents but use high voltages, which complicates their use in hand-held instruments.

Photodiodes provide excellent QE performance but have not been extensively used in scintillation application because the number of photons in a scintillation pulses is often small. APDs, which are photodiodes with gain, are ideal for scintillation detection, as well as the direct detection of x-rays. Impact ionization limits the excess noise performance, $F = 2$, which effects SNR as an equivalent factor-of-2 reduction in the QE, and hole multiplication, which increases at high bias, worsens the excess noise performance, $F > 2$.

SSPM detectors, which are an array of GPD elements, represent an emerging technology that provides the gain and excess noise performance of PMTs. They are robust in that they withstand exposure to large signals, such as room light, while biased. They can be operated as photodetectors down to unity gain by adjusting the bias to a value below the breakdown as an array of photodiodes. The key performance characteristics of SSPM devices are the detection efficiency, the input-referenced dark current, and the excess noise, which is introduced by processes that trigger more GPD elements. The use of standard CMOS technology in the fabrication of SSPM devices facilitates the integration of circuit components on the detector and provides the economy of scale for their production.

ACKNOWLEDGMENTS

The authors would like to thank the people who have contributed to the APD and SSPM development work performed at Radiation Monitoring Devices, Inc., Watertown, MA, including colleagues and collaborators, as well as the following funding agencies that have made this work possible: DTRA, NASA, DNDO, and DOE.

REFERENCES

1. H. Iams and B. Salzberg, The secondary emission phototube, *Proc. Inst. Radio Eng.*, 23: 55–64, 1935.
2. Photomultiplier, Wikipedia, http://en.wikipedia.org/wiki/photomultiplier (retrieved March 2011).
3. H. Kallmann, Lesson in a colloquium in Berlin-Dahlem, reported by Bloch, W, Kannman Elektronen sehen, *Natur und Technik*, 13: 15, July 1947.
4. R. Hofstadter, Alkali halide scintillation counters, *Phys. Rev.*, 74: 100–101, 1948.
5. H. Kallmann, Quantitative measurements with scintillation counters, *Phys. Rev.*, 75: 623–626, 1949.
6. G. F. Knoll, *Radiation Detection and Measurement*, 3rd edn. New York: John Wiley & Sons, 2000.
7. S. R. Cherry, J. A. Sorenson, and M. E. Phelps, *Physics in Nuclear Medicine*. Philadelphia, PA: Saunders, 2003.
8. H. H. Barrett, W. Swindell, and R. Stanton, Radiological imaging: The theory of image formation, detection, and processing, *Phys. Today*, 36: 61–62, 1983.

9. W. R. Leo, *Techniques for Nuclear and Particle Physics Experiments*, 2nd edn. New York: Springer-Verlag, 1994.
10. V. Radeka, Low-noise techniques in detectors, *Annu. Rev. Nucl. Part. Sci.*, 38: 217–277, 1988.
11. R. W. Engstrom, *Photomultiplier Handbook*, Vol. PMT-62. Lancaster, PA: RCA, 1980.
12. B. E. A. Saleh and M. C. Teich, *Fundamentals of Photonics*. New York: John Wiley & Sons, 1991.
13. J. R. Janesick, Charge-coupled CMOS and hybrid detector arrays, *Proceedings of the SPIE*, 5167, pp. 1–18, 2004.
14. J. H. Moore, C. C. Davis, and M. A. Coplan, *Building Scientific Apparatus*. New York: Addison-Wesley Publishing Company, Inc., 1989.
15. Hamamatsu-Photonics, UBA (Ultra Bialkali) SBA (Super Bialkali) Photomultiplier Tube Series, Electron Tube Division (www.Hamamatsu.com), Iwata City, Japan, TPMH1305E01, 2007.
16. J. P. V. S. Cunha, M. Begalli, and M. D. Bellar, High voltage power supply with low power consumption for photomultiplier tubes, in *IEEE Nuclear Science Symposium Conference Record*, Knoxville, TN, Vol. N47-119, pp. 1354–1357, 2010.
17. A. Brunner, R. R. Crittenden, A. R. Dzierba, J. Gunter, R. W. Gardner, C. Hamm, R. Lindenbusch, D. R. Rust, E. Scott, P. T. Smith, C. Steffen, T. Sulanke, and S. Teige, A Cockroft–Walton base for the FEU84-3 photomultiplier tube, *Nucl. Instrum. Methods Phys. Res. Sect. A*, 414: 466–476, 1998.
18. J. L. Wize, Microchannel plate detectors, *Nucl. Instrum. Methods Phys. Res. A*, 162: 587, 1979.
19. Burle Industries, Inc., BURLE Long-Life. MCP Selection Guide, (www.Burle.com—Formerly Galileo), Vol. EP107/MAR05, 2005.
20. Del-Mar-Ventures, Microchannel plates and microchannel plate detectors, http://www.sciner.com/MCP/ (retrieved March 2011).
21. Becker and Hickl-GmBH, The HPM-100-40 hybrid detector application note, http://www.becker-hickl.de/pdf/hpm-appnote03.pdf (retrieved March 2011).
22. A. Fukasawa, J. Haba, A. Kageyama, H. Nakazawa, and M. Suyama, High speed HPD for photon counting, *IEEE Nucl. Sci. Symp. Conf. Rec.*, N05-1: 43, 2006.
23. R. A. La Rue, K. A. Costello, C. A. Davis, J. P. Edgecumbe, and V. W. Aebi, Photon counting III–V hybrid photomultipliers using transmission mode photocathodes, *IEEE Trans. Electron Devices*, 44: 672–678.
24. A. E. Becquerel, Recherches sur les effets de la radiation chimique de la lumiere solair au moyen des courants electriques, *Comptes Rendus de L'Academie des Sciences*, 9: 145–149, 1839.
25. W. Smith, Effect of light on selenium during the passage of an electric current, *Nature*, 7: 303, 1873.
26. S. M. Sze, *Physics of Semiconductors*, 2 edn. New York: John Wiley & Sons, 1981.
27. R. J. McIntyre, Multiplication noise in uniform avalanche diodes, *IEEE Trans. Electron Devices*, 13: 164–168, 1966.
28. R. J. McIntyre, The distribution of gains in uniformly multiplying avalanche photodiodes: Theory, *IEEE Trans. Electron Devices*, ED-19: 702–713, 1972.
29. S. Barber, Photon counting with avalanche photodiodes, *Electr. Eng.*, 56: 63, 1984.
30. R. G. W. Brown, R. Jones, J. G. Rarity, and K. D. Ridley, Characterization of silicon avalanche photodiodes for photon correlation measurements. 2: Active quenching, *Appl. Opt.*, 26: 2383, 1987.
31. R. G. W. Brown, K. D. Ridley, and J. G. Rarity, Characterization of silicon avalanche photodiodes for photon correlation measurements. 1: Passive quenching, *Appl. Opt.*, 25: 4122, 1986.

32. M. Ghioni and G. Ripamonti, Improving the performance of commercially available Geiger-mode avalanche photodiodes, *Rev. Sci. Instrum.*, 62: 163, 1991.

33. G. C. Huth, Large area avalanche photodiodes, *IEEE Trans. Nucl. Sci.*, NS-13: 36, 1966.

34. G. C. Huth, Recent results obtained with high field, internally amplifying semiconductor radiation detectors, *IEEE Trans. Nucl. Sci.*, 13: 36–42, 1966.

35. G. Reiff, M. R. Squillante, H. B. Serreze, G. Entine, and G. C. Huth, Large area silicon avalanche photodiodes: Photomultiplier tube alternate, *Mat. Res. Soc. Symp. Proc.*, 16: 131, 1982.

36. M. R. Squillante, G. Reiff, and G. Entine, Recent advances in larger area avalanche photodiodes, *IEEE Trans. Nucl. Sci.*, NS-32: 563, 1985.

37. R. Farrell, K. Vanderpuye, L. Cirignano, M. R. Squillante, and G. Entine, Radiation detection performance of very high gain avalanche photodiodes, *Nucl. Instrum. Methods Phys. Res. A*, 352: 176–179, 1994.

38. G. Entine, G. Reiff, M. Squillante, H. B. Serreze, S. Lis, and G. Huth, Scintillation detectors using large area silicon avalanche photodiodes, *IEEE Trans. Nucl. Sci.*, 30: 431–435, 1983.

39. R. Farrell, F. Olschner, E. Frederick, L. McConchie, K. Vanderpuye, M. R. Squillante, and G. Entine, Large area silicon avalanche photodiodes for scintillation detectors, *Nucl. Instrum. Methods Phys. Res. A*, A288: 137–139, 1990.

40. N. Z. Hakim, B. E. A. Saleh, and M. C. Teich, Generalized excess noise factor for avalanche photodiodes of arbitrary structure, *IEEE Trans. Electron Devices*, 37: 599, 1990.

41. J. C. Campbell, S. Demiguel, F. Ma, A. Beck, X. Guo, S. Wang, X. Zheng, X. Li, J. D. Beck, M. A. Kinch, A. Huntington, L. A. Coldren, J. Decobert, and N. Tscherptner, Recent advances in avalanche photodiodes, *IEEE J. Sel. Top. Quantum Electron.*, 10: 777–787, 2004.

42. X. G. Zheng, X. Sun, S. Wang, P. Yuan, G. S. Kinsey, J. A. L. Holmes, B. G. Streetman, and J. C. Campbell, Multiplication noise of AlxGa1ÀxAs avalanche photodiodes with high Al concentration and thin multiplication region, *Appl. Phys. Lett.*, 78: 3833–3835, 2001.

43. M. Mcclish, R. Farrell, K. Vanderpuye, and K. S. Shah, A reexamination of deep diffused silicon avalanche photodiode gain and quantum efficiency, *IEEE Trans. Nucl. Sci.*, 53: 3049–3054, 2006.

44. Advanced-Photonix, Noise characteristics of advanced photonix avalanche photodiodes, in *Application Notes: API/NOIS/1291/B*, Camarillo, CA, 1991.

45. R. J. McIntyre, On the avalanche initiation probability of avalanche diodes above the breakdown voltage, *IEEE Trans. Electron Devices*, ED20: 637–641, 1973.

46. R. J. McIntyre, Recent developments in silicon avalanche photodiodes, *Measurement*, 3: 146–152, 1985.

47. S. Cova, A. Longoni, and A. Andreoni, Towards picosecond resolution with single-photon avalanche diodes, *Rev. Sci. Instrum.*, 52: 408–412, 1981.

48. S. Cova, M. Ghioni, A. Lacaita, C. Samori, and F. Zappa, Avalanche photodiodes and quenching circuits for single-photon detection, *Appl. Opt.*, 35: 1956, 1996.

49. S. Vasile, J. S. Gordon, R. Farrell, and M. R. Squillante, Fast avalanche photodiode detectors for the superconducting super collider, in *Semiconductors for Room-Temperature Radiation Detector Applications Symposium*, San Francisco, CA, pp. 537–542, 1993.

50. S. Vasile, P. Gothoskar, D. Sdrulla, and R. Farrell, Photon detection with high gain avalanche photodiode arrays, *IEEE Trans. on Nucl. Sci.*, 45: 720–723, June 1998.

51. J. F. Christian, G. Svolos, A. I. Kogan, F. L. Augustine, M. R. Squillante, and G. Entine, Characterization and modeling of APD pixels made with CMOS technology, in *Nano Materials for Defense Applications*, Maui, HI, 2004.

52. P. Buzhan, B. Dolgoshein, A. Ilyin, V. Kantserov, V. Kaplin, A. Karakash, A. Pleshko, E. Popova, S. Smirnov, Y. Volkov, L. Filatov, S. Klemin, and F. Kayumov, The advanced study of silicon photomultiplier, in *Proceedings of the 7th International Conference on Advanced Technology and Particle Physics (ICATPP-7)*, Como, Italy, pp. 717–728, 2002.

53. P. Buzhan, B. Dolgoshein, L. Filatov, A. Ilyin, V. Kantzerov, V. Kaplin, A. Karakash, F. Kayumov, S. Klemin, E. Popova, and S. Smirnov, Silicon photomultiplier and its possible applications, in *Nuclear Instruments and Methods in Physics Research Section A*, 504: 48–52, 2003.

54. C. J. Stapels, W. G. Lawrence, J. Christian, M. R. Squillante, G. Entine, F. L. Augustine, P. Dokhale, and M. McClish, Solid-state photomultiplier in CMOS technology for gamma-ray detection and imaging applications, in *Nuclear Science Symposium Conference Record*, Fajardo, Puerto Rico, Vol. 5, pp. 2775–2779, 2005.

55. D. Phelan, C. Jackson, R. Redfern, A. P. Morrison, and A. Mathewson, Geiger mode avalanche photodiodes for microarray systems, *Proc. SPIE—Int. Soc. Opt. Eng.*, 4626: 89–97, 2002.

56. A. G. Rochas, M. Gani, B. Furrer, P. A. Besse, R. S. Popovic, G. Ribordy, and N. Gisin, Single photon detector fabricated in a complementary metal–oxide–semiconductor high-voltage technology, *Rev. Sci. Instrum.*, 74: 6263–6270, 2003.

57. F. L. Augustine, J. F. Christian, W. G. Lawrence, and C. Stapels, Single photon detection using Geiger mode CMOS avalanche photodiodes, in *Proceedings of the SPIE International Society for Optical Engineering*, Vol. 6013, Bellingham, WA, 2005.

58. C. Stapels, W. G. Lawrence, M. R. Squillante, G. Entine, F. L. Augustine, and J. Christian, The solid-state photomultiplier for an improved gamma-ray detector, in *IEEE Conference on Technologies for Homeland Security (1105)*, April 26–28, Boston, MA, 2005.

59. W. G. Oldham, R. R. Samuelson, and P. Antognetti, Triggering phenomena in avalanche diodes, *IEEE Trans. Electron Devices*, 19: 1056–1060, 1972.

60. A. L. Lacaita, F. Zappa, S. Bigliardi, and M. Manfredi, On the bremsstrahlung origin of hot-carrier-induced photons in silicon devices, *IEEE Trans. Electron Devices*, 40: 577–582, 1993.

61. A. Lacaita, S. Cova, A. Spinelli, and F. Zappa, Photon-assisted avalanche spreading in reach-through photodiodes, *Appl. Phys. Lett.*, 62: 606–608, 1993.

62. R. Mirzoyan, Low-light level sensor applications and needs, in *Light 07*, Ringberg, Germany, 2007.

63. K. F. Johnson, Extending the dynamic range of silicon photomultipliers without increasing pixel count, *Nucl. Instrum. Methods Phys. Res. Sect. A*, 621: 387–389, 2010.

64. S. M. Ignatov, D. A. Maneuski, V. N. Potapov, and V. M. Chirkin, A scintillation g-ray detector based on a solid-state photomultiplier, *Instrum. Exp. Tech.*, 50: 51, 2007.

65. J. F. Christian, C. J. Stapels, E. B. Johnson, M. McClish, P. Dokhale, K. S. Shah, S. Mukhopadhyay, E. Chapman, and F. L. Augustine, Advances in CMOS solid-state photomultipliers for scintillation detector applications, *Nucl. Instrum. Methods Phys. Res. Sect. A*, 624: 449–458, 2010.

66. M. Ghioni, S. Cova, F. Zappa, and C. Samori, Compact active quenching circuit for fast photon counting with avalanche photodiodes, *Rev. Sci. Instrum.*, 67: 3440, 1996.

67. C. J. Stapels, W. G. Lawrence, F. L. Augustine, and J. F. Christian, Characterization of a CMOS Geiger photodiode pixel, *IEEE Trans. Electron Devices*, 53: 631–635, 2006.

68. C. J. Stapels, M. R. Squillante, W. G. Lawrence, F. L. Augustine, and J. F. Christian, Direct photon-counting scintillation detector readout using an SSPM, *Nucl. Instrum. Methods Phys. Res. A*, 579: 87–90, 2007.

69. Y. Du and F. Retiere, After-pulsing and cross-talk in multi-pixel photon counters, *Nucl. Instrum. Methods Phys. Res. Sect. A*, 596: 396–401, 2008.

70. E. B. Johnson, C. J. Stapels, X.-J. Chen, C. Whitney, E. C. Chapman, G. Alberghini, F. L. Augustine, and J. F. Christian, CMOS solid-state photomultipliers for ultra-low light levels, in *SPIE 2011 Devense: Security + Sensing*, Orlando, FL, 2011.

71. N. J. Cherepy, S. A. Payne, S. J. Asztalos, G. Hull, J. D. Kuntz, T. Niedermayr, S. Pimputkar, J. J. Roberts, R. D. Sanner, T. M. Tillotson, E. v. Loef, C. M. Wilson, K. S. Shah, U. N. Roy, R. Hawrami, A. Burger, L. A. Boatner, W. S. Choong, and W. W. Moses, Scintillators with potential to supersede lanthanum bromide, *IEEE Trans. Nucl. Sci.*, 56: 873, 2009.

72. K. S. Shah, Advances in scintillation and related technologies, in *Presented to DOE NA-22 Program Managers*, Washington, DC, March 18, 2010.

73. E. D. Bourret-Courchesne, G. Bizarri, S. M. Hanrahan, G. Gundiah, Z. Yan, and S. E. Derenzo, BaBrI:Eu^{2+}, a new bright scintillator, *Nucl. Instrum. Methods Phys. Res. Sect. A*, 613: 95–97, 2010.

74. J. Glodo, W. M. Higgins, E. V. D. v. Loef, and K. S. Shah, GdI$_3$:Ce—A new gamma and neutron scintillator, in *2006 IEEE NSS/MIC/RTSD*, San Diego, CA, pp. 1574–1577, 2006.

75. K. S. Shah, *New Scintillation Detectors for PET*. Delft, the Netherlands: Delft University of Technology, 2010.

76. E. V. van Loef, J. Glodo, U. Shirwadkar, N. Zaitseva, and K. S. Shah, Novel organic scintillators for neutron detection, *IEEE Nucl. Sci. Symp. Conf. Rec.*, 1007–1009, 2010 (in press).

77. E. V. van Loef, J. Glodo, U. Shirwadkar, N. Zaitseva, and K. S. Shah, Solution growth and scintillation properties of novel organic neutron detectors, *Nucl. Instrum. Methods A*, 652: 424–426, 2011.

78. P. Schotanus, P. Dorenbos, and V. D. Ryzhikov, Detection of Cds(Te) and ZnSe(Te) scintillation light with silicon photodiodes, *IEEE Trans. Nucl. Sci.*, 39: 546, 1992.

79. V. V. Nagarkar, B. Singh, V. B. Gaysinskiy, S. R. Miller, V. Gelfandbein, H. Bhandari, and M. R. Squillante, Novel synthesis of large area ZnTe:O films for high resolution imaging applications, in *SPIE Medical Imaging Conference*, Orlando, FL, 2011.

80. Y. D. Porter-Chapman, E. D. Bourret-Courchesne, G. A. Bizarri, M. J. Weber, and S. E. Derenzo, Scintillation and luminescence properties of undoped and cerium-doped LiGdCl$_4$ and NaGdCl$_4$, *IEEE Trans. Nucl. Sci.*, 56: 881–886, 2009.

81. F. Olschner, Cremat CSP Application Notes, in (www.cremat.com/CSP_app_notes.htm). Watertown, MA, 2011.

82. Amptek, AE250 Application Notes, in AN250-2, Rev. 3 (http://www.amptek.com/pdf/a250app.pdf). Bedford, MA, 2011.

83. J. B. Birks, Energy transfer in organic phosphors, *Phys. Rev.*, 94: 1567–1573, 1954.

84. J. B. Birks, J. E. Geake, and M. D. Lumb, The emission spectra of organic liquid scintillators, *Br. J. Appl. Phys.*, 14: 141, 1963.

85. E. R. Benton and E. V. Benton, Space radiation dosimetry in low-earth orbit and beyond, *Nucl. Instrum. Methods Phys. Res. Sect. B*, 184: 255–294, 2001.

86. E. R. Benton, *Space Radiation Dosimetry: Lessons Learned and Recommendations*, E. Johnson, Ed. Stillwater, OK: Eril Research and Oklahoma State University, 2006.

87. H. H. Barrett and W. C. J. Hunter, Detectors for small-animal SPECT I, in *Small-Animal SPECT Imaging*, M. A. Kupinski and H. H. Barrett, Eds. New York: Springer, pp. 9–48, 2005.

88. C. J. Stapels, F. L. Augustine, M. R. Squillante, and J. F. Christian, Characterization of CMOS solid-state photomultiplier for a digital radiation rate meter, in *IEEE 2006 Nuclear Science Symposium Conference Record*, 2: 918–922, 2006, San Diego, CA.

89. C. J. Stapels, E. B. Johnson, R. Sia, P. Barton, D. C. Wehe, M. R. Squillante, and J. F. Christian, Digital scintillation-based dosimeter-on-a-chip, in *Nuclear Science Symposium Conference Record*, Honolulu, HI, pp. 1976–1981, 2007.

90. C. J. Stapels, E. B. Johnson, R. Sia, F. L. Augustine, and J. F. Christian, Integrated signal processing of CMOS Geiger photodiode arrays, in *IEEE NSS*, Honolulu, HI, pp. 4586–4590, 2007.
91. C. J. Stapels, E. B. Johnson, S. Mukhopadhyay, P. S. Linsay, E. C. Chapman, and J. F. Christian, Solid state photomultipliers and Geiger photodiodes with integrated readout and signal processing, in *SPIE Silicon Photonics IV*, San Jose, CA, pp. 72200H-72200H-5, 2009.
92. C. Stapels, W. G. Lawrence, M. R. Squillante, G. Entine, F. L. Augustine, and J. Christian, CMOS-based, position-sensitive solid-state photomultiplier, in *IEEE NSS/MIC*, Oct. 26–29, San Juan, Puerto Rico, NM, 2005.
93. M. McClish, P. Dokhale, J. Christian, C. Stapels, E. Johnson, R. Robertson, and K. S. Shah, Performance measurements of CMOS position sensitive solid-state photomultipliers, *IEEE Trans. Nucl. Sci.*, 57: 2280–2286, 2010.
94. P. Dokhale, C. Stapels, J. Christian, Y. Yang, S. Cherry, W. Moses, and K. Shah, Performance measurements of a SSPM-LYSO-SSPM detector module for small animal positron emission tomography, in *Nuclear Science Symposium Conference Record (NSS/MIC), 2009 IEEE*, Orlando, FL, pp. 2809–2812, 2010.
95. T. Frach, C. Degenhardt, B. Zwaans, R. de Gruyter, A. Schmitz, and R. Ballizany, Arrays of digital silicon photomultipliers—Intrinsic performance and application scintillator readout, *IEEE Nucl. Sci. Symp. Conf. Rec.*, 1954–1956, 2010 (in press).
96. J. Zhang, A. M. K. Foundray, P. D. Olcott, and C. S. Levin, Performance characterization of a novel thin position-sensitive avalanche photodiode-based detector for high resolution PET, *IEEE Nucl. Sci. Symp. Conf. Rec.*, 5: 2478–2482, 2005.
97. M. Janecek, J.-P. Walder, P. J. McVittie, B. Zheng, H. v. d. Lippe, M. McClish, P. Dokhale, C. J. Stapels, J. F. Christian, K. S. Shah, and W. W. Moses, A high-speed multi-channel readout for SSPM arrays, *IEEE Trans. Nucl. Sci.*, (submitted), 2011.

14 High-Resolution CdTe Detectors and Their Application to Gamma-Ray Imaging

Tadayuki Takahashi, Shin Watanabe, Shin-nosuke Ishikawa, Goro Sato, and Shin'ichiro Takeda

CONTENTS

14.1 INTRODUCTION

Efforts have long been made to develop room temperature semiconductors with high atomic numbers and wide band gaps. These materials are useful not only in medical and industrial imaging systems, but also in detectors for high-energy astronomy and particle astrophysics in general. Among the range of semiconductor detectors available for gamma-ray detection, cadmium telluride (CdTe) and cadmium zinc telluride (CdZnTe) (CZT) occupy a privileged position due to their high density, the high atomic number of their components, and a wide band gap. A large band-gap energy ($E_{gap} = 1.44\,eV$) allows us to operate these detectors at room temperature [1–4].

The high absorption efficiency of CdTe or CdZnTe comparable with that of NaI and CsI is a very attractive feature. As shown in Figure 14.1, photoelectric absorption is the main process up to 300 keV for these materials, as compared to 60 keV for silicon. Therefore, CdTe and CdZnTe are expected to become more efficient once

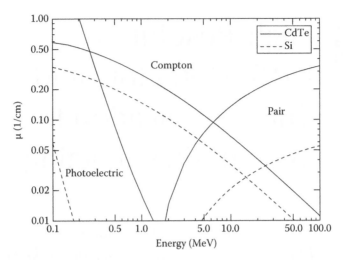

FIGURE 14.1 Linear attenuation coefficients in CdTe and silicon as a function of photon energy. The intensity of photons can be expressed as $I = I_0 \exp(-\mu x)$, where x denotes the path length in centimeters.

the gamma-ray energy exceeds a few hundred kiloelectron volts. Despite long-term efforts to make improvements, high-resolution CdTe and CdZnTe detectors with energy resolution better than a few kiloelectron volts (full width at half maximum or FWHM), however, have only recently become available. In the 1990s, remarkable progress in the technology for producing high-quality single crystals of CdTe by using the traveling heater method (THM) [5] and CdZnTe by the high-pressure Bridgman (HPB) technique [6–8] dramatically changed high-resolution room temperature detectors. The industry has continued such efforts in the 2000s toward producing high-quality CdTe and CdZnTe wafers [9–13]. In addition to the progress made regarding the crystals, various electrode configuration technologies have been proposed and developed to overcome poor carrier transport in the device, as described in this chapter.

To use CdTe and CdZnTe for detecting gamma-rays with energy higher than a few hundred kiloelectron volts, the bias voltage necessary to eliminate the low-energy tail at a level to achieve resolution of a few kiloelectron volts must be higher than a few hundred volts, even with a few millimeters of thickness. If the bias voltage is not high enough, only a fraction of the signal charge generated is induced at the detector electrode. The fraction and resultant pulse height depend on the depth of interaction. This positional dependency produces a shoulder (tailing) in the peaks of gamma-ray lines toward the low-energy region. Increasing the thickness of the CdTe/CdZnTe detector to improve the efficiency of detecting high-energy gamma-rays makes the effects of incomplete charge collection and the low-energy tail more significant. In some cases, the inactive region in the detector volume increases along with thickness, especially for the region near the anode.

Soon after the emergence of HPB-grown CdZnTe, new ideas based on the concept of single charge collection were proposed [14–17]. These ideas included the application of single-carrier detection techniques to semiconductors with low $\mu\tau$ (mobility-lifetime product) values for one type of carrier (typically holes). In this method, an attempt is basically made to change the detectors charge induction property (i.e., weighting potential) in order to improve the charge induction efficiency (CIE) response [3]. Another approach to improve the spectral properties of CdTe detectors is the idea of forming a barrier electrode on the Te-face of the p-type CdTe wafer as an anode [18–20]. A high Schottky barrier between the electrode and CdTe makes the detector operate as a diode (CdTe diode). The significant reduction in the CdTe diode's leakage current allows us to apply high bias voltage for improved CIE without degrading the energy resolution.

Here, we review the achievements made in high-resolution CdTe diode detectors and their application to gamma-ray imaging. Recent progress made in CdTe and CdZnTe detectors has been reported by Takahashi and Watanabe [2], and also by Luke and Amman [3]. The material properties of CdTe and CdZnTe have been reported by Owens and Peacock [4], together with other compound semiconductor materials usable as radiation detectors. Applications to nuclear medicine were reviewed in Refs. [21–24]. The performance of CdTe and CdZnTe onboard astrophysical satellites were described by Limousin et al. [25] for the IBIS instrument onboard INTEGRAL, and by Sato et al. [26] for the BAT instrument used on the Swift gamma-ray burst mission. In the future, the sixth Japanese x-ray satellite, ASTRO-H [27], is planned to be launched in 2014. For this, hard x-ray imager (HXI) [28] as a focal plane detector of the hard x-ray mirror, and soft gamma-ray detector (SGD) [29] are being developed based on the CdTe diode detectors described hereafter.

14.2 HIGH-RESOLUTION CdTe DETECTOR

The technique for growing a large single CdTe crystal with good charge transport properties has been established by the THM [5]. After careful thermal treatment and the selection of proper crystal orientation for the electrode system, the CdTe wafer reportedly exhibits good charge transport properties for both electrons ($\mu_e\tau_e = 1–2 \times 10^{-3}$ cm^2/V) and holes ($\mu_h\tau_h = 1 \times 10^{-4}$ cm^2/V). Electrical resistivity of about $\sim 1 \times 10^9$ Ω cm (p-type) is achieved by using chloride to compensate for native defects. In addition, the detector is free from stability-related problems in the typical electrode configuration with platinum that forms ohmic contacts [5,30]. The wafer's uniform charge transport properties are a very important aspect not only for fabricating large-area strip or pixel detectors, but also for constructing a large-scale gamma-ray camera with many individual detectors. Single crystals 100 mm in diameter are now commercially available for THM-CdTe with high uniformity.

Adopting a Schottky junction, in conjunction with the high quality of THM-CdTe manufactured by Acrorad, dramatically improves the energy resolution of CdTe detectors [18–20]. Since the CdTe material grown by Acrorad has p-type resistivity, a low work function metal, such as indium, can be used to form a Schottky barrier. Moreover, the leakage current is significantly suppressed in reverse bias operation of the In (anode)/CdTe/Pt (cathode) configuration. Thus, the leakage current of a detector with area of 2 mm × 2 mm and thickness of 0.5 mm ($2 \times 2 \times 0.5$ mm^3) is about

0.5 nA with bias voltage of 400 V at 20°C. With this high bias voltage, a CdTe diode with thickness of 0.5–1.0 mm becomes fully active, and even holes generated near the anode face can be completely collected. Cooling the detector further improves the energy resolution.

The theoretical energy resolution of CdTe can be calculated from statistical fluctuations in the number of electron hole pairs and Fano factor (F) [1]. By using $\varepsilon = 4.5$ eV and $F = 0.15$, the theoretical limit (FWHM) is 200 eV at 10 keV, 610 eV at 100 keV, and 1.5 keV at 600 keV [30], provided that we could neglect electronic noise. As shown in Figure 14.2a, a CdTe diode with area of 3 mm × 3 mm and thickness of 1 mm installed in the electronics system manufactured by Amptek [31] had energy resolution of 260 eV (FWHM) at 6.4 keV when operating the detector at −40°C [32,33]. Figure 14.2b shows that the reduction of the low energy tail even in the 662 keV line from ^{137}Cs resulted in resolution of 2.1 keV (0.3%) obtained with a 2 mm × 2 mm and 0.5 mm thick CdTe diode at a bias voltage of 1400 V [30]. In the measurement, the charge signal is integrated in the Clear Pulse CP-5102 charge sensitive preamplifier [34] and shaped by an ORTEC 571 amplifier. Intrinsic resolution by subtracting the contribution of electronic noise was almost as predicted from $\varepsilon = 4.5$ eV and $F = 0.15$ suggesting that the resolution of the CdTe diode with a thickness of 0.5–1 mm almost reaches the theoretical limit for a wide energy range from 10 to 700 keV.

To further reduce the leakage current of Schottky-type detectors, a CdTe diode detector with a guard ring electrode was proposed and tested based on leakage current being proportional to the diode perimeter, not the area or volume [35]. A guard ring structure in the cathode electrode was consequently introduced. This improvement reduced the leakage current by another order of magnitude and enabled us to operate the device at room temperature (i.e., up to ~20°C) with good performance. The typical current at 20°C for this device with an active area of 2 × 2 mm² and a thickness of 0.5 mm was 10 pA at a bias of 400 V [35].

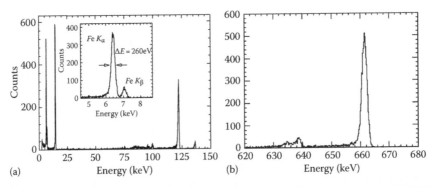

FIGURE 14.2 (a) ^{57}Co spectrum obtained with the CdTe diode. Bias voltage of 800 V was applied. The detector was 3 × 3 × 1 mm³ in size and operating temperature was −40°C. The energy resolution at 6.4 keV was 0.26 keV (FWHM). (From Takahashi, T. et al., *Nucl. Instrum. Methods A*, 541, 332, 2005.) (b) High-energy spectrum of ^{137}Cs above 620 keV with a 2 × 2 mm² CdTe diode 0.5 mm thick. The applied bias voltage was 1400 V and the operating temperature was −40°C. The energy resolution at 662 keV was 2.1 keV (FWHM). (From Takahashi, T. et al., *IEEE Trans. Nucl. Sci.*, 48, 287, 2001.)

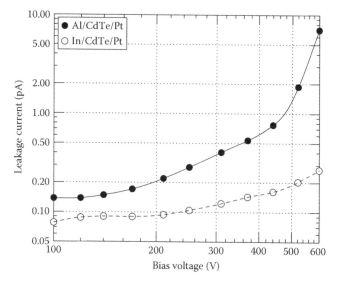

FIGURE 14.3 I–V characteristics at −20°C of CdTe diode detectors with In/CdTe/Pt and Al/CdTe/In electrode configurations, manufactured by ACRORAD. Both detectors had an active area size of $2 \times 2 \, mm^2$ and a thickness of 0.5 mm.

Based on the idea of using a barrier (Schottky) electrode, aluminum recently emerged as an alternative electrode material to indium [36–38]. Because an Al/CdTe/Pt electrode configuration also works as a Schottky diode for p-type CdTe, low leakage current and good energy resolution comparable to those of In/CdTe/Pt detectors can be achieved. Figure 14.3 shows the current–voltage (I–V) characteristics of CdTe diode detectors with the Al/CdTe/Pt and In/CdTe/Pt electrode configurations [39]. As shown in the figure, the lower barrier height afforded by aluminum leads to higher leakage current compared to that of indium. However, an aluminum electrode (which acts as an anode electrode) offers an advantage over an indium electrode because it can be segmented into pixels or strips. Therefore, it is possible to fabricate electron-collecting-type diode pixel detectors utilizing aluminum in pixelated anodes and platinum in the common cathode (Al-pixel/CdTe/Pt) [38]. As shown in Figure 14.4, planar Al/CdTe/Pt detectors with guard-ring electrodes actually offer spectral performance comparable with that obtained with In/CdTe/Pt detectors.

14.3 POLARIZATION PHENOMENON

The CdTe diode detectors show degradation of the spectral properties over time following exposure to high bias voltage, similarly to the so-called polarization phenomenon in semiconductor devices. When CdTe was first utilized for x-ray and gamma-ray detectors, a polarization phenomenon was often observed and considered as attributable to the impurity of crystals [40–43]. Indeed, CdTe detectors based on recent, high-purity crystals manufactured by the THM with the anode and cathode both made of platinum (ohmic devices) do not exhibit the polarization phenomenon,

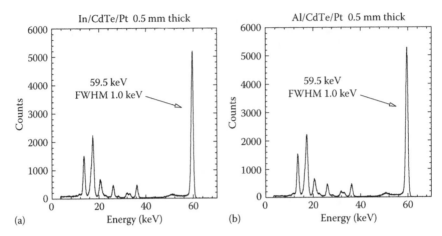

FIGURE 14.4 [241]Am spectrum obtained from CdTe detectors with (a) In/CdTe/Pt and (b) Al/CdTe/Pt electrode configurations with a guard ring structure.

and therefore a series of spectra taken over time does not change. Nevertheless, when Schottky contacts are introduced, we do see a polarization phenomenon at room temperatures and low bias voltages.

As described in our earlier articles (e.g., Takahashi et al. [30]), we found that the polarization phenomenon can be significantly suppressed if we make the detector thin (0.5–1 mm), operating it with high bias voltage at a low temperature of, for example, −20°C. Under these operating conditions, the stability of spectra and absence of polarization are clear for the operation at least a couple of days or more. CdTe diodes can be depolarized by power cycling, that is, turning the bias voltage off and on. It is also generally observed that, when cathode and anode surfaces are irradiated by alpha particles, electron signals rise slower while hole signals rise faster, respectively, as the polarization phenomenon progresses. The fact indicates that negative charges are gradually accumulated inside the detector and the electric field is altered. Together with the temperature dependence, the polarization phenomenon is considered attributable to charge accumulation at deep acceptor levels.

We developed a numerical model and successfully reproduced the changes in spectra over time quantitatively [44], by following the prescription given by Malm and Martini [41]. In the model, a uniform distribution of deep acceptors in the CdTe diode detector was assumed for simplicity. The ionization (detrapping holes) and recombination (trapping holes) processes were formulated. Due to the rectifying property of the Schottky contact, it is considered that the hole density becomes small, and the latter process is suppressed after the bias voltage is turned on. Thus, space charge starts to build up and introduces a change in electric field. Our calculation follows the time evolution of these quantities, and that of corresponding CIE, and a resultant spectrum. As shown in Figure 14.5, the bending in electric field occurs almost at a constant fraction, while the effect in CIE begins with a small change but that produces peak broadening at an early stage of polarization, and proceeds to a drastic decrease causing a peak shift after the time when a region with ~0 strength of

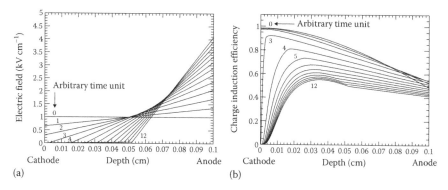

FIGURE 14.5 Simulated time evolution of electric field (a) and CIE (b) during polarization. The detector thickness of 1 mm and the bias voltage of 100 V were assumed.

electric field appears near the cathode. Increase of leakage current can be predicted by that of the electric field at the Schottky barrier (anode side). These features are consistent with what we observe with actual CdTe diode detectors. In this scenario, we can interpret that the high electric field either by high bias voltage or by thin detector material reduces the distortion of the electric field, and diminishes the effect of polarization. Given that the detrapping timescale has an exponential dependence on the temperature, one can expect a much longer detrapping timescale, and thus suppress the polarization phenomenon at lower temperatures.

A controversial issue is the determination of the energy level of the deep acceptors responsible for polarization. Toyama et al. [45] measured the increase of leakage current of an Al/CdTe/Pt detector during polarization. The evaluation of time constant at different temperatures indicated the energy level at 0.69 eV. Another work was done by Cola and Farrella [46] based on the Pockels effect measurement. They derived change in electric field inside an In/CdTe/Pt detector and obtained a similar value (0.62 eV). These results can be ascribed to known cadmium vacancies at the middle band gap [47]. Recently, we measured spectral evolutions of both Al/CdTe/Pt and In/CdTe/Pt detectors for the peak shift time, defined as the time duration taken for the photo peak to shift by 5% from the original position. As shown in Figure 14.6, the peak shift time shows stronger dependence on temperature than the timescales obtained by the other methods. Provided that the peak shift time is proportional to the detrapping time, the slopes correspond to the energy level of 1.1 eV. Another group has reported evaluation of spectral evolution, and the independent data set is consistent with our result [48,49]. We also found a photo-induced polarization phenomenon by illuminating a CdTe diode with infrared light, attributable to the same energy level [50].

14.4 CdTe PIXEL DETECTOR MODULE

For an imaging detector, good energy resolution and the ability to fabricate compact arrays expected from semiconductor detectors are very attractive features compared with inorganic scintillation detectors coupled to either photodiodes or photomultiplier

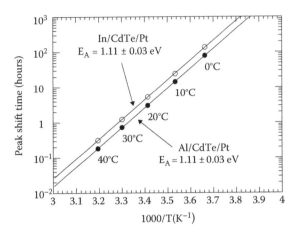

FIGURE 14.6 Peak shift time measured at 59.5 keV peak using In/CdTe/Pt and Al/CdTe/Pt detectors with an area of 4.1×4.1 mm^2 and 0.5 mm thickness. The applied bias voltage was 100 V.

tubes [51]. By forming pixels on the electrode of a large CdTe device, the small capacitance and low leakage current of the pixels help improve energy resolution. Conversely, a large number of channels must be handled in the system, requiring the implementation of very-low-noise analog circuits as a multichannel ASIC (application-specific integrated circuit). Energy resolution of around 1–2 keV would prove challenging if we had to rely on highly compact ASICs based on complementary metal oxide semiconductor (CMOS). Recent advances made in two-dimensional ASIC have led to the development of a fine pitch pixel detector [52–54]. As an application of the CdTe diode, we tried to make a fine-pitch CdTe diode imaging detector based on the ASIC developed by Bonn University [55] and Caltech [56]. In addition to the signals from pixels formed on the anode face, information from the cathode (common electrode) can be used to obtain the depth of gamma-ray interaction. This interaction depth provides correction for the electron path length and trapping, which are important for detectors up to 1 cm thick. He et al. conducted an extensive study on this possibility [57,58].

A large-area CdTe imaging detector with an area greater than a few tens of square centimeters with energy resolution of 1–2 keV (FWHM) is attractive for various applications. We took a modular structure approach in constructing a large-area imaging CdTe detector [59]. Figure 14.7a shows a CdTe pixel detector module consisting of an 8×8 CdTe pixel device, a fanout board, and a 64-channel readout ASIC (VA64TA) [60]. The CdTe device is 13.35 mm $\times$ 13.35 mm in size and 0.5 mm thick. This is a Schottky CdTe diode device that utilizes an indium anode and a platinum cathode. The indium side is used as a common electrode; the platinum side is divided into $8 \times 8 = 64$ pixels. The pixel size is 1.35 mm $\times$ 1.35 mm, with a 50 μm gap between the pixel electrodes. To reduce leakage current, a guard ring electrode 1 mm wide is attached at the outer edge of the detector. A thin layer of gold is evaporated on the platinum side for ensuring good bump-bonding connectivity.

(a)   (b)

FIGURE 14.7 (a) An 8×8 CdTe pixel detector module. The detector had $8 \times 8 = 64$ pixels, each $1.4 \times 1.4\,mm^2$ in size. The detector thickness was 0.5 mm. (b) Sixteen CdTe pixel modules were arranged on a plate to form a large-area CdTe detector.

The substrate of the fanout board is made of a 96% Al_2O_3 ceramic with thickness of 300 µm. The fanout board consists of bump pads, through-holes, and patterns that route signals from the bump pads on both surfaces of the ceramic substrate. Stud bump-bonding technology developed specifically for CdTe detectors [30] is employed to connect each pixel to the bump pads on the fanout board. The studs are made with gold wire and a thin layer of indium printed on top of each stud. The fanout board with the CdTe device was wire-bonded to an ASIC (VA64TA) [61].

Figure 14.8 shows gamma-ray spectra from ^{241}Am and ^{57}Co obtained with the CdTe pixel detector module. The spectra were derived by combining all 64 spectra

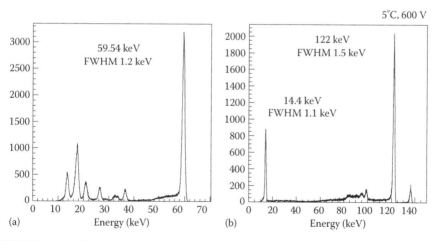

FIGURE 14.8 ^{241}Am (a) and ^{57}Co (b) spectra obtained with a CdTe pixel detector module. Each spectrum was constructed from spectra obtained from all 64 pixels. Bias voltage of 600 V was applied at an operating temperature of 5°C. The energy resolutions were 1.2 and 1.5 keV (FWHM) at 59.5 and 122 keV, respectively.

140 keV   69–80 keV

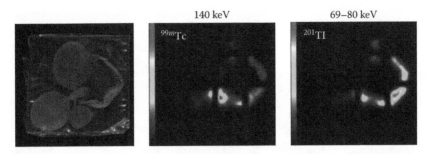

FIGURE 14.9 Gamma-ray image taken from tobacco leaf that absorbed different kinds of radioactive liquid (^{99m}Tc and ^{201}Tl). The leaf was placed above the large-area CdTe detector (Figure 14.7b) with a tungsten collimator. The difference in distributions of ^{99m}Tc and ^{201}Tl can be clearly seen.

from the individual pixels after applying gain corrections. Bias voltage of 600 V was applied at an operating temperature of 5°C. The energy resolutions were 1.2 and 1.5 keV (FWHM) at 59.5 and 122 keV, respectively.

Once CdTe pixel detector modules have been established, a CdTe imager with larger area can be produced by arranging the modules. Figure 14.7b shows a CdTe imager in which $4 \times 4 = 16$ CdTe pixel detector modules are arranged on a plate. The total size of the imager is 5.4×5.4 cm^2. It works as a conventional gamma camera with a collimator attached above the imager. Because the imager has high energy resolution (about 1%), various gamma-ray lines can be easily separated in the spectra obtained. Therefore, the imager is capable of simultaneous multitracer imaging using different gamma-ray lines as demonstrated in Figure 14.9.

14.5 STACKED CdTe DETECTOR

In order to extend the application of CdTe diodes to the detection of megaelectron volt gamma-rays, an effective detector must be more than 1 cm thick. However, it is difficult to obtain good energy resolution and good peak detection efficiency with such a thick CdTe device due to incomplete charge collection. One approach to overcome this difficulty is to stack several tens of thin and fully active CdTe diode detectors, and operate them as a single detector [2,18,62–64]. Figure 14.10 shows such a stacked detector consisting of 40 layers of CdTe diodes each with an area of 21.5×21.5 mm^2 and thickness of 0.5 mm [63]. A gap of only 0.7 mm has been achieved, with the detector housing utilizing a ceramic sheet 0.5 mm thick. The whole detector volume is 9.2 cm^3 and the effective thickness is 2.0 cm, thereby providing efficiency of 20% at 500 keV and 7% at 1 MeV. In the stacked detector, the signal from each layer is processed independently by using an individual analog chain. The gamma-ray spectrum from the detector is obtained by summing the spectra from all layers where pulse heights above a certain threshold are recorded. With this approach, the energy resolution could be maintained at the same level as that of a single layer. The energy resolution obtained with this stacked detector was 1.6%–1.7% (FWHM) for the two peaks of 1.17 and 1.33 MeV in the energy spectrum of gamma-rays from ^{60}Co [63].

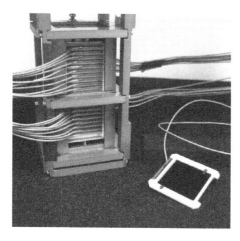

FIGURE 14.10   The 40-layer CdTe stacked detector together with a layer used in the detector. (From Watanabe, S. et al., *Nucl. Instrum. Methods A*, 505, 118, 2003.)

In addition to increased efficiency, the stack configuration with individual readouts provides depth information of interactions. The information obtained is useful for reducing the background, since low-energy gamma-rays can be expected to interact in the upper layers; therefore, low-energy events detected in lower layers can be rejected. Moreover, since the background rate is proportional to the detector volume, low-energy events collected from the first few layers in the stacked detector have a high signal-to-background ratio compared with events obtained from a monolithic detector with a thickness equal to the sum of all layers.

Figure 14.11 shows a CdTe stack detector with four layers, each containing four CdTe pixel modules in a 2×2 configuration [59]. The spectroscopic performance was measured with a ^{133}Ba radioactive source. The spectra from the first, second, third, and fourth layers are presented in Figure 14.12. Photoelectric absorption peaks

FIGURE 14.11   Photo of a CdTe stacked detector with four layers. Each layer consists of $2 \times 2 = 4$ CdTe pixel modules. The layer pitch is 2 mm.

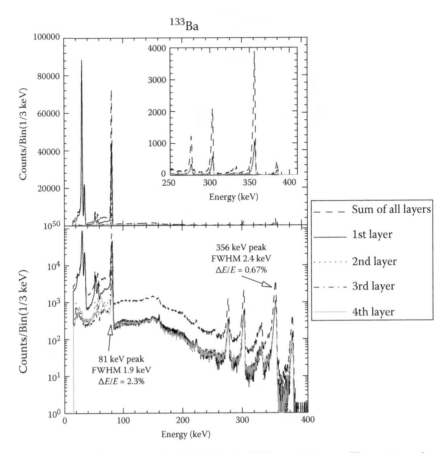

FIGURE 14.12 ^{133}Ba spectra obtained with the CdTe stack detector. The spectrum from each layer and the summed spectrum of all layers is shown in different line styles. The energy resolutions (FWHM) achieved are 1.9 and 2.4 keV at 81 and 356 keV, respectively.

are clearly seen: On the low-energy side, the first layer detects most gamma-rays; on the high-energy side, the peak areas detected in all layers are almost identical. This indicates that stacking detectors improve the detection efficiency for higher-energy gamma-rays. As shown in Figure 14.13, the energy resolution for the 511 keV gamma-ray is 0.9% (FWHM) at an operating temperature of −20°C and under bias voltage of 600 V.

The stacked detector can also be applied to measurement of distance from a gamma-ray source. When a source at a distance of x emits mono-energetic gamma-rays isotropically, the photo peak counts detected in the ith layer (N_i) are given as

$$N_i \propto \exp\left(-\mu(i-1)t\right) \times \left(\frac{x}{x+(i-1)d}\right)^2, \qquad (14.1)$$

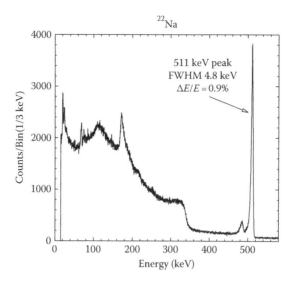

FIGURE 14.13 The 511 keV gamma-ray spectrum obtained with the CdTe stack detector. A [22]Na radioisotope was used. The energy resolution (FWHM) for the 511 keV gamma-ray peak was 4.8 keV (0.9% in $\Delta E/E$).

where
    μ denotes the total photon cross section of detector material
    t denotes the thickness of each layer
    d denotes the pitch between layers

The first term corresponds to the effect of blocking by the upper layers; the second term is introduced by the difference in distance between each layer and the source. Due to the second term, the ratio of photo peak counts obtained in each layer changes with respect to distance from the source. Therefore, once the ratio of each layer count is measured, the distance from the source can be calculated by fitting with the function given in Equation 14.1 [63].

14.6 CdTe DOUBLE-SIDED STRIP DETECTOR

The approach of using a pixelated electrode design is widely adopted to obtain an imaging device based on CdTe and CdZnTe. However, when considering a large imaging sensor with dimensions greater than $10\,cm^2$ and position resolution of a few hundred micrometers as the next target for a "gamma-ray-sensitive" imager, pixel detectors can be very consuming in terms of electronic design. The number of readout channels is N^2 for an $N \times N$ pixel detector. To reduce the number of readout channels, various methods have been proposed. These include ones for coplanar grids [65] and orthogonal strips [66]. An orthogonal strip geometry where the non-collective strip contacts are made capacitive rather than conductive [67] has also been proposed. With these configurations, only $2N$ read out channels are needed for an $N \times N$ pixel detector.

Although several attempts to make a strip electrode on the indium side of the CdTe diode detector have resulted in poor performance, we finally succeeded in making a double-sided strip detector (DSD) as an extension of Al/CdTe/Pt electrode technology [68]. As described in Section 14.2, aluminum acts as a barrier electrode for p-type CdTe, and the resultant detector shows very low leakage current and thus good energy resolution. In the CdTe diode DSD, orthogonal strips on both sides of the detector provide two-dimensional position information. To extract a signal from the strips, we utilized gold stud bump bonding [30] rather than direct wire bonding to the aluminum and platinum electrodes. Several CdTe DSDs have been developed with different dimensions varying the strip pitch from 400 to 60 μm so far.

One of the first prototypes is shown in Figure 14.14, where the dimensions are 2.6 cm × 2.6 cm and the thickness is 500 μm. There are 64 strips at a pitch of 400 μm both on the anode and cathode sides, dividing space into 64 × 64 = 4096 pixels for readout from only 64 + 64 = 128 channels. Two 32-channel ASICs (VA32TAs), which are used for the CdTe pixel detectors, are also employed for readout from the strips on each side [68]. Figure 14.15 shows gamma-ray images obtained with the CdTe DSD. These are shadow images of brass nuts, a brass washer, and a soldering wire with gamma-rays from various radioisotopes: ^{241}Am (60 keV), ^{133}Ba (81 keV), and ^{57}Co (122 keV). The hole of a 2 mm nut and the solder wire 0.6 mm in diameter can be clearly seen. The thin washer becomes transparent as the energy of gamma-ray increases. The DSD provides simultaneous imaging and spectroscopic information with an energy resolution of 2.6 keV (FWHM) for 60 keV gamma-rays.

Figure 14.16 shows spectral performance obtained with another prototype of CdTe DSDs with shorter strips [68]. The detector is 1.3 cm × 1.3 cm in size and 500 μm thick. The electrodes were each divided into 32 strips with a pitch of 400 μm, and read out by a 64-channel ASICs (VA64TA2s) on each side. The detector was

FIGURE 14.14 A CdTe DSD 2.6×2.6 cm² in size and 500 μm thick. The strip pitch was 400 μm with 64 strips formed on each side, and using two VA32TAs for readout on each side.

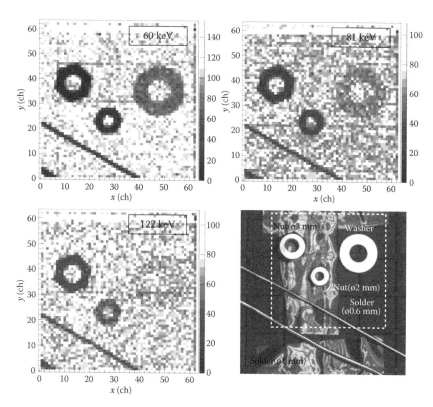

FIGURE 14.15 Shadow images obtained with the CdTe DSD prototype and a photo of the target. Energies of the gamma-rays were 60 keV (^{241}Am), 81 keV (^{133}Ba), and 122 keV (^{57}Co). The pixel size of the images corresponds to a strip pitch of 400 μm.

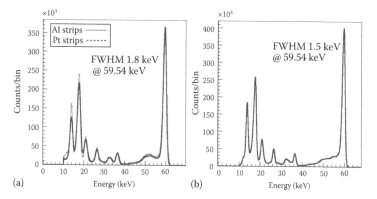

FIGURE 14.16 (a) ^{241}Am spectra obtained using a CdTe DSD 0.5 mm thick. The operation temperature was $-20°$ C and the bias voltage was 500 V. The energy resolution of both the anode and cathode sides was 1.8 keV (FWHM) for 60 keV. (b) A spectrum obtained by averaging anode and cathode signals for each photon, resulting in the improved energy resolution of 1.5 keV (FWHM).

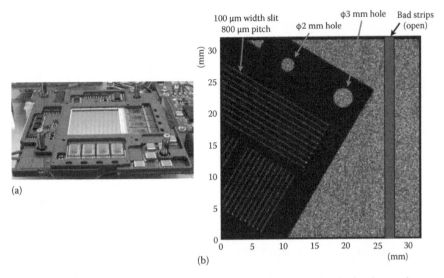

FIGURE 14.17 (a) Photo of a large-area CdTe DSD developed for the hard x-ray imager onboard ASTRO-H. (b) Imaging performance of the large CdTe DSD.

operated at a temperature of −20°C and under a bias voltage of 500 V. The anode spectrum was derived with signals from 30 out of the 32 strips, while the cathode spectrum with signals from all the 32 strips. The energy resolutions of 1.8 keV (FWHM) were obtained for both sides. Improved spectral performance could be expected by averaging the pulse heights from anode and cathode strips. This suppresses the noise components independent between the strips by a factor of up to $\sqrt{2}$. As presented in Figure 14.16b, the averaging operation improved the energy resolution down to 1.5 keV (FWHM).

Based on the series of prototypes, a large-area CdTe DSD has been developed for the hard x-ray imager onboard the ASTRO-H satellite. A photo of the CdTe DSD is shown in Figure 14.17a. The detector size is $32 \times 32\ mm^2$, and the thickness is 0.75 mm. The anode and cathode are each divided into 128 strips with a pitch of 200 μm. As shown in Figure 14.17b, a gamma-ray shadow image was obtained with this detector. The shadow of a tungsten slit with 100 μm width and 800 μm pitch slits, $\phi2$ mm and $\phi3$ mm holes is seen. The 60 keV gamma-rays from ^{241}Am were used for the experiment. The operating temperature was −20°C, and the applied bias voltage was 250 V.

14.7 SEMICONDUCTOR COMPTON CAMERA

Compton cameras are used to reconstruct gamma-ray-emitting source distributions based on Compton scattering information recorded in the detector. Compton scattering typically plays a dominant role in the energy band from a few hundred kiloelectron volts to 10 MeV. It entails elastic collision between an incident photon and an electron in the scattering medium. In conventional Compton cameras, the incident gamma-ray is identified by successive interactions in two detector layers. The ideal case would be that a gamma-ray photon emitted from the source is Compton-scattered in the first layer and then photoabsorbed in the second layer. Once the locations and energy deposits of both interactions are measured, Compton kinematics allows us to calculate the energy and direction (as a cone in the sky) of the incident gamma-ray by using the following Compton equations:

$$E_{in} = E_1 + E_2,$$ (14.2)

$$\cos\theta = 1 - m_e c^2 \left(\frac{1}{E_2} - \frac{1}{E_1 + E_2} \right),$$ (14.3)

where
  E_1 denotes the energy of the recoil electron
  E_2 denotes the energy of the scattered gamma-ray photon
  θ denotes the scattering angle

For every event, a cone can be reconstructed as an opening angle of 2θ. The source is somewhere on the cone surface. Compton cameras offer the advantage of needing only a few photons to recover the position of sources without mechanical collimators in front of the camera. If the direction of a recoil electron can be measured, the Compton cone is reduced to a segment of the cone, the length of which depends on recoil electron measurement accuracy.

Semiconductor imaging detectors are desired for Compton imaging. As expressed in Equations 14.2 and 14.3, the energy and position resolution provided with semiconductors should improve the angular resolution and hence the sensitivity of Compton cameras. From this perspective, several semiconductor-based Compton cameras have been proposed [69,70]. However, most of these cameras have been developed based on combining a semiconductor such as silicon with scintillators. Since CdTe has large atomic numbers (48, 52) and high density (5.8 g/cm³), it offers the potential to replace scintillators and to form a full-semiconductor Compton camera. A semiconductor Compton camera based on HP-Ge has been reported elsewhere [71,72].

Figure 14.18 shows the conceptual design of the Si/CdTe semiconductor Compton camera. It differs from conventional ones in the sense that it uses many layers of silicon and CdTe imaging detectors [73,74]. In principle, each layer could act not only as a scattering part but also as an absorber part. Combining a low-Z material (Si) with a high-Z material (CdTe) is suitable for obtaining high probability of Compton scattering. Importantly, silicon works as a better scatterer than other semiconductor materials with larger atomic numbers in respect of the "Doppler broadening effect" [75]. For low-energy gamma-rays (less than several hundred kiloelectron volts), the task of CdTe is to measure the energy of scattering gamma-rays through photoabsorption. When the energy becomes higher and the effect of Doppler broadening in angular resolution smaller, Compton interactions recorded in CdTe layers could be used to improve detection efficiency [74]. In order for the Compton camera to cover a wide energy band and the wide scattering angle of Compton interaction in the device, the detectors must have low energy thresholds.

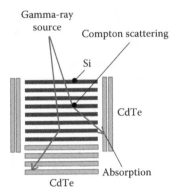

FIGURE 14.18   Conceptual design of the semiconductor Compton camera consisting of layers of thin silicon and CdTe imaging detectors.

As gamma-ray energy increases to the megaelectron volt region, detection becomes more difficult. One attractive idea to improve detection efficiency is a concept called the multiple Compton method [69,74], which utilizes events where an incident photon undergoes multiple Compton scatterings within a detector system. A stack of many thin layers in the Si/CdTe Compton camera can be used to record such events. The order of interactions, necessary to reconstruct energy and direction of the incident photon, can be determined by examining the energy-momentum conservation for all possible sequences. Another important point of this method is that photons need not be fully absorbed and are allowed to escape from the detector system, provided that the photons undergo at least three Compton scatterings, which are sufficient to solve simultaneous equations.

Imaging and spectral performances have been demonstrated for the Si/CdTe Compton camera based on the development of high-resolution CdTe imaging detectors, double-sided silicon strip detectors (DSSDs), and low-noise readout ASICs [61,76–79]. Figure 14.19 shows a prototype consisting of four layers of CdTe pixel detectors placed under four layers of DSSDs. Compton imaging for 60–662 keV gamma-rays was successfully performed. As shown in Figure 14.20 (a), the angular resolution of Compton imaging was about 2° for a point-like radioactive source of 511 keV gamma-rays by using a simple back-projection method. This value is consistent with the theoretical limit due to the Doppler broadening effect. Since the effect becomes more significant for higher Z materials, events where incident gamma-rays are scattered in a CdTe layer and are absorbed in another CdTe layer exhibit larger angular resolution, for example, ~10° [76]. Figure 14.20 (b) shows the effect of background subtraction using the Compton imaging. If we select events in which the Compton cone intersects with the bright spot in the image, most low-energy continuum gamma-rays are eliminated from the spectrum. Figure 14.21 demonstrates the imaging capability for an extended source. A target made of paper is soaked with a liquid radioisotope of ^{131}I and located above the Compton camera at a distance of 3 cm. The C-like shape is properly reconstructed and the 3 mm gap

FIGURE 14.19   Photo of the prototype Si/CdTe Compton camera. Four layers of CdTe pixel detectors were placed under four layers of DSSDs.

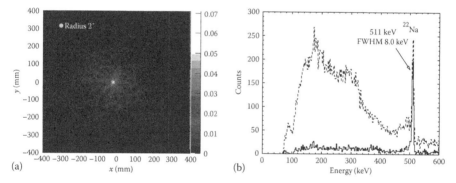

FIGURE 14.20   (a) Image for a point source of 511 keV gamma-rays reconstructed with the Si/CdTe Compton camera. A simple back-projection method was applied. (b) Spectra before and after the background subtraction utilizing the Compton imaging method Takeda et al.

is clearly resolved. Here, we selected a deconvolution algorithm called list-mode maximum likelihood expectation maximization (LM-ML-EM) method [80].

Fine energy resolution enables us to precisely distinguish emission lines from different radioisotopes, and thus is the key for simultaneous imaging of multiple probes described in the next section. As shown in Figure 14.20, the camera has achieved energy resolution of around 1.5% (FWHM) at 511 keV. Distinguishing events in the energy space, the image reconstruction can be performed individually to each radioisotope. Figure 14.22 shows an imaging test of two radioisotopes, ^{133}Ba and ^{22}Na, both placed 30 mm above the camera, and 56° off-axis. The distance between the two sources was 25 mm. The right panel shows energy spectrum taken during the measurement. Multiple lines from the radioisotopes were detected. For

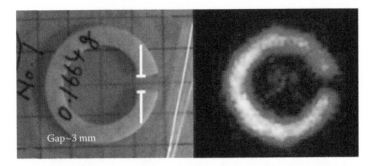

FIGURE 14.21 Image for an extended source of 364 keV gamma-rays (^{131}I) reconstructed with the Si/CdTe Compton camera. The 3 mm gap in the C-shaped target is clearly resolved. The distance between the Compton camera and target was 3 cm. (From Takeda, S. et al., *IEEE Trans. Nucl. Sci.*, 56, 783, 2009.)

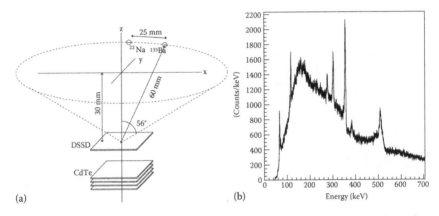

FIGURE 14.22 (a) Experimental setup to demonstrate simultaneous imaging of two radio isotopes with different energies. (b) Energy spectrum taken during the measurement. Multiple lines from ^{133}Ba and ^{22}Na are seen.

the image reconstruction, the energy windows for ^{133}Ba and ^{22}Na were set to 350–360 and 500–511 keV, respectively. Figure 14.23 shows resultant images for ^{133}Ba and ^{22}Na using the simple back-projection method (a), and the LM-ML-EM deconvolution after fifth iteration (b). The auxiliary circles are drawn with a pitch of 10° from zenith (0°) to horizontal (90°) direction. Simultaneous imaging of two radioisotopes was successfully demonstrated. Besides, we indicate that the camera achieved a field of view larger than 100°.

14.8 POSSIBLE APPLICATIONS TO MEDICAL IMAGING

CdTe devices begin to appear in the field of medical imaging. One interesting application is a semiconductor PET using CdTe detectors. The high-energy resolution of CdTe could provide a narrow energy window for coincidence measurement leading

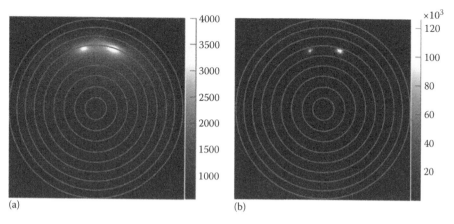

FIGURE 14.23   The two radioactive isotopes, ^{133}Ba and ^{22}Na, are clearly distinguished in the reconstructed images. The simple back-projection method (a) and the LM-ML-EM method (b) were applied.

to a high signal-to-noise ratio. The performance of a prototype system is reported in Yanagita et al. [81]. CdTe-based photon counting detectors have also been utilized for clinical and dental CT applications to reduce x-ray dose to patients [82–85]. We consider that the detector technologies developed for the next generation gamma-ray astrophysics is also attractive for medical imaging. One prospective application is nuclear medicine. In this field, a radiopharmaceutical, a pharmaceutical labeled by radioisotopes, injected into a living object need to be visualized. Because the CdTe imagers and the Si/CdTe semiconductor Compton camera have wide-band imaging capability from a few tens keV to MeV, various radioisotopes can be used including undetectable radioisotopes for current instruments. Moreover, simultaneous tracking of multiple radiopharmaceuticals could be realized by the capability to precisely distinguish emission lines from different radioisotopes. Such a multi-probe tracker has a possibility to differentiate between benign and malignant lesion, and to stage malignancies by analyzing comprehensive information provided by the multiple probes.

For the next generation SPECT cameras, the CdTe strip detectors with a strip pitch of several hundred micrometers could be used as attractive element detectors. In addition to the fine position resolution, even better than that of current commercial instruments for small animal imaging, typically 0.5–0.8 mm [86–88], the fine energy resolution of 1–2 keV enables us to perform simultaneous tracking of several probes labeled by SPECT radioisotopes. These isotopes include ^{201}Tl (71–81 keV), ^{99m}Tc (140 keV), ^{123}I (159 keV), ^{67}Ga (93.3, 185 keV), and ^{111}In (175, 245 keV). The fine energy resolution will also help to reduce the effect of gamma-rays scattered in a target and a collimator, more precisely.

The study for practical applications of Compton cameras to medical imaging has just started. Although the potential performance of this kind of detector was proposed in the early 1970s [89], adequate spatial resolution for medical use has not been achieved by traditional sensor technologies. It is only quite recently that

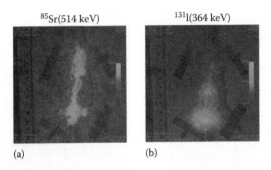

FIGURE 14.24 Results of simultaneous imaging with multiple radiopharmaceuticals by a prototype Si/CdTe Compton camera. The difference in accumulation of ^{85}Sr (a) and iodinated (^{131}I) methylnorcholesterol (b) is clearly seen.

semi-conductor-based Compton cameras has been made available. An application of a Compton camera based on Ge semiconductor detectors operated at liquid nitrogen temperature in the field of molecular imaging has been reported [90].

A high angular resolution Si/CdTe Compton camera, originally developed for gamma-ray observation in space, has demonstrated a good spatial resolution of a few mm when applied to objects as close as several 10 cm. Figure 14.24 shows an experimental result of imaging multiple radio pharmaceuticals injected into a living mouse [91]. The difference in accumulation of ^{85}Sr (514 keV) and iodinated (^{131}I, 364 keV) methylnorcholesterol is clearly seen. In the case of iodinated methylnorcholesterol, we can see three areas of accumulation around the neck, abdomen, and genitals. The accumulations around neck and abdomen are thought to originate from the thyroid and adrenal regions, which is consistent with the product sheet. Intense accumulation was determined around the genitals, which might be from the testicle or prostate. Conversely, the accumulations of ^{85}Sr in new bone growth, such as the skull, spine, lumbar, and femur, were clearly identified. Images taken with a prototype already shows the technical feasibility for simultaneous imaging using multiple radiopharmaceuticals. By using the same prototype, three-dimensional distribution of radio isotopes has been resolved [92]. Future improvements of the Si/CdTe Compton camera in this field are discussed in Takeda et al. [93].

14.9 CONCLUSIONS

The recent advances made in CdTe detectors relative to diode configuration were reviewed. After 10 years of research and development, CdTe diode detectors have now entered the phase of actual application. In particular, the fine energy and position resolution obtained with CdTe pixel detectors or CdTe DSDs are expected to lead to dramatically improved performance in the area of gamma-ray imaging. For astronomical observations, developments of HXI and SGD based on the CdTe diode detectors are being progressed satisfactorily for the ASTRO-H satellite mission. HXI utilizes the large-area CdTe DSD on the focal plane of the hard x-ray mirror, while SGD is composed of 40-layer Si/CdTe Compton cameras, expecting improved sensitivity in the wide energy band. The technology developed here would also offer improved sensitivity in gamma-ray detection for various applications, including nuclear medical imaging, and nondestructive industrial imaging.

ACKNOWLEDGMENTS

This research was conducted in collaboration with T. Mitani, K. Oonuki, K. Tamura, T. Kishishita, M. Ushio, J. Katsuta, H. Odaka, M. Kokubun (Institute of Space and Astronautical Science [JAXA]), K. Nakazawa (the University of Tokyo), H. Tajima (Nagoya University), T. Tanaka (Stanford University), M. Nomachi (Osaka University), Y. Fukazawa (Hiroshima University), N. Kawachi, M. Yamaguchi, (Japan Atomic Energy Agency [JAEA]), K. Arakawa, T. Nakano (Gunma University) and S. Enomoto (Okayama University/RIKEN), and was funded by Japan's Ministry of Education, Science, Sports and Culture through grants-in-aid (13304014, 14079207, 20244017, 21684015). Images in Figures 14.9 and 14.21 were taken in an experiment jointly conducted with the medical department at Gunma University and JAEA. Images in Figure 14.24 were taken in an experiment jointly conducted with RIKEN.

REFERENCES

1. G.F. Knoll, *Radiation Detection and Measurement*, 3rd edn. John Wiley & Sons Inc., New York, 1999.
2. T. Takahashi and S. Watanabe, Recent progress on CdTe and CdZnTe detectors, *IEEE Trans. Nucl. Sci.*, 48: 950–959, 2001.
3. P.N. Luke and M. Amman, Room-temperature replacement for Ge detectors—Are we there yet? *IEEE Trans. Nucl. Sci.*, 54: 834–842, 2007.
4. A. Owens and A. Peacock, Compound semiconductor radiation detectors, *Nucl. Instrum. Methods A*, 531: 18–37, 2004.
5. M. Funaki, T. Ozaki, K. Satoh, and R. Ohno, Growth and characterization of CdTe single crystal for radiation detectors, *Nucl. Instrum. Methods A*, 436: 120–126, 2000.
6. F.P. Doty, J.F. Butler, J.F. Schetzina, and K.A. Bowers, Properties of CdZnTe crystals grown by a high pressure Bridgman method, *J. Vac. Sci. Technol. B*, 10: 1418–1422, 1992.
7. J.F. Butler, C.L. Lingren, and F.P. Doty, $Cd_{1-x}Zn_xTe$ gamma-ray detectors, *IEEE Trans. Nucl. Sci.*, 39(4): 605–609, 1992.
8. C. Szeles, and M.C. Driver, Growth and properties of semi-insulating CdZnTe for radiation detector applications, *Proc. SPIE*, 3446: 1–8, 1998.
9. C. Szeles et al., Development of the high-pressure electro-dynamic gradient crystal-growth technology for semi-insulating CdZnTe growth for radiation detector applications, *J. Electron. Mater.*, 33(6): 742–751, 2004.
10. H. Chen et al., Characterization of large cadmium zinc telluride crystals grown by traveling heater method, *J. Appl. Phys.*, 103: 014903-1-5, 2008.
11. H. Chen et al., High-performance, large-volume THM CdZnTe detectors for medical imaging and homeland security applications, *IEEE Trans. Nucl. Sci.*, 6: 3629–3637, 2006.
12. H. Shiraki, M. Funaki, Y. Ando, S. Kominami, K. Amemiya, and R. Ohno, Improvement of the productivity in the growth of CdTe single crystal by THM for the new PET system, *IEEE Nuclear Science Symposium Conference Record*, Honolulu, HI, pp. 1783–1787, 2007.

13. H. Shiraki, M. Funaki, Y. Ando, A. Tachibana, S. Kominami, and R. Ohno, THM growth and characterization of 100 mm diameter CdTe single crystals, *IEEE Nuclear Science Symposium Conference Record*, Dresden, Germany, pp. 126–132, 2008.

14. P.N. Luke, Single-polarity charge sensing in ionization detectors using coplanar electrodes, *Appl. Phys. Lett.*, 65: 2884–2886, 1994.

15. H.H. Barrett, J.D. Eskin, and H.B. Barber, Charge transport in arrays of semiconductor gamma-ray detector, *Phys. Rev. Lett.*, 75: 156–159, 1995.

16. H.B. Barber, D.G. Marks, B.A. Apotovsky, F.L. Augustine, H.H. Barret, J.F. Butler, E.L. Dereniak et al., Progress in developing focal-plane-multiplexer readout for large CdZnTe arrays for nuclear medicine applications, *Nucl. Instrum. Methods A*, 380: 262–265, 1996.

17. J.F. Butler, Novel electrode design for single-carrier charge collection in semiconductor nuclear detectors, *Nucl. Instrum. Methods A*, 396: 427–430, 1997.

18. T. Takahashi, B. Paul, K. Hirose, C. Matsumoto, R. Ohno, T. Ozaki, K. Mori, and Y. Tomita, High-resolution Schottky CdTe diode for hard x-ray and gamma-ray astronomy, *Nucl. Instrum. Methods A*, 436: 111–119, 1999.

19. T. Takahashi, K. Hirose, C. Matsumoto, K. Takizawa, R. Ohno, T. Ozaki, K. Mori, and Y. Tomita, Performance of a new Schottky CdTe detector for hard x-ray spectroscopy, *Proc. SPIE*, 3446: 29–37, 1998.

20. C. Matsumoto, T. Takahashi, K. Takizawa, R. Ohno, T. Ozaki, and K. Mori, Performance of a new Schottky CdTe detector for hard x-ray spectroscopy, *IEEE Trans. Nucl. Sci.*, 45: 428–432, 1998.

21. H.B. Barber, Applications of semiconductor detectors to nuclear medicine, *Nucl. Instrum. Methods A*, 436: 102–111, 1999.

22. C. Scheiber, CdTe and CdZnTe gamma ray detectors in nuclear medicine, *Nucl. Instrum. Methods A*, 448: 513–524, 2000.

23. D.G. Darambara, State-of-the-art radiation detectors for medical imaging: Demands and trends, *Nucl. Instrum. Methods A*, 569: 153–158, 2006.

24. L. Verger et al., Performance and perspectives of a gamma camera based on CdZnTe for medical imaging, *IEEE Trans. Nucl. Sci.*, 51: 3111–3117, 2004.

25. O. Limousin, C. Blondel, J. Cretolle, H. Dzitko, P. Laurent, F. Lebrun, J.P. Leray et al., The ISGRI CdTe gamma-ray camera first steps, *Nucl. Instrum. Methods A*, 442: 244–249, 2000.

26. G. Sato, A. Parsons, D. Hullinger, M. Suzuki, T. Takahashi, M. Tashiro, K. Nakazawa et al., Development of a spectral model based on charge transport for the swift/BAT 32K CdZnTe detector array, *Nucl. Instrum. Methods A*, 541: 372–384, 2005.

27. T. Takahashi et al., The ASTRO-H mission, *Proc. SPIE*, 7732: 77320Z, 2010.

28. M. Kokubun et al., Hard x-ray imager (HXI) for the ASTRO-H mission, *Proc. SPIE*, 7732: 773215, 2010.

29. H. Tajima et al., Soft gamma-ray detector for the ASTRO-H mission, *Proc. SPIE*, 7732: 773216, 2010.

30. T. Takahashi, S. Watanabe, M. Kouda, G. Sato, Y. Okada, S. Kubo, Y. Kuroda, M. Onishi, and R. Ohno, High resolution CdTe detector and applications to imaging devices, *IEEE Trans. Nucl. Sci.*, 48: 287–291, 2001.

31. Amptek Inc., X-ray and gamma-ray detector, http://www.amptek.com, 2011.

32. T. Takahashi, T. Mitani, Y. Kobayashi, M. Kouda, G. Sato, S. Watanabe, K. Nakazawa, Y. Okada, M. Funaki, and R. Ohno, High resolution Schottky CdTe Diodes, *IEEE Trans. Nucl. Sci.*, 49(3): 1297–1303, 2002.

33. T. Takahashi, K. Nakazawa, S. Watanabe, G. Sato, T. Mitani, T. Tanaka, K. Oonuki, K. Tamura, H. Tajima, T. Kamae, G. Madejski, M. Nomachi, Y. Fukazawa, K. Makishima, M. Kokubun, Y. Terada, J. Kataoka, and M. Tashiro, Application of CdTe for the NeXT mission, *Nucl. Instrum. Methods A*, 541: 332–341, 2005.

34. Cear Pulse Company Ltd., Preamplifiers, http://www2.clearpulse.co.jp/indexEng.html, 2009.
35. K. Nakazawa, K. Oonuki, T. Tanaka, Y. Kobayashi, K. Tamura, T. Mitani, G. Sato, S. Watanabe, T. Takahashi, R. Ohno, A. Kitajima, Y. Kuroda, and M. Onishi, Improvement of the CdTe diode detectors using a guard-ring electrode, *IEEE Trans. Nucl. Sci.*, 51(4): 1881–1885, 2004.
36. H. Toyama, A. Nishihira, M. Yamazato, A. Higa, T. Maehara, R.Ohno, and M. Toguchi, Formation of aluminium Schottky contact on plasma-treated cadmium telluride surface, *Jpn. J. Appl. Phys.*, 43: 6371–6375, 2004.
37. H. Toyama, M. Yamazato, A. Higa, T. Maehara, R.Ohno, and M. Toguchi, Effect of He plasma treatment on the rectification properties of Al/CdTe Schottky contacts, *Jpn. J. Appl. Phys.*, 44: 6742–6746, 2005.
38. S. Watanabe, S. Ishikawa, S. Takeda, H. Odaka, T. Tanaka, T. Takahashi, K. Nakazawa, M. Yamazato, A. Higa, and S. Kaneku, New CdTe pixel gamma-ray detector with pixelated Al Schottky anodes, *Jpn. J. Appl. Phys.*, 46: 6043–6045, 2007.
39. S. Ishikawa, H. Aono, S. Watanabe, S. Takeda, K. Nakazawa, and T. Takahashi, Performance measurements of Al/CdTe/Pt pixel diode detectors, *Proc. SPIE*, 6706: 67060M, 2007.
40. R.O. Bell, G. Entine, and H.B. Serreze, Time-dependent polarization of CdTe gamma-ray detectors, *Nucl. Instrum. Methods*, 117: 267–271, 1974.
41. H.L. Malm and M. Martini, Polarization phenomena in CdTe nuclear radiation detectors, *IEEE Trans. Nucl. Sci.*, 21: 322–330, 1974.
42. P. Siffert, J. Berger, C. Scharager, A. Cornet, and R. Stuck, Polarization in cadmium telluride nuclear radiation detectors, *IEEE Trans. Nucl. Sci.*, 23: 159–170, 1976.
43. M. Hage-Ali, C. Scharager, J.M. Koebel, and P. Siffert, Polarization-free semi-insulating chlorine doped cadmium telluride, *Nucl. Instrum. Methods*, 176: 499–502, 1980.
44. G. Sato, Development and evaluation of CdTe/CdZnTe detectors for space applications, Master's thesis, University of Tokyo, Tokyo, Japan, 2002.
45. H. Toyama, A Higa, M. Yamazato, T. Maehama, R. Ohno, and M. Toguchi, Quantitative analysis of polarization phenomena in CdTe radiation detectors, *Jpn. J. Appl. Phys.*, 45: 8842–8847, 2006.
46. A. Cola and I. Farrella, The polarization mechanism in CdTe Schottky detectors, *Appl. Phys. Lett.*, 94: 102113, 2009.
47. M. Ayoub, M. Hage-Ali, J.M. Koebel, A. Zumbiehl, F. Klotz, C. Rit, R. Regal, P. Fougeres, and P. Siffert, Annealing effects on defect levels of CdTe:Cl materials and the uniformity of the electrical properties, *IEEE Trans. Nucl. Sci.*, 50: 229–237, 2003.
48. Meuris, Study and optimization of the high energy detector in Cd(Zn)Te of the Simbol-X space mission for and gamma-ray astronomy, Ph.D. thesis, Paris Diderot University, Paris, France, 2009.
49. Meuris et al., Characterization of polarization phenomenon in Al-Schottky CdTe detectors using a spectroscopic analysis method, *Nucl. Instrum. Methods A*, submitted for publication, 2011.
50. G. Sato, T. Fukuyama, S. Watanabe, H. Ikeda, M. Ohta, S. Ishikawa, T. Takahashi, H. Shiraki, and R. Ohno, Study of polarization phenomena in Schottky CdTe diodes using infrared light illumination, *Nucl. Instrum. Methods A*, in press, 2011.
51. L. Rossi, P. Fischer, R. Tilman, and N. Wermes, *Pixel Detectors: From Fundamentals to Applications*, From Fundamentals to Applications Series, Springer-Verlag, Berlin, Germany, 2006.
52. M. Chmeissani et al., First experimental tests with a CdTe photon counting pixel detector hybridized with a medipix2 readout chip, *IEEE Trans. Nucl. Sci.*, 51: 2379–2385, 2004.

53. S. Basolo et al., A 20 kpixels CdTe photon-counting imager using XPAD chip, *Nucl. Instrum. Methods A*, 589: 268–274, 2008.

54. F. Harrison, A.E. Bolotnikov, C.M.H. Chen, W.R. Cook, P.H. Mao, and S.M Schindler, Development of a high spectral resolution cadmium zinc telluride pixel detector for astrophysical applications, *Proc. SPIE*, 4851: 823–830, 2003.

55. M. Loecker, P. Fischer, S. Krimmel, H. Krueger, M. Lindner, K. Nakazawa, T. Takahashi, and N. Wermes, Single photon counting x-ray imaging with Si and CdTe single chip pixel detectors and multichip pixel modules, *IEEE Trans. Nucl. Sci.*, 51: 1717–1723, 2004.

56. K. Oonuki, H. Inoue, K. Nakazawa, T. Mitani, T. Tanaka, T. Takahashi, H.C.M. Chen, W.R. Cook, and F.A. Harrison, Development of uniform CdTe pixel detectors based on Caltech ASIC, *Proc. SPIE*, 5501: 218–228, 2004.

57. Z. He, W. Li, G.F. Knoll, D.K. Wehe, J. Berry, and C.M. Stahle, 3-D position sensitive CdZnTe gamma-ray spectrometers, *Nucl. Instrum. Methods A*, 422: 173–178, 1999.

58. F. Zhang, Z. He, G.F. Knoll, D.K. Wehe, and J.E. Berry, 3-D position sensitive CdZnTe spectrometer performance using third generation VAS/TAT readout electronics, *IEEE Trans. Nucl. Sci.*, 52: 2009–2016, 2005.

59. S. Watanabe, S. Takeda, S. Ishikawa, H. Odaka, M. Ushio, T. Tanaka, K. Nakazawa, T. Takahashi, H. Tajima, Y. Fukazawa, Y. Kuroda, and M. Onishi, Development of semiconductor imaging detectors for a Si/CdTe Compton camera, *Nucl. Instrum. Methods A*, 579: 871–877, 2007.

60. H. Tajima, T. Nakamoto, T. Tanaka, S. Uno, T. Mitani, E. do Couto e Silva, Y. Fukazawa, T. Kamae, G. Madejski, D. Marlow, K. Nakazawa, Y. Nomachi, Y. Oakada, and T. Takahashi, Performance of a low noise front-end ASIC for Si/CdTe detectors in Compton gamma-ray telescope, *IEEE Trans. Nucl. Sci.*, 51: 842–847, 2004.

61. T. Tanaka, S. Watanabe, S. Takeda, K. Oonuki, T. Mitani, K. Nakazawa, T. Takashima, T. Takahashi, H. Tajima, N. Sawamoto, Y. Fukazawa, and M. Nomachi, Recent results from a Si/CdTe semiconductor Compton telescope, *Nucl. Instrum. Methods A*, 568: 375–381, 2006.

62. S. Watanabe, T. Takahashi, Y. Okada, G. Sato, M. Kouda, T. Mitani, Y. Kobayashi, K. Nakazawa, Y. Kuroda, and M. Onishi, CdTe stacked detectors for gamma-ray detection, *IEEE Trans. Nucl. Sci.*, 49(3): 1292–1296, 2002.

63. S. Watanabe, T. Takahashi, K. Nakazawa, Y. Kobayashi, Y. Kuroda, K. Genba, M. Onishi, and K. Otake, A stacked CdTe gamma-ray detector and its application as a range finder, *Nucl. Instrum. Methods A*, 505: 118–121, 2003.

64. R. Redus, A. Huber, J. Pantazis, T. Pantazis, T. Takahashi, and S. Woolf, Multielement CdTe stack detectors for gamma-ray spectroscopy, *IEEE Trans. Nucl. Sci.*, 51: 2386–2394, 2004.

65. M.L. McConnel, J.R. Macri, J.M. Ryan, K. Larson, L.-A. Hamel, G. Bernard et al., Three-dimensional imaging and detection efficiency performance of orthogonal coplanar CZT strip detectors, *Proc. SPIE*, 4141: 157–167, 2000.

66. J.R. Macri, B.A. Apotovsky, J.F. Butler, M.L. Cherry, B.K. Dann, A. Drake, F.P. Doty, T.G. Guzik, K. Larson, M. Mayer, M.L. McConnell, and J.M. Ryan, Development of an orthogonal-stripe CdZnTe gamma radiation imaging spectrometer, *IEEE Trans. Nucl. Sci.*, 43: 1458–1462, 1996.

67. G. Montémont, M.C. Gentet, O. Monnet, J. Rustique, and L. Verger, Simulation and design of orthogonal capacitive strip CdZnTe detectors, *IEEE Trans. Nucl. Sci.*, 54: 854–859, 2007.

68. S. Watanabe, S. Ishikawa, H. Aono, S. Takeda, H. Odaka, M. Kokubun, T. Takahashi, K. Nakazawa, H. Tajima, M. Onishi, and Y. Kuroda, High energy resolution hard x-ray and gamma-ray imagers using CdTe diode devices, *IEEE Trans. Nucl. Sci.*, 56: 777–782, 2009.

69. T. Kamae, R. Enomoto, and N. Hanada, A new method to measure energy, direction, and polarization of gamma rays, *Nucl. Instrum. Methods A*, 260: 254, 1987.

70. J.W. LeBlanc, N.H. Clinthorne, C.H. Hua, E. Nygard, W.L. Rogers, D.K. Wehe, P. Weilhammer, and S.J. Wilderman, C-SPRINT: A prototype Compton camera system for low energy gamma ray imaging, *IEEE Trans. Nucl. Sci.*, 45: 943–949, 1998.

71. J.E. McKisson, P.S. Haskins, G.W. Phillips, S.E. King, R.A. August, R.B. Piercey, and R.C. Mania, Demonstration of three-dimensional imaging with a germanium Compton camera, *IEEE Trans. Nucl. Sci.*, 41: 1182–1189, 1994.

72. E.A. Wulf, B.F. Philips, W.N. Johnson, R.A. Kroeger, J.D. Kurfess, and E.I. Novikova, Germanium strip detector Compton telescope using three-dimensional readout, *IEEE Trans. Nucl. Sci.*, 50: 1221–1224, 2003.

73. T. Takahashi, K. Makishima, and T. Kamae, Future hard x-ray and gamma-ray observations, *Astron. Soc. Pac.*, 251: 210–213, 2001.

74. T. Takahashi, K. Nakazawa, T. Kamae, H. Tajima, Y. Fukazawa, M. Nomachi, and M. Kokubun, High resolution CdTe detectors for the next generation multi-Compton gamma-ray telescope, *Proc. SPIE*, 4851: 1228–1235, 2002.

75. A. Zoglauer and G. Kanbach, Doppler broadening as a lower limit to the angular resolution of next generation Compton telescopes, *Proc. SPIE*, 4851: 1302–1309, 2002.

76. K. Oonuki, T. Tanaka, S. Watanabe, S. Takeda, K. Nakazawa, T. Mitani, T. Takahashi, H. Tajima, Y. Fukazawa, and M. Nomachi, Results of a Si/CdTe Compton telescope, *Proc. SPIE*, 5922: 78–88, 2005.

77. S. Watanabe, T. Tanaka, K. Nakazawa, T. Mitani, K. Oonuki, T. Takahashi, T. Takashima, H. Tajima, Y. Fukazawa, M. Nomachi, S. Kubo, M. Onishi, and Y. Kuroda, A Si/CdTe semiconductor Compton camera, *IEEE Trans. Nucl. Sci.*, 52: 2045–2051, 2005.

78. S. Takeda, H. Aono, S. Okuyama, S. Ishikawa, H. Odaka, S. Watanabe, M. Kokubun, T. Takahashi, K. Nakazawa, H. Tajima, and N. Kawachi, Experimental results of the gamma-ray imaging capability with a Si/CdTe semiconductor Compton camera, *IEEE Trans. Nucl. Sci.*, 56: 783–790, 2009.

79. S. Takeda, Experimental study of a Si/CdTe semiconductor Compton camera for the next generation of gamma-ray astronomy, Ph.D. thesis, University of Tokyo, Tokyo, Japan, 2009.

80. S. Wilderman et al., List-mode maximum likelihood reconstruction of Compton scatter camera images in nuclear medicine, *IEEE Nuclear Science Symposium*, Toronto, Ontario, Canada, Vol. 3, pp. 1716–1720, 1998.

81. N. Yanagita et al., Physical performance of a prototype 3D PET scanner using CdTe detectors, *IEEE Nuclear Science Symposium Conference Record*, Honolulu, HI, pp. 2665–2668, 2007.

82. K. Spartiotis et al., X- and gamma ray imaging system based on CdTe-CMOS detector technology, *IEEE Nuclear Science Symposium Conference Record*, Dresden, Germany, pp. 518–522, 2008.

83. J.S. Iwanczyk et al., Photon counting energy dispersive detector arrays for x-ray imaging, *IEEE Trans. Nucl. Sci.*, 56(3): 535–542, 2009.

84. W.C. Barber et al., Large area photon counting x-ray imaging arrays for clinical dual-energy applications, *IEEE Nuclear Science Symposium Conference Record*, Orlando, FL, pp. 3029–3031, 2009.

85. A. Katsumata, K. Ogawa, K. Inukai, T. Nagano, H. Nagaoka, and T. Yamakawa, Initial evaluation of linear and spatially oriented planar images from a new dental panoramic system based on tomosynthesis, *Oral Surg. Oral Med. Oral Pathol. Oral Radiol.*, in press, 2011.

86. Bioscan, NanoSPECT/CT In vivo Preclinical Imager, http://www.bioscan.com/molecular-imaging/nanospect-c, 2011.

87. GE healthcare X-SPECT, https://www2.gehealthcare.com/portal/site/usen/gehchome/, 2011.
88. MILabs, U-SPECT-II & U-SPECT-II/CT, http://www.milabs.com/imaging-solutions/u-spect-ii-ct/
89. R.W. Todd et al., A proposed gamma camera, *Nature*, 251: 132–134, 1974.
90. S. Motomura et al., Multiple molecular simultaneous imaging in a live mouse using semiconductor Compton camera, *J. Anal. At. Spectrom.*, 23: 1089–1092, 2008.
91. S. Takeda et al., Demonstration of in-vivo multi-probe tracker based on a Si/CdTe semiconductor Compton camera, *IEEE Trans. Nucl. Sci.*, submitted for publication, 2011.
92. M. Yamaguchi et al., Spatial resolution of multi-head Si/CdTe Compton camera for medical application, *IEEE Nuclear Science Symposium Conference Record*, Orlando, FL, pp. 4001–4003, 2009.
93. S. Takeda et al., Simulation study of 3-D gamma-ray imager with Si/CdTe semiconductor Compton camera, *IEEE Sensors Application Symposium*, Limerick, Ireland, pp. 170–174, 2010.

15 Recent Advances in Positron Emission Tomography Technology

Farhad Taghibakhsh and Craig S. Levin

CONTENTS

15.1 INTRODUCTION

While common medical imaging techniques such as x-ray radiography, computed tomography (CT), or magnetic resonant imaging (MRI) provide anatomical information of the subject under study, positron emission tomography (PET) generates functional information related to biological activities within living subjects. PET has found various research and clinical applications including cancer studies, pharmaceutical discoveries, neurological and cardiovascular disorders, and even plant studies [1]. Because of major differences in metabolic activities of malignant tumors compared to those of healthy tissue, PET systems are extensively used for detection and treatment in advanced stages of cancer. However, researchers are pushing the technology to employ the method for earlier detection of cancer as well.

PET is known to be the most sensitive medical imaging modality when compared to CT or MRI, for example, for detecting particular cancers [2]. Despite its superior sensitivity, PET has the lowest spatial resolution among the aforementioned modalities due to a number of factors including the physics of 511 keV photon detection and positron annihilation, imaging time, and health-related limitations of the amount of radioactive tracer used. Active fields of research in technology of PET related to early cancer detection include improving contrast and spatial resolution and combining PET with other imaging modalities such as MRI. Because PET does not provide an anatomical frame of reference for its functional images, almost all commercial PET systems today are combined with a CT scanner. Fusion of PET with MRI is considered to be further advantageous in early cancer detection mainly because of the superior MRI soft tissue contrast resolution, as well as the possibility to combine PET with the functional capabilities of MRI (fMRI) [3].

Traditionally, PET scanners are made for whole body imaging and are based on photomultiplier tubes (PMTs). Recent advances in the development of compact radiation detectors and solid-state CMOS compatible photomultiplying photodetectors have helped researchers innovate detector designs that not only are capable of generating high resolution PET images, but also provide the possibility of integrating PET scanners with MRI because of their inherent magnetic field insensitivity (unlike PMTs). Such detectors have particularly found applications in dedicated imagers, that is, small PET imagers dedicated to specific organs such as brain or breast, which have higher sensitivity and resolution than large whole body PET scanners.

This chapter updates the readers with some of the most recent advances in PET detector design and instrumentation for high resolution molecular imaging applications, provides information on new concepts in scintillation light detection for PET, and points to some future directions in technology of PET detectors and instrumentation. We begin with a quick review of PET detector physics and the necessity of resolving depth of interaction (DOI) within detectors for high resolution applications. Various methods of resolving interaction position in segmented and monolithic scintillators are explained, compared, and contrasted. Different solid-state photodetectors are introduced; however, the focus is on newly introduced multipixel Geiger-mode avalanche photodiodes (APDs) (known as silicon photomultipliers, or SiPMs), and their applicability to dedicated imagers. Today, technology may limit application of time-of-flight (TOF) for dedicated PET imaging; however, transient response of SiPMs could be even faster than PMTs and therefore provides the opportunity for implementing TOF dedicated PET imaging. Such concepts will be addressed.

15.2 PHYSICS OF PET AT A GLANCE

PET is a nuclear imaging technique that is based on the imaging of positron annihilation events inside a patient's body. Positrons are emitted from traces of radiolabeled biochemical agents that are injected into the body and are distributed according to

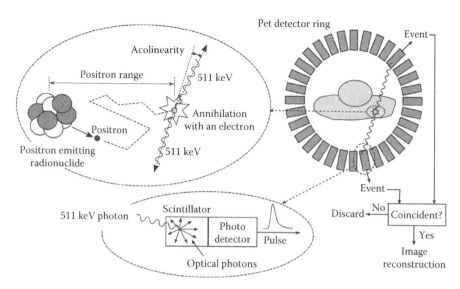

FIGURE 15.1   Schematic diagrams illustrating emission and annihilation of a positron and coincident detection of annihilation photons in a PET detector ring using scintillator-based radiation detectors which are commonly used in PET.

their biochemical properties. Basic physics and operation of PET is illustrated in Figure 15.1. When a radioactive atom on a particular molecule decays, a positron is ejected from the nucleus; the positron eventually annihilates with a nearby electron resulting in emission of two 511 keV photons departing from the annihilation site separated by an angle of almost 180°. A PET imager consists of a set of radiation detectors around the body (or part of the body) to be imaged. If both annihilation photons are detected within a specific time window (coincident detection), the information is used to estimate the location of the annihilation within a specific distance from the radioactive tracer known as positron range. Thus, the intensity of voxels in a PET image relates to the concentration of the radioactive tracer in the corresponding part of the body.

Positron range and photon acolinearity are the two major factors adversely affecting the accuracy of estimating the location of the radioactive agent. Positron range is the distance a positron travels (because of its nonzero kinetic energy) from the nucleus until it annihilates with an electron. At the time of annihilation, a positron may still have a left over kinetic energy that results in a finite deviation from 180° (due to conservation of momentum) in the angle at which annihilation photons depart from each other; this is called acolinearity error. The magnitude of these errors depends on the radioactive material used. For [18]F (the positron emitting isotope of fluorine widely used for PET applications), the positron range is ~0.1 mm FWHM and the acolinearity error is 0.24° full width at half maximum (FWHM) [4]. While it is possible to correct for the positron range to restore the image resolution in certain conditions, the effect of acolinearity error increases as the distance between detectors are increased. As a result of this distance dependence, photon acolinearity is one of the main factors limiting spatial resolution in clinical PET scanners with large detector rings, while it may not be significant in dedicated imagers or small animal imagers.

15.3 FACTORS AFFECTING SPATIAL AND CONTRAST RESOLUTION OF PET

Different methods have been developed to estimate the location of the annihilation using the signals generated by PET detectors. The line connecting two detectors that have detected a coincident event, or the line of response (LOR), ostensibly passes through the annihilation site. Many LOR counts are needed (typically on the order of 10^9) to reconstruct a PET image using various filtered back-projection algorithms.

Scattered events and random counts that result in false counts in LORs (Figure 15.2) are among other causes of image blurring and loss of spatial resolution.

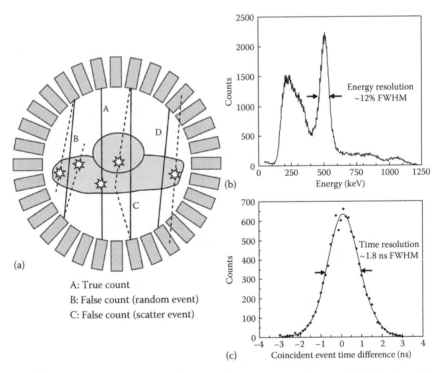

FIGURE 15.2 (a) Diagram of various possible event types generated in PET rings. The "true" events yield the highest image quality and accuracy. Random and scatter coincidences yield loss in image quality and accuracy. (b) A typical energy spectrum of ^{22}Na positron source obtained from a PET detector showing energy resolution of about 12% around 511 keV. The better the energy resolution, the more effectively scattered events are rejected. (c) A typical histogram of event time differences illustrating timing performance of the PET detector with a time resolution of about 1.8 ns. The better the timing resolution, the less random events are counted.

Because a 511 keV photon loses part of its energy if scattered while traveling in the tissue, the capability of PET detectors to accurately measure the energy of detected photons (energy resolution) helps in distinguishing a detected scattered event from a true event. Random counts on the other hand are generated when two 511 keV photons resulting from two different annihilation events are detected almost simultaneously. The better the PET detectors can resolve coincident events (higher timing resolution), the more effectively they can reject random counts.

Quantitatively, the signal-to-noise ratio (SNR) in the PET image is related to the noise equivalent count (NEC) rate which depends on true, scattered, and random counts as defined in the following equation:

$$NEC = \frac{T^2}{T+S+R},$$ (15.1)

where T, S, and R represent true, scattered, and random count rates, respectively. It is thereby clear how detector performance parameters such as energy resolution (for rejecting scattered events) and timing resolution (for rejecting random events) affect the overall quality of a PET image.

Parallax error (Figure 15.2) is a severe cause of image blurring in systems in which the detectors are close to the body. Examples include dedicated PET imagers and systems with small bore diameter such as animal scanners. In animal scanners, spatial resolution degradation due to parallax error becomes more severe as distance from the center of field of view (FOV), incorporation of DOI, or the position along the scintillator crystal where the annihilation photon is absorbed inside the PET detector mitigates the effect of parallax error and therefore improves image resolution [5].

15.4 TIME OF FLIGHT PET

TOF is used to further improve the localization of the annihilation site in PET systems that have sufficiently high timing resolution capability. In TOF-PETs, the time difference in the detection of the two photos of a coincident event, Δt, is related to the difference in the distances the two annihilation photons travel before they are detected (Figure 15.3). If Δx is the distance from the annihilation site to the center of FOV, the following equation relates Δx to Δt, in which c is the speed of light:

$$\Delta x \approx \frac{1}{2} c \cdot \Delta t.$$ (15.2)

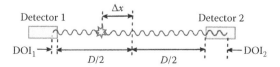

FIGURE 15.3 Estimation of the position of the annihilation point using TOF technique.

Because of limited timing resolution, the estimated position will have a roughly Gaussian distribution around the annihilation site. Therefore, the estimated annihilation site is limited to a certain part of the LOR weighted by a probability density function, as opposed to being uniformly distributed along the full length of the LOR as is the case in conventional PET systems. Compared to conventional PET, TOF-PET benefits from various advantages including reduced dose (or imaging time) to arrive at a certain image quality, and improved contrast and spatial resolution for a given radioactive dose or imaging time.

Equation 15.2 approximates Δx because it does not take into account the effect of DOI or variations of index of refraction along the path, but it clearly indicates that better timing resolution, leading to better accuracy of estimated position of annihilation. However, because the difference in the distances the two annihilation photons travel also depends on DOI in the two detectors detecting the photons (Figure 15.3), incorporating DOI is expected to improve the effectiveness of TOF-PET [6]. The error in positioning of the annihilation site is approximately 1.5 cm FWHM for every 100 ps FWHM of timing resolution (considering $c \sim 30$ cm/ns). Therefore, subnanosecond timing resolution is required for effective implementation of TOF-PET.

15.5 SCINTILLATOR-BASED RADIATION DETECTORS

Generally, the radiation detectors that are used in PET consist of a scintillation material (scintillator) coupled to a photodetector (Figure 15.1), exhibiting an indirect detection scheme. The crystalline or noncrystalline scintillator absorbs the high-energy radiation and generates optical photons that are subsequently converted to an electrical signal by the photodetector representing the energy of the absorbed radiation. Because of the two-step process of converting radiation to electrical signal, various properties of the scintillator and the photodetector are involved in determining the characteristics of the radiation detector such as energy and timing resolution.

An ideal photodetector for PET applications is one that has high gain, high speed, high quantum efficiency (QE), low noise, low power, is temperature insensitive, compact, preferably insensitive to magnetic fields, and, also very importantly, is inexpensive. PMTs have been traditionally used for PET applications and even to date, all clinical whole body PET systems are based on PMTs. Advances in fabrication of high QE, high gain and fast solid-state photodetectors has opened new horizons for PET applications. Some examples of these detectors and devices are shown in Figure 15.4.

15.5.1 SCINTILLATORS

Scintillator materials must meet various key requirements to be considered suitable for PET applications. Important features are high atomic number (high Z) for good stopping power, high photon light yield for high SNR and high QE, short decay time for high timing resolution, and high energy resolution [4]. Other optical characteristics to be considered include low absorption at the scintillation light wavelength, and a wavelength matched with that of the photodetector response [7].

Scintillators used in PET are usually crystals because of their superior optical properties (such as transparency, photon yield, and decay time) compared to

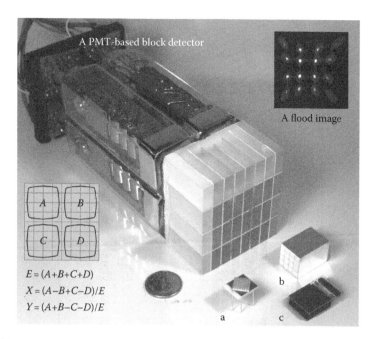

FIGURE 15.4   Image of a clinical PET block detector based on four PMTs coupled to an array of scintillator crystals compared to (a) a position sensitive avalanche photodiode (PS-APD) and a high resolution array of $1 \times 1 \times 1\,mm^3$ LYSO crystals, (b) an array of $2 \times 2 \times 20\,mm^3$ LYSO crystals made for, and (c) a 4×4 array of silicon photomultiplier photodetectors. Anger Logic formulas (lower left corner) are commonly used to resolve photon events occurring in individual elements of a crystal array using event positioning calibrations from flood histogram images (top right corner).

noncrystalline scintillators such as ceramics. Despite having poor stopping power, however, ceramic scintillators have recently been considered for PET applications due to attractive features such as better compatibility of emission spectra with those of silicon photodetectors, possibility of precisely controlled impurity concentration, less expensive fabrication processes, and the possibility of manufacturing complex shaped scintillators [8,9]. In contrast to ceramic scintillators that are highly transparent, one of the main challenges of developing ceramic scintillators is transparency [10].

Tables 15.1 and 15.2 list some common crystalline and ceramic scintillators, and compare their properties. It is worth noting that while the listed ceramic scintillators are not considered suitable for TOF-PET applications due to their poor timing performance, applicability of crystalline scintillators to TOF-PET mostly depends on their initial rate and decay time.

15.5.2   PHOTOMULTIPLIER TUBES

PMTs are the workhorses of PET systems today; they posses many of the aforementioned features except high QE (~25% for a generic PMT), compactness, and

TABLE 15.1
Properties of Common Crystalline Scintillators Used/Researched for PETs

Crystal Scintillators	Atten. Length (cm)	Luminosity (Photons/MeV)	Initial Rate (Luminosity/ns)	Decay Time (ns)	Timing Resolution	Energy Resolution (%)	λ (nm)
$Bi_4Ge_3O_{12}$ (BGO)	1.1	8,200	37	60	3–6 ns	12	505
BaF_2	2.3	11,800	2,266	0.8	200–500 ps	10	175,300
Gd_2SiO_5:Ce (GSO)	1.5	10,000	232	15	1 ns	9	430
Lu_2SiO_5:Ce (LSO)	1.2	25,000	676	37	225 ps to 1.2 ns	10	420
$LaCl_3$:Ce	2.9	50,000	2,500	20	218 ps	3	330
$LaBr_3$:5%Ce	2.2	60,000	4,000	15	260–315 ps	3	358
$CeBr_3$	2.2	68,000	4,000	17	129 ps	3	370
LuI_3: 2%Ce	1.8	100,000	4348	23	125 ps	4	472

Sources: Knoll, G.F.: *Radiation Detection and Measurements*, 3rd edn. 2000. Copyright Wiley-VCH Verlag GmbH & Co. KGaA.; Birowosuto, M.D.. *Novel γ-Ray and Thermal-Neutron Scintillators: Search for High-Light-Yield and Fast-Response Materials*, Vol. 2007, IOS Press, Amsterdam, the Netherlands, 2007.

TABLE 15.2

Properties of Some Ceramic Scintillators Investigated for PET

Ceramic Scintillators	Density (g/cm³)	Luminosity (Photons/ MeV)	Transparency	Decay Time (ns)	Energy Resolution (%)	λ (nm)
Lu_2O_3:Eu	9.4	27,000	Transparent (++)	$\sim 10^6$	12	610
Lu_2SiO_5:Ce	7.4	16,000	Translucent (+)	40	10	420
$Lu_2Si_2O_7$:Ce	6.2	10,000	Translucent (+)	40	9	390
$SrHfO_3$:Ce	7.6	20,000	Translucent (+)	20	10	410
$BaHfO_3$:Ce	8.4	44,000	Translucent (+)	20	3	400
LuAG:Ce	6.7	29,000	Transparent (++)	71	8.6	550

Sources: Van Loef, E.V. et al., Recent advances in ceramic scintillators, *Materials Research Society Symposium Proceedings*, Vol. 1038, p. 1038-O06-02, 2008; Hull, G. et al., *Proc. SPIE*, 6706, 670617, 2007.

magnetic sensitivity requirements. The main advantages of PMTs are very high gain ($\sim 10^6$) and almost noiseless output signal, resulting in high SNR and leading to simple electronic processing requirements and lower overall system complexity. While bulkiness, magnetic sensitivity, and high voltage requirement which are inherent to PMTs limit their application for some particular detector designs, recent development of ultrahigh QE PMTs (such as Hamamatsu R7600U-200 with $QE=42\%$ at 380 nm) and position sensitive multianode PMTs (such as Hamamatsu H9500 16×16 channel flat panel PS-PMT) have addressed the main shortcomings of the traditional vacuum tube photodetector. Reported improvements in energy resolution ($\sim 20\%$ improvement [11]) and spatial resolution of detectors made using these PMTs, coupled with their outstanding characteristics such as high SNR, low temperature sensitivity, and reasonable price indicate that PMTs will sill continue to be the photodetector of choice for large systems such as whole body PET in applications not involving MRI.

15.5.3 AVALANCHE PHOTODIODES AND POSITION SENSITIVE APDS

While silicon APDs have high QE of 60%–80% (much more than generic and ultrahigh QE PMTs), their low gain of ~ 100 and slow response time requires complex electronics to achieve reasonable energy and timing resolution for PET applications. These issues have limited the applicability of APDs to small systems mainly because of the large number of channels of electronics required for large systems. Extensive research and development work on APDs have led to the successful commercialization of various high resolution PET systems for small animal imaging [12], as well as some prototype systems for clinically dedicated imaging applications [13].

The invention of position sensitive APDs (PS-APDs) addressed the main limiting factors of wide spread use of APDs. PS-APDs are large area photodiodes that provide four outputs for positioning the interaction, as well as an additional common output for more accurate timing and energy measurements. This detector configuration considerably reduces the number of required electronic channels, and thus, paves the way for high resolution PET systems at reasonable costs. Although it is possible to correct for position dependencies in the energy and timing response of PS-APDs in order to optimize performance, their overall slow response time limits their applications to non-TOF PET.

15.5.4 Silicon Photomultipliers

SiPMs have been the center of much attention in the last few years because their favorable characteristics such as high gain (~10^6) and fast response time (~100 ps) are similar to those of PMTs. The device consists of a number of microscopic Geiger mode APDs, or microcells, connected in parallel, where each microcell fires a current up on photon excitation (Figure 15.5). Ironically, because the photocurrent of a microcell is independent of the absorbed photons, SiPMs are not really photo-multiplying devices. This is why they are also called multipixel photon counting (MPPC) devices, or Geiger mode APDs (G-APDs or GM-APDs) depending on the manufacturer.

In SiPMs, photodetection efficiency (*PDE*) is analogous to *QE* in PMTs and APDs. *PDE* is defined in Equation 15.3 in which, *FF* is the geometric fill factor (the ratio of active area of microcells to total area of microcells), *QE* is the quantum efficiency of the material, and P_G (between 0.5 and 1.0) is the probability of a Geiger discharge upon absorption of a photon in the active area of the device:

$$PDE = FF \times QE \times P_G. \tag{15.3}$$

Since *QE* is defined by the semiconductor used (silicon in SiPMs), and P_G is dependent on bias voltage as well as fabrication process technology used, one can appreciate the important role of *FF* (dependent on device structure and design) in determining *PDE*. Therefore, maximizing *FF* is one of the challenges in design and fabrication of SiPMs (see Section 15.5.4.3).

As silicon photodetectors, SiPMs have several advantages over APDs, namely higher gain, higher speed, and lower operating voltage; their disadvantages are small

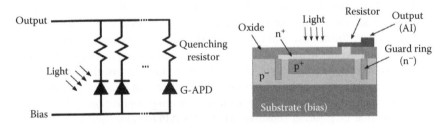

FIGURE 15.5 Silicon photomultiplier consisting of an array of Geiger mode APDs and passive quenching resistors. The microstructure of a single G-APD on p-type substrate is shown on the right.

PDE of ~25% resulting from their small active area fill factor; and smaller dynamic range which does not appreciably affect their application for PET.

Compared to PMTs however, SiPMs provide compactness and magnetic field insensitivity at the cost of much higher noise (higher dark count that adversely affect their SNR), as well as undesirable temperature dependent characteristics. Several novel photodetector technologies have been developed based on SiPMs to address specific shortcomings of the device. It is expected that the following technologies individually or in combination, will lead to widespread use of SiPMs in future PET systems.

15.5.4.1 CMOS Compatible SiPM

Some SiPMs are fabricated using custom silicon processing facilities, as is the case for APDs. However, adoption of the well-established CMOS technology in manufacturing of SiPMs resulted in considerable cost reduction at the expense of finite degradation in device quality. Examples of CMOS compatible SiPMs are the SensL devices that are relatively inexpensive but may not be suitable for TOF applications due to degraded timing performance. Additionally, the technology provides the possibility of integrating CMOS processing circuitry with the photodetector for further performance optimization (see Section 15.5.4.4).

15.5.4.2 Position-Sensitive SiPMs

Position sensitive SiPMs (PS-SiPMs) were developed to resolve individual elements of a scintillator crystal array using fewer numbers of photodetectors, thereby reducing the number of electronic processing channels. Generally, PS-SiPMs are fabricated by interconnecting individual SiPMs using a resistive charge sharing network, very similar to the resistive network used for PS-PMTs [14], that provides four output signals for positioning the event, as well as a fifth optional common signal that may be used for energy and timing resolution independent of the four positioning signals similar to PS-APDs already discussed. A technology recently developed by RMD Inc., Watertown, MA, incorporates fabrication of position sensitive SSPM (PS-SSPM) using an array of microcells each connected to the resistive network allowing the capability to resolve very fine segmented crystals (0.5 mm pitch) without the need for light sharing [15]. Testing results indicate energy resolution of 13%–14% and timing resolution of 1–2 ns FWHM using a LYSO crystals of $1 \times 1 \times 20 \, mm^3$, and spatial resolution of 70 μm using a 15 μm diameter pulsed laser beam [15]. RMD Inc. has also used a continuous resistive sheet, as opposed to a resistive network, to fabricate novel PS-SSPM devices. While fabrication process of such devices might have fewer steps compared to those made using resistive networks, their timing performance is expected to be limited to the distributed RC time constant of the resistive sheet especially if the sheet resistance is high. Compared to PS-APDs, PS-SiPMs and PS-SSPMs have the advantages of high gain, low bias voltage, and fast response.

15.5.4.3 Back-Illuminated SiPM

The small *PDE* of SiPMs is mainly due to small fill factor. Integration of quenching resistors in each microcell of a planar configuration results in excessive dead space

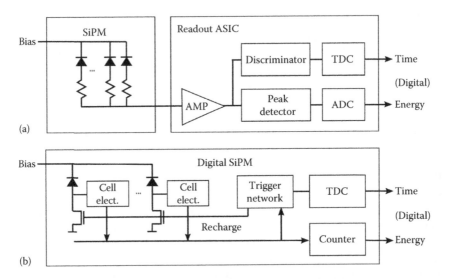

FIGURE 15.6 Comparison between a conventional SiPM photodetector and readout ASIC (a), and an all-in-one digital SiPM (b). The digital SiPM uses active quenching to boost its speed, as opposed to passive quenching used in conventional SiPMs. The digital SiPM also integrates processing electronics that yield digital time and energy information.

and reduction of SiPM active area fill factor. Illuminating a thinned device from the back provides almost 100% fill factor and restores the *PDE* of SiPM up to 80% (similar to that of APDs) [16]. However, the fabrication process may not necessarily be CMOS compatible [17].

15.5.4.4 Digital SiPM

This is a novel technology recently developed by Philips which integrates an SiPM with processing circuits at microcell level to produce an all-in-one digital sensor that considerably reduces the need for external processing circuitry. Each device contains an integrated counter and an integrated time to digital converter (TDC) to provide energy and arrival time information (Figure 15.6), respectively [18]. In these devices, each detected photon is directly converted into a digital signal in each of the Geiger-mode cells of the sensor. Integration of the TDC with the digital sensor provides excellent timing resolution of 153 ps FWHM, while the reported energy resolution is 10.4% FWHM when the sensor is coupled to a $4 \times 4 \times 22 \, mm^3$ LYSO crystal [19]. The sub-200 ps timing resolution for digital SiPM suggests that the detector is promising for time of flight PET applications.

15.6 SEMICONDUCTOR RADIATION DETECTORS

Although semiconductor detectors that directly convert the radiation energy into measurable electrical charge have been widely exploited in high energy physics, these detectors have recently attracted much attention in the PET research

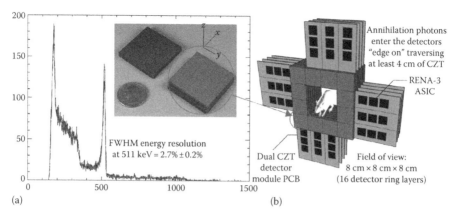

FIGURE 15.7 (a) An example of an energy spectrum obtained from CZT detectors with high energy resolution of 2.7% FWHM at 511 keV. The CZT detectors shown in the inset are being used for (b) a high resolution high sensitivity animal scanner using an edge-on configuration of detectors. (From Gu, Y. et al., *Phys. Med. Biol.*, 56, 1563, 2011.)

community as an alternative to conventional indirect (scintillator-based) radiation detectors [20,21].

The main advantage of direct detectors is their high energy resolution that allows effective filtering of scattered radiation for reconstructing sharp images. The other important advantage of semiconductor radiation detectors is their inherent DOI capability, due to the spread of charge as it drifts toward the collecting electrodes. However, semiconductors such as cadmium zinc telluride, or CZT, have poor transport properties that results in inferior timing resolution compared to their indirect radiation detector counterparts. They also have lesser stopping power for 511 keV photos compared to LYSO, because of smaller effective atomic number, which requires use of thicker detectors to achieve sufficient sensitivity.

As an example of a novel configuration for a high resolution small animal scanner, 40 mm × 40 mm × 5 mm CZT detectors are arranged in such a way that the incoming radiation enters from the side of the detector (parallel to the electrode planes), increasing the chance of absorption through the bulk of the semiconductor (Figure 15.7). Initial evaluation of the detector performance indicates peak energy, coincident timing, and DOI resolution of 2.7%, 15 ns, and 1 mm FWHM, respectively [22].

15.7 PET DETECTORS WITH DOI CAPABILITY

DOI detectors are one of the mainstream research and development topics in PET detectors because of their essential importance in the development of high resolution PET systems. Among various methods proposed or implemented for extraction of DOI information (Figure 15.8), multilayer crystals are probably the most heavily investigated. The method involves use of different scintillator crystals with different scintillation light decay time characteristics known as a Phoswich (contraction of phosphor-sandwich) configuration, intentional misalignment (including offset and

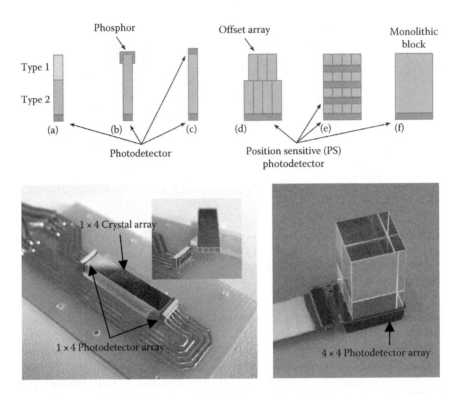

FIGURE 15.8 Diagrams of various methods of extracting DOI information using different detector configurations: (a) Phoswich with two different crystals and single-ended readout. (b) Phosphor-coated scintillator and single-ended readout. (c) Dual-ended readout. (d) Offset crystal array. (e) Stack of crystal array and photodetectors. (f) Monolithic detectors. Pictures show (bottom left) 1×4 array of $1 \times 1 \times 20\,\text{mm}^3$ LYSO crystals in dual-ended readout configuration with Hamamatsu 1×4 MPPC arrays, and (bottom right) a $12.6 \times 12.6 \times 20\,\text{mm}^3$ monolithic LYSO crystal coupled to a SensL 4×4 SiPM array.

rotation) between crystal layers, or a combination of the two [23]. Different crystal decay times are detected from the pulse-shape, and misaligned crystals appear in a different position in the flood image (if the absorption event occurs in an offset crystal, the scintillation light is shared between the two crystals on the bottom, creating a dot between them in the flood image). Although the multilayer detector design is a cost effective solution, especially when the crystal array is read from one side only, this method provides discrete DOI information limited to the number of crystal layers used (~5 mm for three layers [23]). This design might be adequate for small detector rings for animal scanners [24]; however, light losses at crystal interfaces, as well as the complexity of the crystal array, are considered to be the major drawbacks of the multilayer crystal detector designs.

One method of improving the DOI resolution in multilayer crystal detectors is to couple each crystal layer to a separate photodetector. In this method, the DOI resolution and photon collection efficiency improves at the cost of increased photodetectors and electronics. In order to obtain continuous DOI information from monolayer crystal arrays, the dual-ended readout method was developed [25]. This method which extracts DOI by comparing the relative strengths of the signals obtained from the two ends of the crystal, might not be considered cost-effective because it requires a separate set of photodetectors and additional electronic processing circuitry. However, dual-ended readout can provide high DOI resolution down to 2.0 mm [26] and has been implemented in Clear-PEM, a dedicated positron emission mammography prototype system [27].

Recently, novel methods have been reported for continuous DOI information using single-ended readout of monolayer crystal arrays. One approach involves use of a phosphor coating on one end of the crystal array, and extracting DOI from pulse shapes resulting from the different timing characteristics of the light re-emitted by the phosphor layer. DOI resolution of ~8.0 mm in LYSO crystals of 20 mm length have been reported using this method [28]. In another approach, triangular reflectors are used between individual crystals allowing the scintillation light to be shared between two or more crystals depending on the DOI. In this scheme, DOI is extracted from the pattern of detected signals among the elements of the photodetector array coupled to the crystal array. An early detector design based on this concept used light sharing between only a pair of crystals [29], and an advanced design recently reported 2D light sharing among all crystals of the array in X and Y directions resulting in an improved DOI resolution of ~2 mm [30].

As an alternative to segmented crystal arrays, a cost effective solution for DOI detectors is provided by a monolithic crystal coupled to an array of photodetectors on one side. Monolithic crystals intrinsically provide DOI information as the result of light spread on the photodetector array allowing extraction of DOI from the pattern of detected signals among the photodetector array elements. Various methods have been employed to estimate the 3D position of interactions from detector responses (Figure 15.9). The methods are categorized into two major groups: methods that need training data, and direct methods that do not need to be trained. The former uses statistical methods to estimate the position of interaction from the pattern of the signals in the photodetector array. Those statistical methods include maximum likelihood estimation (MLE) methods [31,32] and neural network–based methods [33]. Direct methods are based on nonstatistical calculations [34,35] such as the amount of light received by the photodetector through the solid angle of the light cone. Although monolithic crystals are cost effective both in terms of scintillator manufacturing and required number of photodetectors/processing channels, from a spatial resolution point of view, they suffer from resolution degradation in the X–Y plane close to the walls of the scintillator block (generally about half of the resolution in the center of the block), as well as low DOI resolution [36].

15.8 MRI COMPATIBLE PET DETECTORS

The challenges related to developing detectors for simultaneous PET-MRI multimodality imaging are not limited only to fabrication of photodetectors insensitive to magnetic fields (including photodetectors such as APDs and SiPMs that have been in use for years). In addition to the static magnetic field of 1.5 T up to 11.7 T, serious challenges to be addressed include, but not limited to the following:

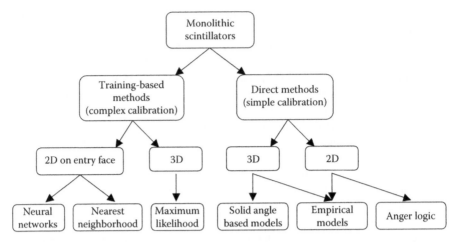

FIGURE 15.9 Diagram depicting the two main event positioning methods (training-based and direct) recently exploited in research and development of PET detectors using monolithic scintillators. The third layer from the top describes the positioning capability of each method and the bottom layer shows the positioning algorithm used. (From Moore, S.K. et al., Maximum-likelihood estimation of 3D event position in monolithic scintillation crystals: Experimental results, *IEEE Nuclear Science Symposium Conference Record*, Honolulu, HI, pp. 3691–3694, 2007; Ling, T. et al., *Phys. Med. Biol.*, 52, 2213, 2007; Bruyndonckxa, P.P. et al., *Nucl. Instrum. Methods Phys. Res. A*, 571, 182, 2007; Li, Z. et al., 3D nonlinear least squares position estimation in a monolithic scintillator block, *IEEE Nuclear Science Symposium Conference Record*, Orlando, FL, pp. 2654–2657, 2009; Taghibakhsh, F. et al., Novel methods of resolving energy and 3D position of interactions in monolithic scintillator plates, *IEEE Nuclear Science Symposium Conference Record*, Knoxville, TN, pp. 2549–2552, 2010.)

- Gradient fields of 20–100 mT/m, and radiofrequency pulses of 64–500 MHz which result in interference and generation of eddy currents in electronics
- Geometric constraints for fitting the PET detector ring(s) inside the MR bore
- Attenuation correction for PET-MRI in absence of CT

In general, therefore, PET-MRI detector blocks should be of compact size, have minimum electrical wiring to minimize the effects of electromagnetic (EM) interference, and preferably be EM shielded. In 2007, Siemens introduced its prototype BrainPET (Figure 15.10) and later commercialized the first brain PET insert for simultaneous clinical PET-MRI. BrainPET consists of 192 detector blocks arranged in 6 rings, each block consisting of a 12×12 array of $2.5 \times 2.5 \times 20\,mm^3$ LSO crystals coupled to a 3×3 matrix of APDs. The insert, with dimensions of 35.5 cm diameter and 19.25 cm axial, achieves spatial resolution of 2.1 mm in the center of its FOV [37].

A prototype small animal PET-MRI scanner has also been developed based on LSO and PS-APDs, in which PS-APDs inside the MRI bore are coupled to scintillator

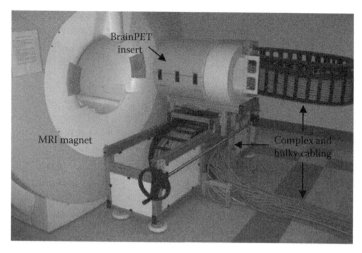

FIGURE 15.10   MRI BrainPET insert prototype built by Siemens. Large cables are used to connect the PET insert to the PET data acquisition and processing system residing outside of the MR room. (From Schmand, M. et al., J. *Nucl. Med.*, 48(suppl. 2), 45, 2007.)

arrays using optical fibers. The prototype system, which has a 35 mm diameter and 12 mm axial FOV, achieves spatial resolution of 1.2 mm [38].

Other detector designs based on SiPMs have also been reported recently. In one approach to TOF-PET/MR, a 33 mm × 33 mm detector block made from a 4 × 4 × 22 mm³ LYSO crystal array is coupled to a matrix of 64 SiPMs readout by custom-developed ASICs, achieving energy resolution of 15% and subnanosecond timing resolution of 650 ps [39].

In a radically novel approach to minimizing EM interference, optical coupling is utilized between PET detectors in the MR bore and processing electronics outside of the MR room using up to 20 m long optical fibers (Figure 15.11). In this method, cross-strip capacitive multiplexing (passive network) is employed to directly couple

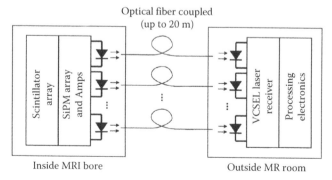

FIGURE 15.11   Schematic diagram of fiber-based optical coupling between PET detectors inside of an MRI bore and data acquisition and processing system outside of the MR room. Optical fibers are lighter and more compact than coaxial cables and are not susceptible to electromagnetic interferences. See Ref. [41] for details.

SiPMs to 850 nm wavelength nonmagnetic VCSEL laser diodes. Coincident energy and timing resolution (FWHM) of 15.5% and 1.3 ns were measured, with no significant degradation between co-axial and optical coupling chains [40].

15.9 DEDICATED PET SYSTEMS

Although almost all PET systems today are whole body PET systems, many groups from clinics to research labs and industry have voiced the need for dedicated PET imagers. Compared to whole body PET, dedicated PET systems are smaller in size, with fewer detectors arranged in such a way that they are dedicated to the imaging of a particular part of the body. Examples are breast, prostate, and brain imaging.

Dedicated PET imaging is one the most active fronts of research and development in the field of PET, mainly because of its potential for detecting cancer at earlier stages than whole body PET systems. Technically, because the detectors of dedicated PET imagers are much closer to the patient's body, they can have higher sensitivity. However, the main challenge in dedicated PET imagers is the need for high spatial resolution. Financially, these systems are also attractive because their small number of detectors makes them more affordable to develop or to buy and operate. Currently, there is one breast dedicated imaging system available in the market: Naviscan, which is based on PMTs. This system is very similar to conventional mammography imagers (Figure 15.12), but uses much less breast compression, making the procedure essentially pain-free for women [41].

Several other breast imaging prototype systems based on semiconductor photodetectors are being developed to further improve the performance of breast-dedicated

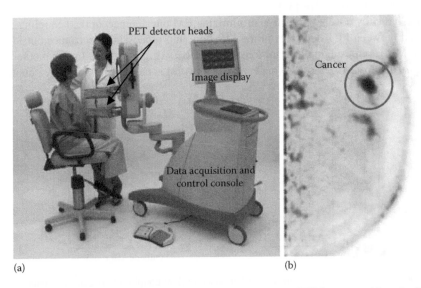

(a)   (b)

FIGURE 15.12 Naviscan positron emission mammography (PEM) scanner (a), and a PEM image (b) spotting a cancer lesion in the breast. The system has a spatial resolution of ~2 mm. (From Berg, W.A. et al., *Breast J.*, 12(4), 309, 2006.)

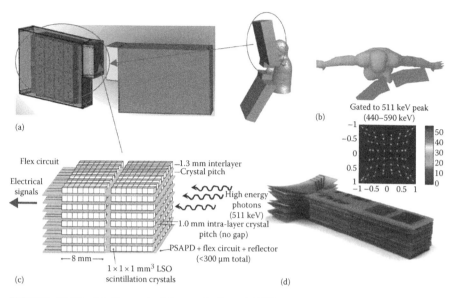

FIGURE 15.13 (a) Concept of breast-dedicated PET system with 9 cm × 16 cm detector heads and data acquisition electronics. (b) Two different potential breast imaging orientations for a patient standing/sitting upright. (c) Schematic diagram of a module comprising layers of 1 mm³ LYSO scintillator arrays coupled to PS-APDs mounted on flex circuits. (d) Image of assembled PET detectors and a flood image showing the capability of resolving events positioned in 1 mm³ scintillators. See Ref. [44] for details.

imaging beyond what Naviscan system offers. One example is ClearPEM, a prototype breast imaging system based on APDs and a dual-ended readout configuration to achieve 2.0 mm DOI resolution. ClearPEM targets an image resolution of 1.2 mm to detect cancer lesions smaller than 3 mm [27,42]. In a different approach to ultrahigh resolution breast imaging, a prototype PET system is being developed at Stanford based on stacks of PS-APDs and arrays of 1 mm³ scintillators to achieve submillimeter spatial resolution of 0.835 mm FWHM [42]. In this design (Figure 15.13), PS-APD readout channels are multiplexed in order to reduce the highly increased complexity of the 3D detector, without any appreciable degradation in detector performances [43].

15.10 SUMMARY

With development of SiPMs as a cost-effective solution for compact and high performance photodetection, it is expected that advancements in PET, particularly in spatial and contrast resolution, will continue through incorporation of DOI and FOF. Although systems based on TOF are already available, TOF resolution must be enhanced for TOF to have a considerable impact on clinical imaging performance. This will require development of faster scintillators and photodetectors, where the greatest obstacle perhaps is the cost. Therefore, exploring cost-effective solutions for these technologies must be regarded as high priority.

ACKNOWLEDGMENT

Authors would like to thank Alex Grant for his careful review of this manuscript, and providing constructive critiques.

REFERENCES

1. M.R. Kiser, C.D. Reid, A.S. Crowell et al., Exploring the transport of plant metabolites using positron emitting radiotracers, *HFSP Journal*, 2(4): 189–204 (2008).
2. K. Kinkel, Y. Lu, M. Both et al., Detection of hepatic metastases from cancers of gastrointestinal tract by using non-invasive imaging methods (US, CT, MR imaging, PET): A meta analysis, *Radiology*, 224: 748–756 (2002).
3. H. Zaidi, M.L. Montandon, A. Alavi, The clinical role of fusion imaging using PET, CT, and MR imaging, 3(3): 275–291 (2008).
4. S.R. Cherry, M. Dahlbom, PET: Physics, instrumentation, and scanners. In *Molecular Imaging and Its Biological Applications*, M.E. Phelps (ed.), 1st edn., Springer, New York (2004).
5. Y. Yang, Y. Wu, J. Qi et al., A prototype PET scanner with DOI-encoding detectors, *Journal of Nuclear Medicine*, 49(7): 1132–1140 (2008).
6. V. Spanoudaki, C.S. Levin, Investigating the temporal resolution limits of scintillation detection from pixellated elements, *Physics in Medicine and Biology*, 56(7): 735–765 (2011).
7. G.F. Knoll, *Radiation Detection and Measurements*, 3rd edn., Wiley, New York (2000).
8. E.V. Van Loef, Y. Wang, J. Glodo et al., Recent advances in ceramic scintillators, *Materials Research Society Symposium Proceedings*, Vol. 1038, p. 1038-O06-02 (2008).
9. M.D. Birowosuto, *Novel γ-Ray and Thermal-Neutron Scintillators: Search for High-Light-Yield and Fast-Response Materials*, Vol. 2007, IOS Press, Amsterdam, the Netherlands (2007).
10. G. Hull, J.J. Roberts, J.D. Kuntz et al., Ce-doped single crystal and ceramic garnet for γ-ray detection, *Proceedings of the SPIE*, 6706: 670617 (2007).
11. R. Pania, M.N. Cintib, R. Scafec et al., Energy resolution measurements of LaBr3:Ce scintillating crystals with an ultra-high quantum efficiency photomultiplier tube, 610(1): 41–44 (2009).
12. M. Bergeron, J. Cadorette, J. Beaudoin et al., Performance evaluation of the LabPET APD-based digital PET scanner, *IEEE Transactions on Nuclear Science*, 56(1): 10–16 (2009).
13. R. Bugalho, B. Carriço, C.S. Ferreira et al., Characterization of avalanche photodiode arrays for the ClearPEM and ClearPEM-sonic scanners, *Journal of Instrumentation*, 4: 1–12 (2009).
14. P.D. Olcott, J.A. Talcott, C.S. Levin et al., Compact readout electronics for position sensitive photomultiplier tubes, *IEEE Transactions on Nuclear Science*, 52(1): 21–27 (2005).
15. M. McClish, P. Dokhale, J. Christian et al., Characterization of CMOS position sensitive solid-state photomultipliers, *Nuclear Instruments and Methods in Physics Research Section A*, 624(2): 492–497 (2010).
16. C.J. Staples, P. Barton, E.B. Johnson et al., Recent developments with CMOS SSPM photodetectors, *Nuclear Instruments and Methods in Physics Research Section A*, 610(1): 145–149 (2009).
17. C. Merck, R. Eckhardt, R. Hartmann et al., Back illuminated drift silicon photomultiplier as novel detector for single photon counting, *IEEE Nuclear Science Symposium Conference Record*, San Diego, CA, pp. 1562–1565 (2006).

18. T. Frach, G. Prescher, C. Degenhardt et al., The digital silicon photomultiplier—Principle of operation and intrinsic detector performance, *IEEE Nuclear Science Symposium Conference Record*, Orlando, FL, pp. 1959–1965 (2009).

19. C. Degenhardt, G. Prescher, T. Frach, The digital silicon photomultiplier—A novel sensor for the detection of scintillation light, *IEEE Nuclear Science Symposium Conference Record*, Orlando, FL, pp. 2383–2386 (2009).

20. P. Vaska, D.H. Kim, S. Southekal et al., Ultra-high resolution PET: A CZT-based scanner for the mouse brain, *Journal of Nuclear Medicine*, 50: 293 (2009).

21. Y. Yin, S. Komarov, H. Wu et al., Characterization of highly pixelated CZT detectors for sub-millimeter PET Imaging, *IEEE Nuclear Science Symposium Conference Record*, Orlando, FL, pp. 2411–2414 (2009).

22. Y. Gu, J.L. Matteson, R.T. Skelton et al., Study of a high-resolution, 3D positioning cadmium zinc telluride detector for PET, *Physics in Medicine and Biology*, 56: 1563–1584 (2011).

23. S.-J. Hong, S.-I. Kwon, M. Ito et al., Concept verification of three-layer DOI detectors for small animal PET, *IEEE Transactions on Nuclear Science*, 55(3): 912–917 (2008).

24. S. Yamamoto, M. Imaizumi, T. Watabe et al., Development of a Si-PM-based high-resolution PET system for small animals, *Physics in Medicine and Biology*, 55: 5817–5831 (2010).

25. W.W. Moses, S.E. Derenzo, C.L. Melcher, R.A. Manente, Room temperature LSO pin photodiode PET detector module that measures depth of interaction, *IEEE Transactions on Nuclear Science*, 42: 1085–1089 (1995).

26. Y. Shao, K. Meadors, R.W. Silverman, R. Farrell, L. Cirignano, R. Grazioso, K.S. Shah, S.R. Cherry, Dual APD array readout of LSO crystals: Optimization of crystal surface treatment, *IEEE Transactions on Nuclear Science*, 49: 649–654 (2002).

27. E. Albuquerquea, V. Bexigaa, R. Bugalhob et al., Experimental characterization of the 192 channel Clear-PEM frontend ASIC coupled to a multi-pixel APD readout of LYSO: Ce crystals, *Nuclear Instruments and Methods in Physics Research Section A*, 598(3): 802–814 (2009).

28. H. Du, Y. Yang, J. Glodo et al., Continuous depth-of-interaction encoding using phosphor-coated scintillators, *Physics in Medicine and Biology*, 54: 1757–1771 (2009).

29. R.S. Miyaoka, T.K. Lewellen, H. Yu, Design of a depth of interaction (DOI) PET detector module, *IEEE Transactions on Nuclear Science*, 45(3): 1069–1073 (1998).

30. M. Ito, J.-S. Lee, M.-J. Park, Design and simulation of a novel method for determining depth-of-interaction in a PET scintillation crystal array using a single-ended readout by a multi-anode PMT, *Physics in Medicine and Biology*, 55: 3827–3841 (2010).

31. S.K. Moore, W.C.J. Hunter, L.R. Furenlid et al., Maximum-likelihood estimation of 3D event position in monolithic scintillation crystals: Experimental results, *IEEE Nuclear Science Symposium Conference Record*, Honolulu, HI, pp. 3691–3694 (2007).

32. T. Ling, T.K. Lewellen, R.S. Miyaoka, Depth of interaction decoding of a continuous crystal detector module, *Physics in Medicine and Biology*, 52: 2213–2228 (2007).

33. P.P. Bruyndonckxa, C. Lemaitrea, D. Schaart et al., Towards a continuous crystal APD-based PET detector design, *Nuclear Instruments and Methods in Physics Research A*, 571: 182–186 (2007).

34. Z. Li, P. Bruyndonckx, M. Wedrowski, G. Vandersteen, 3D nonlinear least squares position estimation in a monolithic scintillator block, *IEEE Nuclear Science Symposium Conference Record*, Orlando, FL, pp. 2654–2657 (2009).

35. F. Taghibakhsh, S.G. Cuddy, J.A. Rowlands, Novel methods of resolving energy and 3D position of interactions in monolithic scintillator plates, *IEEE Nuclear Science Symposium Conference Record*, Knoxville, TN, pp. 2549–2552 (2010).

36. W.C.J. Hunter, H.H. Barrett, L.R. Furenlid, Calibration method for ML estimation of 3D interaction position in a thick gamma-ray detector, *IEEE Transactions on Nuclear Science*, 65(1): 189–196 (2009).
37. M. Schmand, Z. Burbar, J. Corbeil et al., BrainPET: First human tomograph for simultaneous (functional) PET and MR imaging, *Journal of Nuclear Medicine*, 48(suppl. 2): 45(2007).
38. C. Catana, Wu Y., Judenhofer, M.S. et al., Simultaneous acquisition of multislice PET and MR images: Initial results with a MR-compatible PET scanner, *Journal of Nuclear Medicine*, 47: 1968–1976 (2006).
39. T. Solf, V. Schulz, B. Weissler et al., Solid-state detector stack for ToF-PET/MR, *IEEE Nuclear Science Symposium Conference Record*, Orlando, FL, pp. 2798–2799 (2009).
40. P.D. Olcott, H. Peng, C.S. Levin, Novel electro-optically coupled MR compatible PET detectors, *IEEE Nuclear Science Symposium Conference Record*, Dresden, Germany, pp. 4640–4645 (2008).
41. W.A. Berg, I.N. Weinberg, D. Narayanan et al., High-resolution fluorodeoxyglucose positron emission tomography with compression ("positron emission mammography") is highly accurate in depicting primary breast cancer, *The Breast Journal*, 12(4): 309–323 (2006).
42. E. Albuquerque, F.G. Almeida, P. Almeida et al., An overview of the Clear-PEM breast imaging scanner, *IEEE Nuclear Science Symposium Conference Record*, Dresden, Germany, pp. 5616–5618 (2008).
43. A. Vandenbroucke, A.M.K. Foudray, P.D. Olcutt, C.S. Levin, Performance characterization of a new high resolution PET scintillation detector, *Physics in Medicine and Biology*, 55: 5895–5911 (2010).
44. F.W.Y. Lau, A. Vandenbroucke, P.D. Reynolds et al., Analog signal multiplexing for PSAPD-based PET detectors: Simulation and experimental validation, *Physics in Medicine and Biology*, 55(23): 7149–7174 (2010).

Index